PHYSICS RESEARCH AND TECHNOLOGY

QUASICRYSTALS: TYPES, SYSTEMS, AND TECHNIQUES

PHYSICS RESEARCH AND TECHNOLOGY

Additional books in this series can be found on Nova's website under the Series tab.

Additional E-books in this series can be found on Nova's website under the E-books tab.

MATERIALS SCIENCE AND TECHNOLOGIES

Additional books in this series can be found on Nova's website under the Series tab.

Additional E-books in this series can be found on Nova's website under the E-book tab.

PHYSICS RESEARCH AND TECHNOLOGY

QUASICRYSTALS: TYPES, SYSTEMS, AND TECHNIQUES

BETH E. PUCKERMANN
EDITOR

Nova Science Publishers, Inc.
New York

For permission to use material from this book please contact us:
Telephone 631-231-7269; Fax 631-231-8175
Web Site: http://www.novapublishers.com

LIBRARY OF CONGRESS CATALOGING-IN-PUBLICATION DATA

Quasicrystals : types, systems, and techniques / [edited by] Beth E. Puckermann.
p. cm.
Includes index.
ISBN 978-1-61761-123-0 (hardcover)
1. Quasicrystals. I. Puckermann, Beth E.
QD926.Q375 2009
530.4'1--dc22

2010027150

Published by Nova Science Publishers, Inc. † New York

CONTENTS

PREFACE

Quasicrystals are metallic alloys that exhibit atomic scale order, but not periodic order. Atomic scale properties of these materials are different from single crystalline material, for example, extraordinary mechanical properties, electrical and thermal transport properties, and electronic structure. This book presents topical research in the study of quasicrystals, including vacancies in quasicrystals; the formation of quasicrystals in bulk metallic glasses and their effects on mechanical behavior; the electrical transport observed in Al-Pd-Mn quasicrystals; logarithmic periodicity in quasicrystals; and positron annihilation studies of quasicrystals.

Chapter1- The critical electronic states originating from the quasiperiodic long-range atomic order make its electrical properties quite unusual. Critical states decay as power-law and hence the temperature dependence of conductivity is expected to follow a power-law i.e. $\sigma \propto T^a$. In fact this behavior has been nicely evidenced in some non-magnetic quasicrystals like Al-Pd-Re and Al-Cu-Ru. In this series of Al-based stable quasicrystals Al-Pd-Mn system appears to be the most interesting. It has been reported to contain magnetic Mn sites. Due to its magnetic character, Al-Pd-Mn quasicrystals are expected to show interesting electrical transport properties, which is of basic importance from the "effect of magnetic scattering on the transport of localized electric systems" point of view. We have reviewed and tried to put forward an approach, which describes all the features of the electrical transport observed in Al-Pd-Mn quasicrystals.

To investigate the important role of Mn in these alloys, we changed the concentration of magnetic Mn in $Al_{70}Pd_{20}Mn_{10}$ quasicrystal following three approaches, (i) by exchanging the concentration of Pd and Mn atoms (ii) by replacing Mn by other transition metals like Fe, Cr, Co and Ni and (iii) by replacing aluminum(Al) by boron(B). Studies on these three different compositions have been described in Part-I, Part-II and Part-III. All these stable quasicrystal compositions were investigated using low-temperature (down to 2K) magnetic measurements, embodying magnetization-vs-temperature (M-T) and magnetization-vs-field (M-H) measurements and low-temperature zero-field and in-field magneto-transport measurements using four- probe method.

The σ-T variation of nearly all the studied samples show qualitatively the same behavior i.e. while cooling a σ-T minima is followed by a σ-T maxima. The observed features are all in accordance with the other observations. In the existing literature the occurrence of conductivity minima has been attributed to anti weak localization effect appearing due to strong spin-orbit scattering of conduction electrons by Pd. However, based on strong arguments as corroborated by other magnetic and magneto-transport measurements, we have shown that although the observed σ-T variation is due to weak-localization the minima may not be due to spin-orbit scattering. We have concluded that the electronic transport in Al-Pd-Mn quasicrystals is fully dominated by quantum interference effects including electron-phonon scattering and Kondo-type magnetic scattering as two competing random dephasing processes. The σ-T minima appear as a result of these two competing scattering processes. The observed σ-T maxima is due to maxima in the spin-flip scattering rate which is expected while spin-flip scattering of electrons is from a system of interacting moments.

Chapter2-'Logarithmic periodicity' refers to three features in quasicrystals: firstly the ideal structure is uniquely icosahedral and infinitely extensive; secondly, the diffraction patterns contain corresponding orders that are geometrically spaced; and thirdly the mathematical description of electronic states is by special Fourier transforms in logarithmic order. This periodicity is driven by the low enthalpy in the subcluster. The model differs from most mathematical models because the three dimensional tiles share edges not faces. Experimental evidence of several types supports the model, beyond its conceptual simplicity. The principal three sources are: electron diffraction; electron microscopy; and diffraction simulations. The variety of properties and predictions are consistent with available experimental data in the binary quasicrystals such as Al_6Mn. Though logarithmic periodicity describes ideal solids with perfect icosahedral symmetry, the structure is defective in realization. While the defects should be expected in rapidly cooled and metastable solids, they imply uncertainties that require further refinement. If dendritic crystal growth depends on deposition of supercluster planar quads, the higher the order, the more nearly icosahedral.

Chapter3- Positron annihilation studies of quasicrystals (QC's) and their related materials (crystalline approximants) are reviewed. We describe why a positron, anti-particle of electron, is suitable for probing vacancies locally in aperiodic QC's. A series of positron annihilation spectroscopy is then briefly outlined. Positron lifetime spectroscopy reveals high concentration of structural vacancies more than 10^{-4} in atomic concentration for QC's and crystalline approximants studied. Chemical environments around the structural vacancies are investigated by coincident Doppler broadening spectroscopy. In addition, the concentration of structural vacancies is discussed based on the positron diffusion data obtained by a variable-energy slow positron beam. Besides the structural vacancy, we refer to other two kinds of vacancies: thermally formed high-temperature vacancy and electron-irradiation induced vacancy. Finally, the structural phase transition in QC's probed through the local atomic and electronic structures around structural vacancies is presented.

Chapter4- The structures of many quasicrystals have still remained unknown since the publication of the first icosahedral quasicrystal in rapidly solidified Al-Mn alloys in 1982. The main obstacle is that the quasicrystals always contain defects and it is difficult to synthesize high quality single crystals which are needed for a good structure determination by single crystal X-ray diffraction. In most cases, quasicrystals coexist with several complex quasicrystal approximants. These approximants have similar local atomic structures as the quasicrystals and many of them also contain defects that make diffraction spots from quasicrystals and different approximants overlapped and the whole diffraction pattern blurred. Meanwhile, some less complicated approximants in the same system can be synthesized as large single crystals with fewer defects, and their atomic structures can be determined. Due to the similar local atomic structures, a quasicrystal and its approximants always show similar intensity distribution and phase relationships for the strong reflections in reciprocal space.

Thus, the structure factors with both amplitudes and phases can be calculated from a known approximant for strong reflections and after re-indexing them, they can be used to calculate a 3D electron density map for more complex approximants by inverse Fourier transformation. The structure model can be deduced from this 3D electron density map since the strongest reflections mainly determine the atomic positions in a structure. In principle, the perfect quasicrystal structure model can be obtained by this approach. The strong reflections approach avoids a direct structure determination from quasicrystals containing defects but takes the advantage of using the common features of quasicrystals and approximants. The model deduced from this approach will be an ideal model for the quasicrystal, free of defects.

Chapter5- The dominant cluster in the Ti/Zr-based quasicrystals is a Bergman-type cluster possessing a large number of tetrahedral interstitial sites; this makes these quasicrystals attractive as potential hydrogen storage materials. This paper summarizes our recent research results on the hydrogen absorption and desorption properties of the Ti-Zr-Ni and Ti-Hf-Ni quasicrystals and related amorphous or crystal phases produced by a combination of mechanical alloying and subsequent annealing. The effects on the microstructures and hydrogenation properties of the substitution of Zr for either Ti or Hf in alloys based on the $Ti_{45}Zr_{38}Ni_{17}$ compositions are investigated. Comparisons between results reported for samples prepared by rapid quenching or annealing are also made.

Chapter6- The annealing of bulk metallic glasses (BMGs) at elevated temperatures usually leads to partial or full crystallization. The crystallization in several systems starts with the formation of metastable quasicrystalline (QC) particles and then the material can be regarded as a composite of QC and amorphous phases. The appearance of QC particles significantly affects the mechanical properties of BMGs. In this chapter, the morphology, structure and chemical composition of QC particles formed during heat-treatment of BMGs are reviewed according to the relevant literature. Special attention is paid to the influence of the formation of QC particles on the mechanical behavior at room and high temperatures. It was found that during heat-treatment of a commercial ZrTiCuNiBe BMG above the glass transition temperature nanosized spherical QC particles containing smaller grains were formed. Depending on the annealing temperature the volume fraction of the QC phase varied between 25 and 37%. The QC particles contain Ti, Zr and Ni in high concentration, while the amorphous matrix is enriched in Be. The high temperature viscosity increases mainly due to the hard QC particles but there is also a slight contribution from the compositional changes of the supercooled liquid matrix. The bending strength measured at room temperature decreases in consequence of QC formation, most probably mainly due to the loss of free volume in the amorphous matrix.

Chapter7- The atomic structure of the 2-fold decagonal Al-Ni-Co quasicrystal surface has been investigated using scanning tunneling microscopy (STM). Decagonal quasicrystals are made of pairs of atomic planes with pentagonal symmetry periodically stacked along a 10-fold axis. It is, therefore, expected that the 2-fold surfaces exhibit a periodic direction along the 10-fold axis, and an aperiodic direction perpendicular to it. The surface shows rough and cluster-like structures at low annealing temperatures ($T<1000K$), whilst annealing to temperatures in excess of 1000K results in the formation of step-terrace structures. The surface consists of terraces separated by steps of heights 1.9, 4.7, 7.8, and 12.6 Å. Ratios of step heights can be properly assigned to different τ powers, suggesting a well defined quasiperiodic long-range order. At the annealing temperature ($1100K < T < 1150K$), atomically resolved STM images of the 2-fold plane reveal atomic rows along the 10-fold direction with a periodicity of 4 Å. The spacing between the parallel rows is aperiodic, with distances following a Fibonacci sequence. We found that the quasiperiodic order in the sequence of atomic rows is destroyed by the presence of phason defects. Above the heating temperature of 1200K, formation of second phase structures was observed. The formation of a second phase could be associated with the preferential evaporation of Al at the elevated temperature.

Chapter8- The atomic structure of the 2-fold decagonal Al-Ni-Co quasicrystal surface has been investigated using scanning tunneling microscopy (STM). Decagonal quasicrystals are made of pairs of atomic planes with pentagonal symmetry periodically stacked along a 10-fold axis. It is, therefore, expected that the 2-fold surfaces exhibit a periodic direction along the 10-fold axis, and an aperiodic direction perpendicular to it. The surface shows rough and cluster-like structures at low annealing temperatures (T<1000K), whilst annealing to temperatures in excess of 1000K results in the formation of step-terrace structures. The surface consists of terraces separated by steps of heights 1.9, 4.7, 7.8, and 12.6 Å. Ratios of step heights can be properly assigned to different τ powers, suggesting a well defined quasiperiodic long-range order. At the annealing temperature (1100K < T < 1150K), atomically resolved STM images of the 2-fold plane reveal atomic rows along the 10-fold direction with a periodicity of 4 Å. The spacing between the parallel rows is aperiodic, with distances following a Fibonacci sequence. We found that the quasiperiodic order in the sequence of atomic rows is destroyed by the presence of phason defects. Above the heating temperature of 1200K, formation of second phase structures was observed. The formation of a second phase could be associated with the preferential evaporation of Al at the elevated temperature.

Chapter9- The microstructural characteristics of a commercial quaternary AA8090 (Al-2%Li-1.2%Cu-0.8%Mg, by wt.%) alloy reinforced with 15 vol.% SiCp has been examined in detail. The composite material in the form of plate sheets with the thickness about 1600 μm was thinned to electron beam transparent (~ 20 nm thickness) using mechanical polishing and ion beam milling to carry out microscopy observations. In the alloy matrix (α - Al) the presence of δ'-precipitates ($L1_2$ structure, lattice parameter a = 0.401 nm) as tiny spheres of about 50 – 100 nm in size has been delineated. The presence of icosahedral quasicrystalline phase has also been observed in the matrix. In general, a lamellae structure of δ'-precipitate with the layer thickness of about 250 nm has been revealed on the grain boundaries. Adjacent to δ'-precipitate, a prominent region of precipitate free zones with a thickness between 65 – 85 nm is present at the boundaries. The distribution of SiCp in α-Al matrix is uniform with a clear interface exhibiting some dislocations.

Chapter10- The shapes of icosahedral quasicrystalline (IQC) particles in Al-Mn-Be-(Cu) alloys were determined in samples subjected to very wide range of cooling rates: from around 10^6 K/s in very thin melt-spun ribbons down to below 100 K/s in permanent copper dies. Accordingly, the sizes of quasicrystalline particles ranged from few tenths of nanometres up to more than 100 μm. As a consequence, different methods were employed to properly characterize their shapes: projections of quasicrystalline particles using transmission electron microscopy (TEM), cross-sections of IQCs on metallographic polished surfaces, observation of deep etched samples and extracted particles in a scanning electron microscope (SEM). Despite of different sizes and shapes it was discovered that two the most important features are common to all of them:

preferential growth in the three-fold directions

tendency for faceting and adopting the shape of pentagonal dodecahedron.

The evolution of quasicrystalline shapes from apparently spherical particles to very large and highly branched dendrites is systematically presented. Special attention was devoted to the correct interpretation of quasicrystal shapes obtained from 2D-metallographic cross-sections.

In: Quasicrystals: Types, Systems, and Techniques
Editor: Beth E. Puckermann, pp. 1-45
ISBN 978-1-61761-123-0

Chapter 1

DOMINANCE OF MAGNETIC SCATTERING IN $AL_{70}PD_{20+X}MN_{10-X}$ (X = 0, 1 AND 2), $AL_{70}PD_{20}MN_8(TM)_2$ (TM=FE, CR, CO AND NI) AND $AL_{70-X}B_X PD_{20}MN_{10}$ (X = 0, 0.5,1, 2 AND 4) STABLE ICOSAHEDRAL QUASICRYSTALS

Archna Sagdeo[1] and N.P.Lalla[2]

[1]Raja Ramanna Center of Advanced Technology, Rajendera Nagar, Indore, India

[2]UGC-DAE Consortium for Scientific Research, University Campus, Khandawa Road, Indore, India

ABSTRACT

The critical electronic states originating from the quasiperiodic long-range atomic order make its electrical properties quite unusual. Critical states decay as power-law and hence the temperature dependence of conductivity is expected to follow a power-law i.e. $\sigma \propto T^a$. In fact this behavior has been nicely evidenced in some non-magnetic quasicrystals like Al-Pd-Re and Al-Cu-Ru. In this series of Al-based stable quasicrystals Al-Pd-Mn system appears to be the most interesting. It has been reported to contain magnetic Mn sites. Due to its magnetic character, Al-Pd-Mn quasicrystals are expected to show interesting electrical transport properties, which is of basic importance from the "effect of magnetic scattering on the transport of localized electric systems" point of view. We have reviewed and tried to put forward an approach, which describes all the features of the electrical transport observed in Al-Pd-Mn quasicrystals.

To investigate the important role of Mn in these alloys, we changed the concentration of magnetic Mn in $Al_{70}Pd_{20}Mn_{10}$ quasicrystal following three approaches, (i) by exchanging the concentration of Pd and Mn atoms (ii) by replacing Mn by other transition metals like Fe, Cr, Co and Ni and (iii) by replacing aluminum(Al) by boron(B). Studies on these three different compositions have been described in Part-I, Part-II and Part-III. All these stable quasicrystal compositions were investigated using low-temperature (down to 2K) magnetic measurements, embodying magnetization-vs-temperature (M-T) and magnetization-vs-field (M-H) measurements and low-temperature zero-field and in-field magneto-transport measurements using four- probe method.

The σ-T variation of nearly all the studied samples show qualitatively the same behavior i.e. while cooling a σ-T minima is followed by a σ-T maxima. The observed features are all in accordance with the other observations. In the existing literature the occurrence of conductivity minima has been attributed to anti weak localization effect appearing due to strong spin-orbit scattering of conduction electrons by Pd. However, based on strong arguments as corroborated by other magnetic and magneto-transport measurements, we have shown that although the observed σ-T variation is due to weak-localization the minima may not be due to spin-orbit scattering. We have concluded that the electronic transport in Al-Pd-Mn quasicrystals is fully dominated by quantum interference effects including electron-phonon scattering and Kondo-type magnetic scattering as two competing random dephasing processes. The σ-T minima appear as a result of these two competing scattering processes. The observed σ-T maxima is due to maxima in the spin-flip scattering rate which is expected while spin-flip scattering of electrons is from a system of interacting moments.

1. Introduction

Electronic properties of quasicrystals are expected to be quite unusual. This expectation is due to their critical electronic states [1,2], which are neither exponentially localized like those in disordered materials nor extended like that of crystalline materials. Critical states decay as power-law and hence the temperature dependence of conductivity is expected to follow power-law i.e. $\sigma \propto T^a$ [3,4]. In fact this behavior has been observed in some structurally high quality non-magnetic quasicrystals like Al-Pd-Re and Al-Cu-Ru [4,5 ,6],. In the series of these Al-based stable quasicrystalline alloys, Al-Pd-Mn system, due to the presence of Mn, appears to be the most interesting. It has been reported that only few percent of the total Mn sites are magnetic [28,32,37,46], ,], whose concentration increases with increasing defect [47]. Due to its magnetic character, Al-Pd-Mn quasicrystals are expected to show interesting electrical transport properties, which are of basic importance from "transport in magnetic materials" point of view [7]. The aspect of electrical transport in Al-Pd-Mn quasicrystal has been sparsely studied. We have tried to put forward a universal approach, which describes all the features of the electrical transport in Al-Pd-Mn quasicrystals.

The electrical conductivity σ of icosahedral quasicrystals is unusually sensitive to slight changes in their composition. Such changes shift the position of E_F, which results in a change of the DOS (E_F) and hence a change in σ. Thus, if we dope these quasicrystals by suitable dopants or by internal exchange of relative compositions of their ingredients, we can manipulate the Fermi-level and thus the electrical transport properties. The electronic structure calculations in the case of i-Al-Pd-Mn have shown the presence of structure-induced pseudo-gap at E_F [8]. The sp-d hybridization between Al sp and Mn d states is found to be an important factor in the formation of the pseudo-gap [9,10]. The position and density of Mn states near E_F and hence its magnetic properties are very sensitive to the structural and / or chemical environment of the Mn in the alloy [11]. Looking at the important role of Mn in these alloys, if we change the composition of $Al_{70}Pd_{20}Mn_{10}$ icosahedral quasicrystal by changing the Mn concentration, we can manipulate the DOS of the system. This in turn will affect its magnetic properties. Therefore to investigate the role of magnetic properties of Al-

Pd-Mn quasicrystals on its electrical transport behavior we have changed the concentration of Mn in $Al_{70}Pd_{20}Mn_{10}$ quasicrystalline alloy following three different approaches, (i) by exchanging the concentration of Pd and Mn atoms (ii) by replacing 2% of Mn by other transition metals like Fe, Cr, Co and Ni and (iii) by replacing boron in the place of aluminum. These are separately described in Part I, Part II and Part III.

Part I deals with $Al_{70}Pd_{20+x}Mn_{10-x}$ quasicrystalline samples in which concentration of Pd and Mn have been exchanged, Part II deals with the $Al_{70}Pd_{20}Mn_{10}$ quasicrystalline samples in which 2% of Mn is replaced by other transition metals like Fe, Cr, Co and Ni and finally Part III deals with $Al_{70-x}B_xPd_{20}Mn_{10}$. In these three parts we have studied the magnetic and transport properties of these quasicrystalline alloys.

Prior to 1987, all the known quasicrystals were thermodynamically metastable, exhibiting significant structural disorder, as manifested in the broadening of the X-ray diffraction lines. It was argued that this disorder might inhibit some possible novel intrinsic physical properties of quasicrystals. It was therefore of great importance when the first thermodynamically stable icosahedral alloys Al-Cu-TM (M= Fe, Ru, Os) were discovered [12] as they posses a high degree of structural perfection comparable to that found in the periodic alloys. These stable icosahedral phases were found to be chemically ordered and having face centered icosahedral (FCI) structure [12].

In the series of investigations on the formation and stability of the Al-Cu-TM quasicrystals, it was found that quasicrystals of this series form at the compositions with valence electron numbers (e/a) in the vicinity of 1.75. On the basis of this empirical rule, new FCI phases of Al70Pd20Mn10 and Al70Pd20Re10 were discovered [13]. In addition, it was also found that the icosahedral phase forms in a wide composition range, consisting of the simple icosahedra (SI) and FCI phases in different composition regions in the Al-Pd-TM system. The stable icosahedral quasicrystals, which was found in the Al-Pd-Mn alloy system [13,14] are in general free of atomic disorder and phason strains [15]. Structural studies with different techniques show that this phase forms in a perfect icosahedral state [16,17]. Therefore this system is ideal for studying the effects of magnetic impurities on the unusual electrical transport of quasicrystals. Very good structural quality and several centimeters large single grained quasicrystals of Al-Pd-Mn have been obtained by Bridgemann and of Czochralski [18,19] techniques. The most interesting fact about Al-Pd-Mn quasicrystals is that it melts directly into liquid without involving any crystalline phase, which makes the preparation of large single grained quasicrystals easier.

1.1. Phase Diagram

The phase diagram of melt quenched Al-Pd-Mn alloys has been well investigated by Tsai *et al.* [20] and is illustrated in Figure 1. The broken line shows a composition line with e/a = 1.75. It can be noted that the formation composition range of the icosahedral phase (i-phase) elongates along the line of e/a = 1.75, indicating that the electronic structure plays an important role in the formation of the i-phase. The i-phase with FCI structure forms in an wide composition range of 15 to 25 at.% Pd and 7 to 15 at.% Mn. In lower Pd and Mn regions free FCC Al phase exists as contamination.

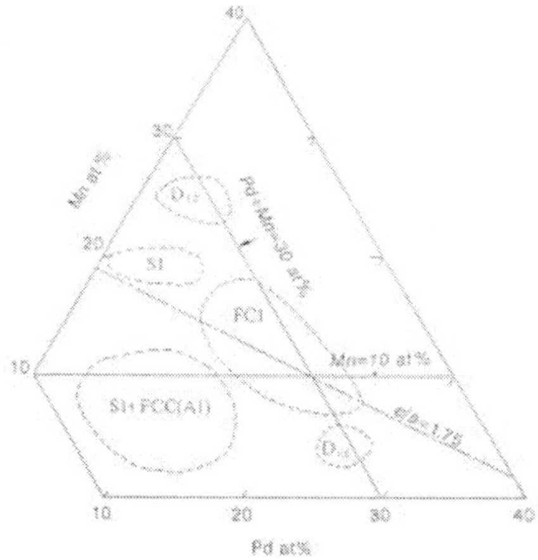

Figure 1. Phase diagram of Al-Pd-Mn alloys.

While an ordered cubic phase occurs in higher Pd and Mn regions, and a decagonal phase exists at the Pd-rich (~25 at.%) and Mn-poor (<5 at. %) regions. It is also expected that a D-phase would be observed in a Pd-poor region with Mn near 20 at. %. Thus, the global concentration of Pd + Mn = 30 at. % seems to be very critical for the formation of quasicrystalline phases. A slight departure from the optimized stoichiometry results in multiphase materials and / or non-perfect icosahedral order.

1.2. Magnetic Properties

Most of the Al-based quasicrystalline alloys show diamagnetic behavior and hence its magnetic susceptibility remains temperature-independent. For instance, diamagnetic behavior is found in icosahedral Al-Cu-Fe and Al-Pd-Re [21] quasicrystals. In contrast, the existence of Curie terms and hence of localized moments, has been reported in stable Al-Pd-Mn quasicrystals. Several experimental studies on polycrystalline [22,23,24] as well as on single grained quasicrystals [25,26,27,28] have shown that the magnetic susceptibility of these Al-Pd-Mn quasicrystals follows the Curie-Weiss law over a wide temperature range with a small and negative value of Curie temperature θ_c, suggesting that the Mn-Mn exchange interaction in these quasicrystals is anti-ferromagnetic in nature [23]. The specific-heat measurements [23,28] suggest that these magnetic moments are coupled through RKKY type of indirect exchange interaction. The observation of small magnetization as compared to the total Mn concentration in these studies [22,23,24,26,27,28] along with the small value of Curie constant C, indicates a very low concentration of moments (~1% or even less) of all the Mn atoms being involved. There exist few other studies [29,30], according to which, Curie-Weiss law cannot account for the susceptibility data in case of Al-Pd-Mn quasicrystals. Their susceptibility as a function of 1/T, exhibits a continuous curvature. According to them the Kondo coupling between the localized moments and the conduction electron spins can explain the anomalous temperature dependence of the susceptibility. Few other studies [23,28] have shown the presence of spin-glass type of transition at low temperatures, indicating high degree of frustration present in the system. Nimori *et al.* [31] gave the cluster-glass type of picture of Al-Pd-Mn quasicrystals. They have suggested the formation of ferrimagnetic clusters and have argued that the observed small magnetization as a result of this ferromagnetic ordering. Besides this, the study of magnetic properties of liquids in Al-Pd-Mn systems reveals a very strong increase of the susceptibility on melting [32]. These features were ascribed to the presence of magnetic moments on a large fraction of the Mn atoms in the liquid state. A similar change is observed in neutron scattering measurements, which indicate that localized magnetic moments appear on melting, and disappears in the solid state [32].

The origin of magnetism in Al-Pd-Mn quasicrystals has been widely debated. The moment formation has been shown to be affected by the pseudo-gap in the electronic density of states (DOS) at the Fermi level and also by the hybridization between Al *s-p* and Mn *d* states [33,34,35,36,37]. Spin-polarized band structure calculations on quasicrystalline Al-Pd-Mn models show that the formation of magnetic moments on Mn atoms is extremely sensitive to their local DOS and occurs only on a few Mn sites, which supports the observed low moment concentration [36].

1.3. Electrical Conductivity

In the series of Al-based stable quasicrystalline alloys, Al-Pd-Mn appears to be the most interesting. It has been investigated for its interesting electronic transport properties and reports have been published [22,24,27,38,39,40,41,42,43,44,45,46]. The resistivity ρ(T) of the i-AlPdMn represents a special class amongst quasicrystals. It displays a maximum in ρ(T) between 40K-130K [22,24,40,42,43] and sometimes an additional minimum at lower temperatures between 4K-25K [40,42,43] Such a peculiar behavior of ρ(T) in Al-Pd-Mn quasicrystals is not yet well understood. The presence of Pd atoms along with Mn in this quasicrystalline system has made it more complicated and at the same time interesting also. There are many earlier studies [22,24,38,39,40,41], which, indicate that Pd plays the dominant role in the transport mechanism of Al-Pd-Mn quasicrystals. These studies have shown that the σ-T behavior can be well explained by strong spin-orbit scattering in the presence of Quantum Interference Effects (QIE's) giving rise to the weak anti-localization effect. As will be discussed in detail in following sections, these interpretations do not appears to be appropriate.

However, because of the presence of magnetic moments of Mn atoms, Al-Pd-Mn quasicrystals are magnetic in nature and the role of magnetic Mn cannot be neglected while explaining its transport properties. It may be the most dominant factor. K. Saito [27] *et al.* and S. Matsua [41] have shown that the ρ(T) data gives a poor fit to the weak localization theory including only spin-orbit scattering. But the σ (T) data can be well accounted, below the σ-T minima temperature only, by weak localization theory, if spin scattering along with spin-orbit scattering is considered as the dephasing mechanism. Moreover, there exist few other experimental studies [22,24,29,40,42,46,47] that indicate a one-to-one correspondence between the temperature T_m of the ρ(T) maximum and the concentration of the magnetic Mn moments, suggesting that ρ(T) maximum could be purely a magnetic effect.

The occurrence of ρ(T) minimum has been attributed to the magnetic scattering, i.e., Kondo-effect, since below ~10K, ρ(T) rises very steeply following a lnT dependence [24,40,,48]. Occurrence of negative magneto-resistance [40] at 4.2K also indicates the presence of Kondo effect. C. R. Wang *et al.* [24] have shown that in the ρ(T) curve, T_{min}, which corresponds to the temperature of ρ(T) minimum, increases with increasing Mn concentration. This indicates that the increase in the ρ(T) or the occurrence of minimum at low temperatures, originates from Kondo effect. The signature of the Kondo like effect at 4.2K is observed by tunneling spectroscopy also [49].

Despite these studies, it is still not very clear as to what is the actual mechanism that is playing role behind this peculiar electrical transport behavior in Al-Pd-Mn quasicrystals. Is it so that maxima and minima have separate origins or both have a common reason of its occurrence? In the present review we have tried to establish a single mechanism, which could explain the σ(T) behavior in the entire studied temperature range.

2. Synthesis and Characterization Details

All the three different alloy series of Al-Pd-Mn quasicrystals i.e. $Al_{70}Pd_{20-x}Mn_{10+x}$ (or x= 0, -1 and-2), $Al_{70}Pd_{20}Mn_{10-x}(TM)_x$ (x=0 and 2 for TM = Fe, Cr, Co and Ni) and $Al_{70-x}B_xPd_{20}Mn_{10}$ (for x=0.5, 1, 2 and 4) were prepared using RF-induction melting. The melting was done under argon atmosphere in boron nitride crucible kept in a quartz jacket. Melting losses during preparation are found to be less than 0.5%. The ingots of as melted alloys were annealed at 800°C in vacuum (with partial Ar- atmosphere). For this the ingots were wrapped in Mo foil and sealed in a quartz ampoule containing partial pressure of argon and kept for annealing at 800°C for 120 hours. The annealed ingots were cut into slices and subjected to structural characterization using powder x-ray diffraction (XRD) and back-scattered electron imaging using SEM. Compositional analysis of the alloys were carried out during SEM using EDAX.

The σ-T measurements (1.4-300K) at zero and 8-Tesla and the magneto-resistance (Δρ(B)/ρ) measurements (0-8 Tesla) at 1.4K, 3K, 6K, and 20K were performed using conventional D.C. four-probe method. Magnetic measurements, M-H (0-4Tesla) and χ-T (5-300K) at 50-Oe and 1000-Oe magnetic field were carried out on all the studied samples using SQUID magnetometer. In the following we will describe the results obtained on three different types of Al-Pd-Mn quasicrystals in Part I , Part II and Part III.

3. Part I

This part of the paper deals with the studies related to the synthesis and structural characterization and magnetic and electrical transport properties measurements of Al-Pd-Mn icosahedral quasicrystalline alloys, in which we have exchanged the relative concentrations of Pd and Mn in $Al_{70}Pd_{20}Mn_{10}$ quasicrystalline alloy.

3.1. Results and Discussions

3.1.1. Structural Characterization

Figure 2 shows the typical XRD patterns of (a) unannealed $Al_{70}Pd_{20}Mn_{10}$ and (b) annealed $Al_{70}Pd_{20-x}Mn_{10+x}$ quasicrystalline samples. The diffraction peaks of all the samples were indexed based on Elser's 6-index system [50]. Looking into the indexed pattern it is found that all the samples are single-phase. The absence of Z- contrast in the back-scattered electron imaging in SEM, as shown in Figure 3, reveals that the samples are single phase. The EDAX results have been summarized in Table-1. It clearly exhibits that the compositions of the alloys are very close to the compositions intended to make. The differences are well within the typical error of EDAX analysis.

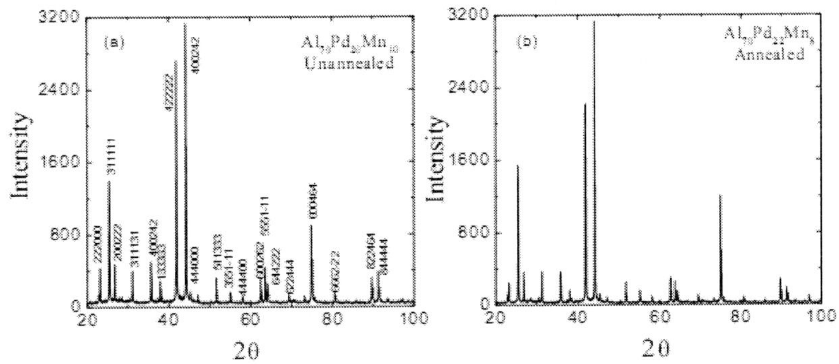

Figure 2. Powder x-ray diffraction patterns of (a) Unannealed Al70Pd20Mn10 and (b) Annealed Al70Pd22Mn8, quasicrystalline samples. These patterns are typical to Al-Pd-Mn quasicrystals. The indices have been given based on Elser's 6-Dimensional indexing scheme.

Figure 3. A typical back-scattered SEM micrograph of unannealed Al70Pd20Mn10 quasicrystalline alloy.

Table 1. Table depicting the composition of Al70Pd20-xMn10+x, samples obtained using EDAX

Observed Composition (At.%)	x = 0 Unannealed	x = 0 Annealed	x = -1 Annealed	x = -2 Annealed
Al	69.8	69.9	69.8	69.6
Pd	19.9	21.0	22.1	22.4
Mn	10.2	9.1	8.1	7.9

3.1.2. Magnetic Characterization

Magnetization measurements were carried out on all the studied samples. Figure 4(a) and (b) depict χ-T and M-H data of the annealed Al70Pd20Mn10 sample. The observed χ-T and M-H behavior is typical of all these samples. The observed zero-field cooled (ZFC) and field-cooled (FC), χ-T done at 50Oe is identical to that of the Al70Pd20Mn10 single crystal data [31]. The observed bifurcation in ZFC and FC χ-T data indicates the presence of interacting moments within the sample. The (χ- χ0)-1-vs- T plot as shown in the inset of Figure 4(a), clearly shows two distinct slopes. In the literature such a features has been attributed to the presence of two types of moments [51,52] in the sample, the one, which are nearest neighbor exchange interacting, and the other, which is paramagnetic type. Following the (χ- χ0)-1 vs-T plot it appears that a group of moments, most probably the interacting ones, starts freezing below 198K forming small clusters [31]

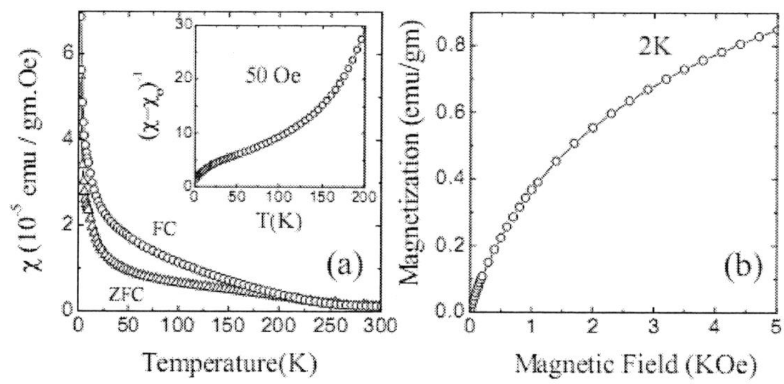

Figure 4. (a) Temperature dependence of susceptibility (χ) (at 50 Oe) for annealed Al70Pd20Mn10 sample. The insets show FC data in the form (χ- χ0)-1 (104gm Oe / emu) vs. T plot. (b) M-H curve at 2K.

The linear feature indicating paramagnetic behavior with weak antiferromagnetic interaction, associated with the single free moments, dominates the χ-T behavior below 90K. The non-saturating trend of the M-H curve also indicates the presence of antiferromagnetic interaction within the sample. Another feature, which is distinctly observed for all the samples, is the down turn of the $(\chi-\chi_0)^{-1}$-vs-T plot below 35K. The paramagnetic regime of the χ-T data, the data above 250K, of the four samples were fitted for Curie-Weiss law, given by equation 3.1

$$\chi = \chi_0 + \frac{C}{T-\theta_c} \qquad (3.1)$$

The refined parameters C and θ_c are tabulated in Table-2. Since the M-H curves shows non-saturating trend even up to the field of 5-Tesla and a temperature of 5K, an estimate of saturation magnetization M_s for each sample, was done by plotting M vs. 1/H curve and then extrapolating the data to infinite field [23]. This value of M_s is used to estimate the spin S ($S=3K_BC/M_sg\mu_B-1$) of the moments, see Table-2. It shows a clear correlation between decreasing magnetization with decreasing total Mn content of the samples. A comparison of the magnetization values of unannealed and annealed $Al_{70}Pd_{20}Mn_{10}$ samples exhibits that after annealing magnetization decreases.

Table 2. Refined values of physical parameters involved in Curie-Weiss law

X	χ_{FC} at 5K (emu /gm. Oe)	χ_0 (emu /gm. Oe)	C (emu.K/ gm.Oe)	θ_c (K)	Spin (S)	P_{eff} (μ_B)	Magnetic Mn (%)
0 (unann)	7.9 E-4	1.4E-6	1.3E-4	238.9	1.97	5.94	0.28
0 (ann)	4.4 E-5	6.5E-7	6.0E-5	198.1	0.37	2.74	1.52
-1 (ann)	3.5E-5	4.5E-7	2.0E-5	197.1	0.32	2.64	0.55
-2 (ann)	1.0E-5	-2.4E-7	1.2E-4	13.7	7.32	16.6	0.02

This observation is in accordance with the results reported by Scheffer *et al.* [53]. Assuming all the Mn to be magnetic, each with $5\mu_B$, Curie constant ($C = N\mu_{eff}^2\mu_B^2/3K_B$) for $Al_{70}Pd_{20}Mn_{10}$ was calculated to be 68×10^{-4}emu.K/gm.Oe, whereas the experimentally measured value for annealed $Al_{70}Pd_{20}Mn_{10}$ was found to be only 6×10^{-5}emu.K/gm.Oe. Comparing the values of spin and the Curie constants, as found from the experimental data, with that of the theoretically expected ones, it is found that only 1.52% of the Mn sites are magnetic in the case of annealed $Al_{70}Pd_{20}Mn_{10}$. Magnetic Mn% of all other samples are given in Table-2. This estimate is in accordance with the spin-polarized band-structure calculation [37], other bulk magnetization measurements [28,36] and also the results coming from microscopic magnetic-probes, like neutron scattering [32]. Keeping in view the divergence of the FC χ-T data with respect to ZFC data at 244K and the presence of very dilute magnetic moments, it appears that the free moments are existing in the form of clusters (quasicrystals do contain icosahedral clusters of Mn atoms) of just few, two or three moments [28]. On lowering the temperature the moments within the cluster, get ordered with an effective finite moment. Looking at the enhanced bifurcation between the ZFC and FC χ-T-data taken using a field of just 50 Oe, it appears that inter-cluster interaction is very weak and can be taken as nearly paramagnetic [31]. The occurrence of such a bifurcation in the ZFC and FC χ-T data is a typical character of cluster-glass type magnetic structure [31,54].

3.1.3. Conductivity Vs. Temperature (σ-T)

Figure 5 depicts the σ-T curves for unannealed $Al_{70}Pd_{20}Mn_{10}$ and annealed $Al_{70}Pd_{20}Mn_{10}$, $Al_{70}Pd_{21}Mn_9$ and $Al_{70}Pd_{22}Mn_8$ quasicrystalline samples respectively.

The σ-T variations of all the studied samples are qualitatively the same. Each shows a pair of minima and maxima. This result is consistent with other reports [22,24,27,40]. In order to elucidate the possible origin of the observed transport behavior, the σ-T measurements of all the samples were also carried out in the presence of magnetic field of 8-Tesla. It can be seen that corresponding to each composition, the σ-T maxima reduces and shifts to higher temperature on application of external magnetic field, see corresponding insets of Figure 5. The reduction in peak value basically means the occurrence of positive magneto-resistance. A summary of all these results is presented in Table-3, together with the corresponding magnetization values. Table-3 depicts that ρ_{RT} increases systematically with decreasing total magnetization. Another definite correlation is found in the temperature (T_{max}) of σ-T maxima and the magnetization, where, T_{max} increases with increasing total magnetization.

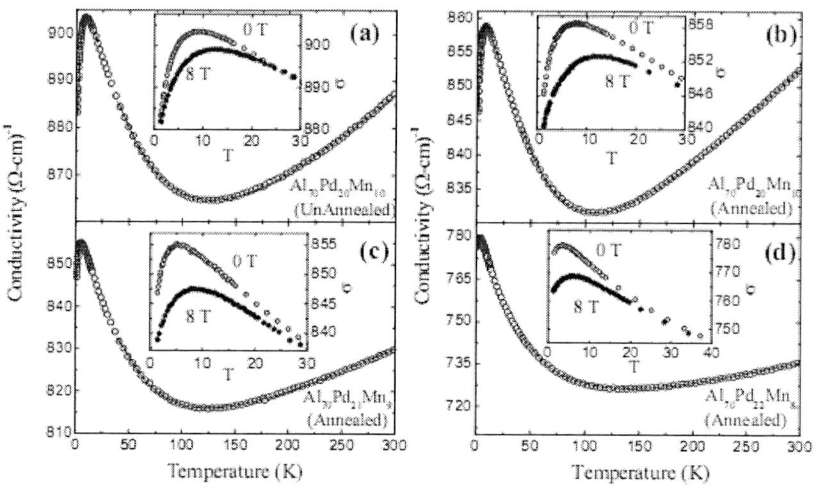

Figure 5. (a)-(d) σ -T curves for unannealed Al70Pd20Mn10 and annealed Al70Pd20-xMn10+x, x = 0, -1 and -2 quasicrystalline alloys. The circles are the observed data points and the lines are the fit to the data. The insets highlight the low temperature σ -T behavior in the zero-field (open circles) and in the 8-Tesla field (solid circles). High temperature shift of σ-T maxima on application of field can be seen.

Table 3. Table showing the Pre and Post minima slopes and shift in Tmax on application of 8-Tesla field for Al70Pd20-xMn10+x

X	χ at 5K (emu /gm. Oe)	ρ_{RT} (μΩ-cm)	T_{min} (K)	T_{max} (K)		Pre min. slope (Ω-cm. K)$^{-1}$	Post min. slope (Ω-cm.K)$^{-1}$
				zero field	in field		
0 (unann)	7.9E-4	1126	126.5	9.10	12.91	0.186	-0.63
0 (ann)	4.4E-5	1172	110.6	8.21	12.04	0.148	-0.49
-1 (ann)	3.5E-5	1205	124.2	5.82	9.40	0.098	-0.63
-2 (ann)	1.0E-5	1359	138.9	3.55	6.44	0.082	-1.00

3.1.3.1. σ -T Minimum

The occurrence of conductivity minima is in accordance with the other observations [22,24,27,40] where it has been attributed to anti weak localization effect appearing due to strong spin-orbit scattering of conduction electrons by Pd atoms. However, following points show that although the observed σ-T variation is due to weak-localization but the minima may not be due to spin-orbit scattering. These points also clarify the possible scattering mechanism.

(1) First of all we will qualitatively compare the results of resistivity measurements on non-magnetic $Al_{70}Pd_{20}Ru_{10}$ quasicrystal [55]. Despite its structural similarity with the Al-Pd-Mn based quasicrystals, the results of electrical transport are completely different. Although the Al-Pd-Ru quasicrystal has nearly same value of room temperature resistivity, the same value of σ-T slope and has similar concentration of Pd, presence of which has been attributed to cause strong spin-orbit scattering in the case of Al-Pd-Mn, but no σ-T minima is observed in this case. This observation indicates that the origin of σ-T minima even in the case of Al-Pd-Mn may lie somewhere else.

(2) Secondly, according to weak-localization including spin-orbit scattering, the temperature dependent part of the conductivity Δσ(T) is given by [56]

$$\Delta\sigma(T)=\frac{e^2}{4\pi^2\hbar}\left(\frac{1}{\sqrt{D}}\right)\left\{3\sqrt{\frac{1}{\tau_i}+\frac{A}{\tau_{so}}}-3\sqrt{\frac{A}{\tau_{so}}}-\sqrt{\frac{1}{\tau_i}}\right\}$$

where, τ_i and τ_{so} are the electron-phonon (e-ph) inelastic scattering and spin-orbit scattering times respectively. At high temperatures, i.e. when $\tau_i \ll \tau_{so}$ and in dirty metallic limit, $\tau_i = C/T^2$, for e-ph interaction

$$\Delta\sigma(T)=\frac{e^2}{2\pi^2\hbar}\left(\frac{T}{\sqrt{DC}}\right),$$

and the slope will be given by

$$\frac{d(\Delta\sigma(T))}{dT}=\frac{e^2}{2\pi^2\hbar}\left(\frac{1}{\sqrt{DC}}\right) \tag{3.2}$$

Now, as temperature decreases τ_i increases as C/T^2 and a minima in σ-T occurs. Below the minima temperature, i.e. when $\tau_i >> \tau_{so}$

$$\Delta\sigma(T)=-\frac{e^2}{4\pi^2\hbar}\left(\frac{T}{\sqrt{DC}}\right)$$

and the slope is given by

$$\frac{d(\Delta\sigma(T))}{dT}=-\frac{e^2}{4\pi^2\hbar}\left(\frac{1}{\sqrt{DC}}\right) \tag{3.3}$$

From the equations 3.2 and 3.3 it is clear that σ(T) slope below minima will be just half and opposite in sign to the σ(T) slope above minima.

But, the experimentally observed σ-T curves, either being described in the present communication or shown earlier in other reports [27,40], exhibit that the average post-minima σ-T slope (slope at temperatures below T_{min} but above T_{max}) is in general more than thrice the corresponding pre-minima slope (slope at temperatures above T_{min}), see Table-3. This observation, even qualitatively, is not commensurate with the expectations from weak anti-localization theory (due to spin-orbit scattering) [56]. The experimentally observed pre-minima slopes (between 240K-300K) are almost linear; if at all curved, it has slight positive curvature. This means that in the present case, from 240K to 300K, weak-localization is being dephased by inelastic scattering which has a temperature dependence given by $\tau_i = C.T^{-p}$, with $p \cong 2$ (e-ph interaction in dirty-metallic limit).

This argument very clearly indicates that in the present case σ-T minima might be originating due to dephasing of weak-localization caused by an altogether different type of scattering mechanism and not due to spin-orbit scattering.

(3) As is obvious from magnetic measurements, all Al-Pd-Mn samples are magnetic and hence a possible origin may lie in its magnetic properties as well. Magnetic moments are known to cause spin-flip scattering of conduction electrons [57]. If we first incorporate a temperature independent magnetic (spin-flip) scattering due to random magnetic fields of paramagnetic Mn-ions, together with spin-orbit scattering, then the most general expression for the temperature dependent part of the conductivity $\Delta\sigma(T)$ will be given by expression 3.4

$$\Delta\sigma(T) = \frac{e^2}{4\pi^2\hbar}\left(\frac{1}{\sqrt{D}}\right)\left\{3\sqrt{\frac{1}{\tau_i}+\frac{2}{\tau_s}+\frac{A}{\tau_{so}}} - 3\sqrt{\frac{A}{\tau_{so}}} - \sqrt{\frac{1}{\tau_i}+\frac{2}{\tau_s}}\right\} \qquad (3.4)$$

Here if τ_s is being considered to be temperature independent then as temperature decreases, τ_i increases and at a particular temperature say T_1, $\tau_i > (\tau_s/2)$ i.e. below this temperature the random dephasing created by (e-ph) inelastic scattering becomes smaller than that of the spin-flip scattering. Since spin-flip scattering time τ_s has been considered to be temperature independent, the total random dephasing becomes temperature independent. The temperature dependence of τ_i becomes insignificant. A constant rate of random dephasing, given by $2/\tau_s$ prevails below that temperature and hence the expression 3.4 can be rewritten as

$$\Delta\sigma = \frac{e^2}{4\pi^2\hbar}\left(\frac{1}{\sqrt{D}}\right)\left\{3\sqrt{\frac{2}{\tau_s}+\frac{A}{\tau_{so}}} - 3\sqrt{\frac{A}{\tau_{so}}} - \sqrt{\frac{2}{\tau_s}}\right\} \qquad (3.5)$$

It can be seen that none of the terms are temperature dependent. Thus, whenever such a situation occurs, $\Delta\sigma$ will become temperature independent i.e. flat with respect to temperature. Two types of situations may arise, (i) when T_1 occurs above the $\Delta\sigma$(T) minima (i.e. $\tau_s < \tau_{so}$) and (ii) when T_1 occurs below $\Delta\sigma$(T) minima (i.e. $\tau_s > \tau_{so}$). These situations are schematically shown in the following Figure 6. In situation (i) no minima is observed at all and the flattening starts above minima (the flat dashed line above minima) whereas in situation (ii) flattening starts after minima (the flat dashed line shown below minima), as shown in Figure 6. The solid line shows the σ-T variation in the situation of complete absence of spin-flip scattering. But experimentally, neither any type of flattening effect is observed nor the observed post minima slope is half of the pre-minima slope as shown by the solid line in Figure 6. This indicates that the experimentally observed minimum is not at all due to spin-orbit scattering and if spin-flip scattering is present, it is not temperature independent. It could have been simple spin-disorder effect but scattering rate due to simple spin-disorder effect decreases with decreasing temperature. Thus, these arguments suggest that spin-flip scattering rate should be increasing with decreasing temperature and probable such effect is the Kondo-type scattering. This concludes that the above assumption of temperature independent magnetic scattering is not adequate.

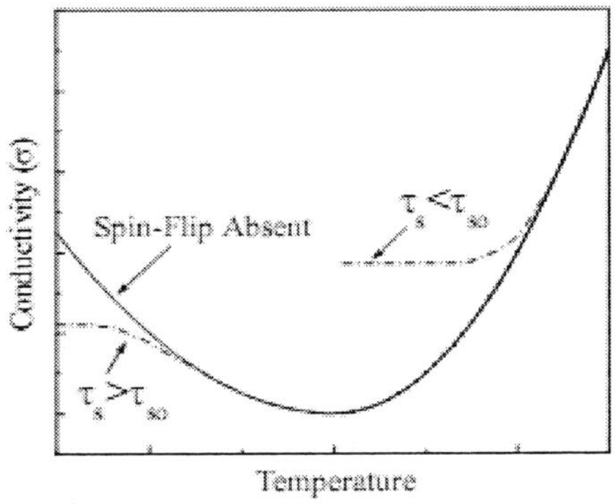

Figure 6. Temperature variation of conductivity for different conditions of τs and τso.

(4) From the steep rise of post-minima σ-T variation, it is obvious that it is being dominated by a random dephasing mechanism, which is strongly temperature dependent. The dephasing rate increases continuously with decreasing temperature. Presumably this may be due to Kondo-like spin-flip scattering. It should be noted that only recently Prejean *et al.*, 2002 [46] and Dolinsek et.al, 2002 [45] have also argued that the σ-T minima (or ρ-T maxima) is probably due to magnetic scattering and not due to spin-orbit scattering effect. The polarization of conduction electrons about the magnetic Mn has been realized in NMR experiments and the magnetic properties are attributed to the Kondo-effect [30,29]. The screening of moments is well known [57] in the Kondo-effect. Thus, from the above-described four points, we can at least qualitatively say that the occurrence of σ-T minima cannot be attributed to weak anti-localization effect appearing due to strong spin-orbit scattering of conduction electrons by Pd atoms, but it may be due to the presence of Kondo-type magnetic scattering.

3.1.3.2. σ -T Maximum

The observed σ-T maximum is in accordance with the other reports [24,40]. Occurrence of such a σ-T maximum has been reported for single grained quasicrystals as well [42]. Akiyama *et al.*, [40] and Wang *et al.* [24], based mainly on the occurrence of linear behavior with negative slope in the ρ-ln(T) plot , have attributed the occurrence of ρ-T minima (or σ-T maxima) to the presence of Kondo scattering. Conventionally the occurrence of linear region with negative slope in ρ-ln(T) plot does indicate the presence of Kondo scattering but our in-field σ-T data completely negate this. Had the σ-T maxima been appeared due to conventional type of Kondo effect, it should have shifted to lower temperature with a negative magneto-resistance on application of field. But the present observation of in-field σ-T is just opposite. It shifts to high temperatures with a positive magneto-resistance.

3.1.3.3. Possible Origin of Observed σ -T Behavior

The presently synthesized Al-Pd-Mn quasicrystals whose room-temperature resistivities (see Table-3) are much above the Mooij's criteria and have resistivities ratios R ($\rho(T_{max})/\rho(300)$) ranging from 1.1-1.3, are far from the proximity of MI- transition [58,59] and lie well within the bad metallic regime, where σ-T variation is simply accounted by quantum interference effects, like weak-localization, even up to high temperatures. In fact the presence of weak-localization effect in these alloys is already experimentally well established, even at temperatures as high as 200K [60]. In the weak-localization regime the Kondo-type scattering would show its presence in a rather unconventional way. In this case the spin-flip scattering will basically cause random dephasing of electron waves moving on time reversed paths and hence decrease the effective back-scattering of electrons. This effect, below a certain temperature (below which $\tau_i > \tau_{sf}$), causes increase in conductivity rather than increase of resistivity, which is usually observed in low resistivity amorphous / crystalline magnetic alloys in the presence of Kondo scattering. When external field is applied the spin-flip rate is suppressed, hence dephasing is suppressed, and resistance increases. This gives rise to a positive magneto-resistance proportional to $+H^2$ [61,62], as is observed in the present case at relatively high temperatures. On the other hand, in conventional Kondo alloys (where weak-localization effect is absent), resistance increases as a result of direct consequence of spin-flip scattering and $-H^2$ dependent magneto-resistance is observed. Details of magneto-resistance are discussed in the section 3.3.4.

Keeping in view the above discussed facts regarding the occurrence of σ-T maximum and minimum along with the observed facts like, (a) the occurrence of cluster like (group of interacting moments) magnetic behavior in this alloy below 240K and (b) the occurrence of correlation between T_{max} and the total magnetization of the samples (see Table-3) as observed for concentrated Kondo systems like Ag-Mn and Au-Mn [63,64], it appears that the observed σ-T behavior for Al-Pd-Mn quasicrystalline alloys may be a case of Kondo scattering of conduction electrons by a system of interacting moments rather than from a single localized moment. Monod *et al.* [65] and Mothe *et al.* [66] have shown that interacting (RKKY-type interaction) magnetic impurities with arbitrary net spin value, will show a maxima of scattering rate at Kondo temperature T_K and hence a maxima in ρ-T variation. This T_K shifts to lower temperatures, with increase in coupling strength between the moments [66,67]. Such maxima have been observed in Ag-Mn and Au-Mn alloys [63,64]. In the case of direct consequence of Kondo scattering on electronic transport, i.e. when $\rho \propto \tau^{-1}_{sf}$, as in the case of conventional Kondo systems, the occurrence of maxima in Kondo-scattering rate τ^{-1}_{sf}, will cause a ρ-T maxima, whereas, in the case of weak-localized systems, the spin-flip scattering will cause random dephasing of electron waves and hence the maxima in Kondo-scattering rate τ^{-1}_{sf} will produce a ρ-T minima (or σ-T maxima).

In the light of the above discussion, it appears that, in the present case the total dephasing rate is dominated by Kondo like spin-flip scattering [68,69,70] only and is given by equation 3.6.

$$\tau^{-1}_{l} = \tau^{-1}_{i} + 2.\tau^{-1}_{sf} \qquad (3.6)$$

where, τ_i is inelastic scattering time for electron-phonon scattering ($\tau_i = C.T^{-2}$) and τ_{sf} is the spin-flip scattering time. The spin-flip scattering rate τ_{sf}^{-1} is taken to be

$$\tau_{Sf}^{-1} = R_{sf} \frac{\pi^2 S(S+1)}{\pi^2 S(S+1) + \ell n^2 (T/T_K)} \tag{3.7}$$

where, R_{sf} is given as $c/(2\pi N(E_F))$ equation 3.7 of spin-flip rate has been successfully used for describing the electrical transport in weak-localized systems with Kondo impurities [70,71]. Here T_K is the Kondo-temperature, S is the spin of the flipping moment, c is the concentration of magnetic scatterers, and $N(E_F)$ is the density of states at the Fermi-level. It should be noted that equation 3.7 gives a maximum in scattering rate at $T=T_K$. This maxima is equal to $R_{sf.}$ It has been experimentally found that equation 3.7 is applicable at temperatures above as well as below T_K [70,71]. Including the scattering rate as given in equation 3.6 the conductivity at any temperature between 1.4-300K will be given by $\sigma=\sigma_0 + \Delta\sigma$ (T) [68,69,70,71] where

$$\Delta\sigma(T) = \frac{e^2}{2\pi^2\hbar}\left[\sqrt{\frac{1}{D\tau_i} + \frac{2}{D\tau_{sf}}}\right] \tag{3.8}$$

Here, D and temperature dependence of τ_i remain same between 1.4-300K but the temperature dependence of τ_{sf} may change. As temperature decreases the τ^{-1}_i decreases but τ^{-1}_{sf} increases. The temperature at which $\tau^{-1}_i < 2\tau^{-1}_{sf}$, commencement of σ-T minima takes place. From the above equation 3.8, in conditions of either $T>>T_K$, $T<<T_K$ or the effective spin $S<<1$, it can be shown that $\Delta\sigma(T) \propto \pm \ln^{-1}(T)$ at temperatures $T>>T_K$, and $\Delta\sigma(T) \propto \pm \ln(T)$ at temperatures $T<<T_K$, see Annexure-I. The occurrence of this behavior is vividly clear in the σ-vs-$\ln^{-1}$(T) and the σ vs. ln(T) plots for unannealed $Al_{70}Pd_{20}Mn_{10}$ alloy, shown in Figure 7. This behavior is same for all the quasicrystalline alloys studied in the present communication.

We have fitted the observed σ-T data based on equation 3.7 using least square refinement of the parameters. The fits to the data are shown as continuous curves in Figure 5. The fitting parameters are σ_0, D, τ_i, Rsf, S and T_K. As is obvious from equation 3.8, D and τ_x (x= i, sf) are multiplied to each other, hence during refinement, both will get refined on cost of each other and hence the product $D.\tau_x$ will remain uniquely defined for a given sample. Since in our case, D is not an independently measured parameter, we have summarized in Table-4, the scattering parameters in terms of $(1/D.\tau_x)$, which is expressed in units of (h/4e). cm^{-2}. This directly gives the scattering–field $B_x = h/4eD\tau_x$ [70,71].

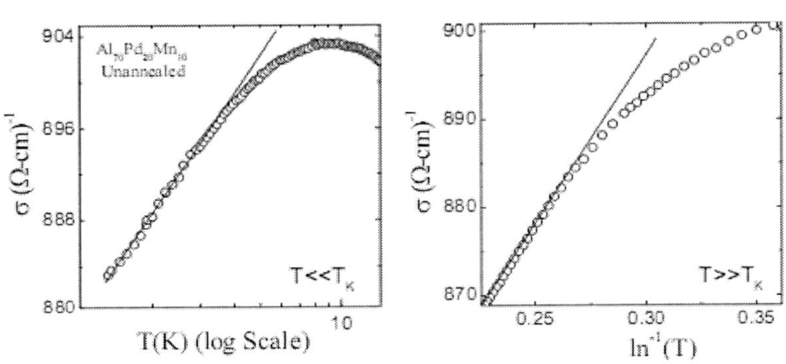

Figure 7. Figure depicting $\Delta\sigma(T) \propto \pm$ ln-1(T) at temperatures T>>TK, and $\Delta\sigma(T) \propto \pm \ln(T)$ at temperatures T<<TK. Straight line shows the presence of linear regions.

In Table-4 we have presented σ_0, Bi, B_{sf}(max), S and T_K. B_{sf} (max) is defined, as the maximum scattering field and is given as ($h.R_{sf}/4eD$). The fitted parameters are found to be physical. The first best fitted temperature range i.e. from 40K to 290K, is more or less same for all the compositions. The quality of fits appears to justify the validity of the proposed model.

Table 4. Refined values of physical parameters involved in Weak-localization including spin-flip scattering for Al70Pd20-xMn10+x

X	σ_0 $(\Omega\text{-cm})^{-1}$	$B_i\,xT^2$ ($h/8\pi e$) (cm^{-2})	High temperature range (40K-300K)			Low temperature range (1.4K-20K)		
			S	T_K (K)	B_{sf}(max) ($h/8\pi e$) (cm^{-2})	S	T_K (K)	B_{sf}(max) ($h/8\pi e$) (cm^{-2})
0 (unan)	815	2.88E8	0.095	28.25	1.61E13	0.38	9.09	2.32E13
0 (ann)	803	1.49E8	0.057	32.79	5.07E12	0.34	7.78	9.41E12
-1 (ann)	791	8.91E7	0.053	31.74	5.60 E12	0.41	5.26	1.23E13
-2 (ann)	707	4.67E7	0.031	43.37	3.33E12	0.52	3.31	1.56E13

Figure 8 presents the temperature-dependence of spin-flip scattering-field $B_{sf,}$ which has been calculated, using parameters R_{sf}, D, S and T_K, obtained from refinement of the σ-T data. It depicts that at each temperature the spin-flip scattering-field is nearly proportional to the magnetization of the samples, see inset of Figure 8.

The refined parameters do not fit to the data below 40K. According to the refined values of T_K there should have been maxima at around 30K, but the experimentally observed maxima occur at lower temperatures, i.e. effective T_K has shifted to lower value. This shift may be attributed to the changing magnetic state of the sample as indicated by the feature in $(\chi\text{-}\ \chi_0)^{-1}$-vs-T plot below 35K. Such a feature may appear due to emergence of a ferromagnetic interaction below 35K, which would push T_K to lower temperatures [66,67]. Due to change of the magnetic state, the parameters R_{sf}, S and T_K are expected to change and hence the low temperature region of the σ-T data were fitted with the same equation 3.8, but keeping the parameters σ_0, D, and τ_i to its values, which were obtained by the refinement of the σ-T data in high temperature range. The parameters R_{sf}, S and T_K were let free to get refined to new values. The fitting was good only in the temperature range from 1.5K-20K. The fitting of σ-T data in the range 1.5K-20K can be seen in Figure 9. As is obvious from Figure 9, these fits very well account for the observed σ-T maxima and hence we attribute them to Kondo-maxima.

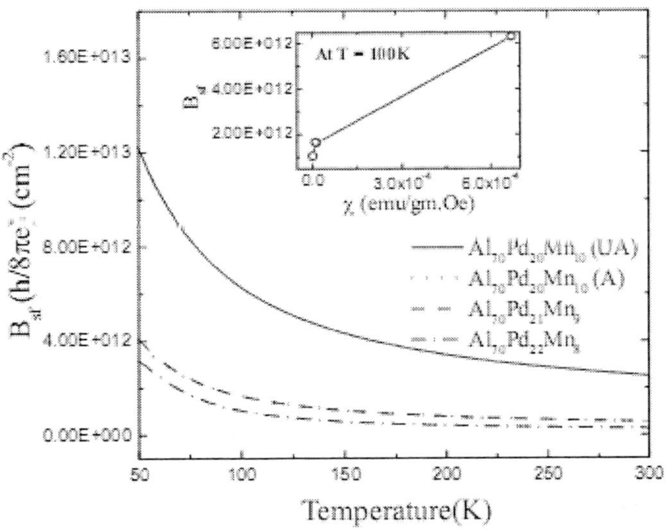

Figure 8. Temperature dependence of spin-flip scattering field for four quasicrystalline samples. These have been derived after refinement of the σ-T data. The curves corresponding to annealed Al70Pd20Mn10 and annealed Al70Pd21Mn9 are almost matching. Inset shows an increasing trend of spin-flip scattering field with the total magnetization of the sample.

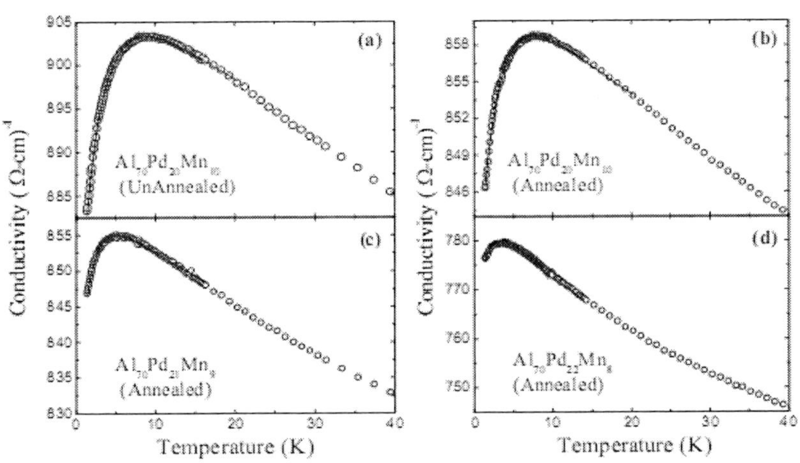

Figure 9. σ-T curves in the temperature range of 1.4-40K showing the occurrence of σ-T maxima for (a) unannealed Al70Pd20Mn10, and annealed (b) Al70Pd20Mn10, (c) Al70Pd21Mn9, and (d) Al70Pd22Mn8 quasicrystalline samples. This also highlights the quality of the fits to the data.

The temperature range from 20-40K may correspond to a region of changing magnetic state, where perhaps no unique value of parameters R_{sf}, S and T_K are defined. It should be noted that the temperature range of 20-40K is coincident with the features below 35K in the $(\chi-\chi_0)^{-1}$-vs- T plot. The values of R_{sf}, S and T_K obtained from the refinement of low-temperature data are also given in Table-4. It can be seen that at low-temperatures, values of S corresponding to each composition has increased and the Kondo temperature T_K has decreased. The increase in the spin value (see Table-4) effectively means the emergence of ferromagnetic interaction between the moments. Increase in interaction between the moments will make the spin-fluctuations rather slow and hence spin-flip scattering will decrease. Effectively this means that the T_K would decrease [66,67]. That is what has actually been observed. We have also estimated the value of $N(E_f)$ using the relation $R_{sf}=c/2\pi hN(E_f)$. The $N(E_f)$ for alloys with different Mn concentration has been found to be in between 0.5 to 1 states/atom. eV. This value matches quite satisfactorily with the value obtained from band structure calculation of higher approximants of Al-Pd-Mn quasicrystals [8].

It can be noted that the value of spin obtained from the refinement of σ-T data is much lower as compared to the value of spin obtained from the magnetization measurements. This may be due to the fact that the spin value obtained from σ-T data, is the net spin of a cluster as seen by the conduction electrons during its spin-flip scattering from the cluster in zero-field. Since the zero-field order of moments within a cluster appears to be a result of frustrated interaction, as reflected by the non-saturating type M-H curve, see Figure 4(b), the net spin will be the result of nearly random arrangement of moments hence low net spin value would result. The measured spin is the spin value of almost one single moment because it has been obtained using saturation magnetization value measured after applying high fields, at which inter moment interactions get suppressed and only single moment character remains.

The σ-T maxima invariably shift towards high temperatures showing positive magneto-resistance, on application of sufficiently high magnetic field, see insets of Figure 5. This is so because σ-T maxima occur due to decrease in spin-flip rate at T_K. At high-fields the spin-flip rate gets further suppressed and hence shifts the σ-T maxima at temperatures above T_K. Thus the observed maxima are appearing basically due to decrease in dephasing rate below T_K. It is not a direct but indirect consequence of Kondo-scattering occurring in weak-localized system. Low- field variation of σ-T is expected to be different than the high-field ones and is discussed in the following section-3.1.4.

3.1.4. Magneto-Resistance

The magneto-resistance ($\Delta\rho(B)/\rho$) data for all the samples is presented in Figure 10. Magneto-resistance at 20K and 6K are positive for all the samples and it follows a nearly H^2 behavior as expected [61,62,68,69,72,73] from weak-localization in high temperature limit. At high temperatures, with increase in the field, spin-flip is suppressed and hence in the present case of weakly localized system, dephasing is suppressed. Decrease in dephasing rate causes increase of resistance giving rise to a positive magneto-resistance. It can be noticed that for unannealed $Al_{70}Pd_{20}Mn_{10}$ sample, magneto-resistance increases while cooling from 20K to 6K. But on further cooling to 3K, magneto-resistance decreases and finally becomes negative at lower temperatures. It remains negative for relatively lower magnetic fields, shows a shallow minima around 2.5-Tesla and finally becomes positive at around 5-Tesla. Magneto-resistance at 1.4K invariably shows a negative component for all the studied samples except for $Al_{70}Pd_{22}Mn_8$ alloy. The crossover from negative to positive is also common to all of them, which show negative magneto-resistance. A comparison of magneto-resistance data of all the samples reveals that magneto-resistance has a definite correlation with the sample magnetization.

The negative component increases with increasing magnetization, see Figure 11(a). Largest negative magneto-resistance has been observed for the unannealed $Al_{70}Pd_{20}Mn_{10}$ sample, which has the largest magnetization also. In another sense when chance of correlation between moments is high, negative component of magneto-resistance is also high. But at a fixed field it decreases with increasing temperature, see Figure 11(b). At 1.4K the largest positive magneto-resistance is observed for the sample with the lowest magnetization; see

Figure 11(c). As stated above, at high-temperature the magneto-resistance is invariably positive. In both conditions, when magnetization (which is proportional to moment concentration) is low enough or temperature is high enough, the correlation will be weak and hence positive magneto-resistance contribution will dominate.

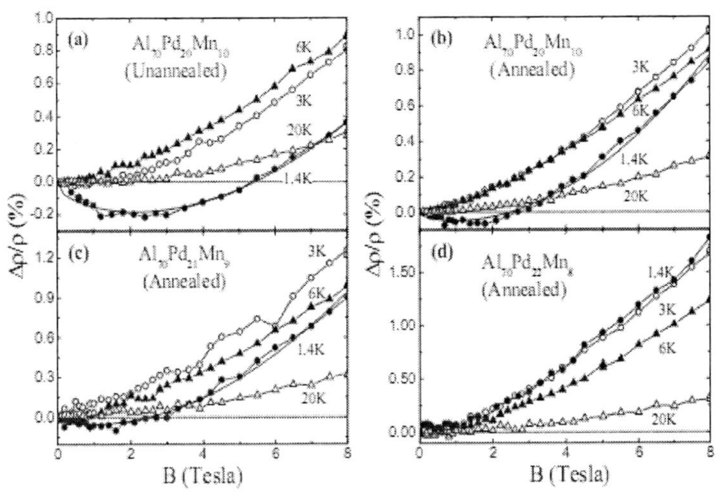

Figure 10. Magneto-resistance curves of (a) unannealed Al70Pd20Mn10, and annealed (b) Al70Pd20Mn10, (c) Al70Pd21Mn9, and (d) Al70Pd22Mn8 quasicrystalline samples. Continuous curves with 1.4K data are fits to the data based on eqn.3.9, please see text in section 3.4.

Thus, at any temperature magneto-resistance will have two components one positive, with $+H^2$ dependence and the other negative, with $-H^p$ dependence and hence the total magneto-resistance can be expressed as follows

$$\frac{\Delta\rho(B)}{\rho} = C_1 H^2 - C_2 H^p \qquad (3.9)$$

$-H^p$ dependence has been guessed from the nature of observed curvature of the negative part of the magneto-resistance curves. Here C_1 and C_2 are constants denoting the fractions of (+ve) and (-ve) magneto resistance contributions respectively. The magneto-resistance curves with both the components are very well fitted with the equation 3.9 with values of p~0.4 to 0.6. In the following, based on experimentally observed facts, we will try to give a possible explanation of the origin of the observed negative magneto-resistance. Looking into the dominant role of quantum interference effects in these samples, the explanation uniquely demands the presence of interacting moments.

Figure 12 shows σ-T data in the temperature range of 1.4K to 40K for $Al_{70}Pd_{20}Mn_{10}$ sample at zero (O), 2.5 Tesla (□) and 8.0 Tesla (Δ) fields. The in-field data has been taken at 2.5Tesla because it corresponds to the maximum negative magneto-resistance value for $Al_{70}Pd_{20}Mn_{10}$ sample. The three data were refined using equation 3.8, keeping σ_0, D, τ_i fixed, as refined in high temperature range and R_{sf}, S and T_K left free for refinement. The refinement is shown as continuous line in Figure 12. From refinement it is found that effective spin value increases from 0.36 to 0.42 to 0.51 with increase of applied external field from zero to 2.5 Tesla to 8.0 Tesla.

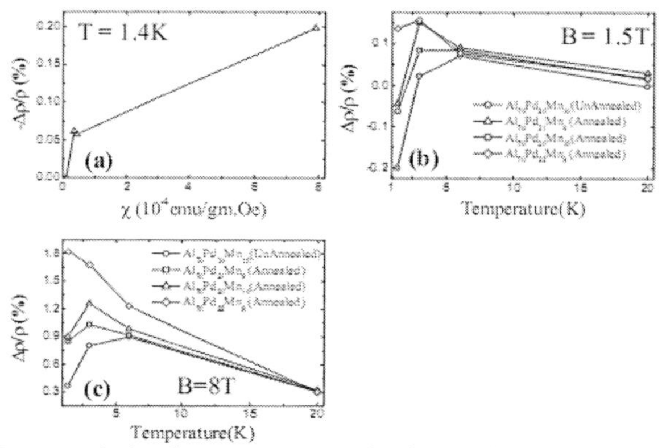

Figure 11. Plots exhibiting correlation between magnetization and (a) negative-magneto-resistance, (b) its temperature dependence (c) Correlation between the magnetization and positive magneto-resistance of samples studied in the present investigation.

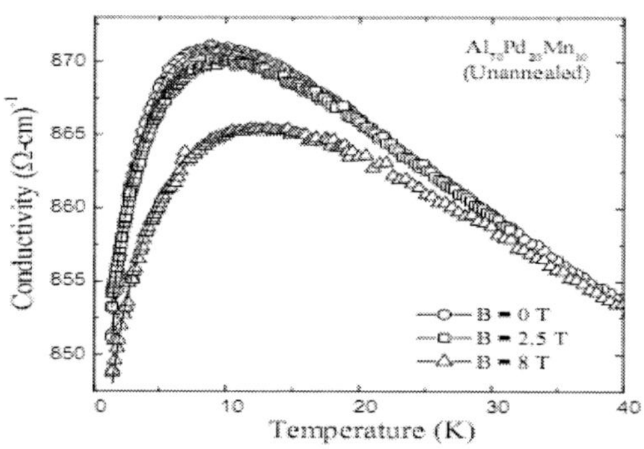

Figure 12. σ-T curve for an unannealed Al70Pd20Mn10 sample at zero, 2.5 Tesla and 8.0 Tesla fields. The lines represent to the fits corresponding to eqn. 3.8.

This is quite expected in the present case of interacting moments where S stands for the net spin of the system of interacting moments (cluster). On application of field the alignment between the moments within the cluster improves and hence the net spin S also improves. Increase in S will try to increase the scattering cross-section, see equation 3.8. But for a fixed spin, increasing field will directly decrease the spin-flip scattering rate by suppressing the spin-fluctuations. Thus, in the presence of external field the net spin-flip rate will be resultant of the two effects. In the present case it so appears that at lower fields indirect increase in spin-flip rate due to rise in S, on application of field, surpasses the direct effect of spin-flip suppression due to the applied field. Thus the net effect of field results in increase of spin-flip rate and decrease of resistance. The rate of increase of net spin S of the cluster with applied field will depend on the intra-cluster coupling strength. Since the magnetization will also be proportional to coupling strength, it will show a correlation with the magnetization and that is what has been experimentally observed, see Figure 11(a). As the field is increased further the net spin S of the cluster will saturate but the direct suppression of spin-fluctuation continues and hence net spin-flip rate starts decreasing with field. As discussed above, this will cause a net positive magneto-resistance. Thus at each field value the net magneto-resistance can be given as in equation 3.9. The observed nature of field dependence of negative magneto-resistance i.e. $\Delta\rho/\rho \propto -H^p$ (with $p < 1$) is yet not clear.

4. Part II

In Part I we have tried to prove that electronic transport in Al-Pd-Mn quasicrystals is fully dominated by quantum interference effects including electron-phonon (e-ph) scattering as the inelastic and Kondo-type spin-flip scattering as the random dephasing processes. The minimum in the σ-T is appearing as a result of these two competing scattering processes. The observed maximum in σ-T is due to the occurrence of maximum in the spin-flip scattering rate, which is expected when spin-flip scattering of electrons is caused by a system of interacting moments in the place of isolated moments. To strengthen these conclusions further in this part we have studied the structural, magnetic and transport properties of Al-Pd-Mn icosahedral quasicrystalline alloys, in which we have replaced 2% of Mn by other transition metals of the same series, like Fe, Co, Cr and Ni. For this purpose we have prepared a series of $Al_{70}Pd_{20}Mn_{10-x}(TM)_x$ samples with x=0 and 2 for TM = Fe, Co, Cr and Ni. The reason for substituting Mn by other transition metals is that by doing so it will be possible to keep the Pd concentration fixed in the sample and hence the contribution of spin-orbit scattering, if at all effective, will remain same. On the other hand if the electronic transport in this system is affected by magnetic properties, then the partial substitution of Mn by other transition metals will cause changes in electronic transport properties. Thus these substitutions will help in clarifying the role of magnetic scattering over spin-orbit scattering.

Table 5. Table depicting the composition of Al70Pd20Mn8TM2 samples obtained using EDAX

Observed Composition (At.%)							
	Al	Pd	Mn	Fe	Cr	Co	Ni
$Al_{70}Pd_{20}Mn_{10}$	69.81	19.94	10.25	-	-	-	-
$Al_{70}Pd_{20}Mn_8Fe_2$	69.32	19.85	8.62	2.21	-	-	-
$Al_{70}Pd_{20}Mn_8Cr_2$	69.97	20.24	7.93	-	1.86	-	-
$Al_{70}Pd_{20}Mn_8Co_2$	70.26	19.76	7.73	-	-	2.25	-
$Al_{70}Pd_{20}Mn_8Ni_2$	68.91	20.42	8.48	-	-	-	2.18

4.1. Results and Discussion

4.1.1. Structural Characterization

Figure 13 shows the XRD patterns of $Al_{70}Pd_{20}Mn_{10-x}(TM)_x$ with x = 0 and 2 for TM = Fe, Cr, Co and Ni, quasicrystalline samples. The diffraction peaks of all the samples were indexed based on Elser's 6-index system [50]. A careful analysis of these XRD patterns, indicates the presence of small amount of Al_3Pd_2 phase in transition metal doped samples. This was found to be rather high with the Cr doped sample. For this case it was estimated to be ~ 2 Wt. %. An arrow marked at ~ 42.84°, corresponds to the diffraction peak of the Al_3Pd_2 phase. Back-scattered electron imaging in SEM, showed the presence of a white looking localized regions, which correspond to Al_3Pd_2, as shown in Figure 14. As is clear from the image that these white spots are quite widely separated, 50-100μm apart and effectively harmless for transport measurements. This is automatically confirmed by the occurrence of a σ-T behavior (Figure 16) exactly similar to the undoped sample, which does not contain any such parasitic phase. Please compare with Figures 3 and 5. The EDAX results have been summarized in Table-5. It clearly reveals that the compositions of the alloys are very close to the initial compositions before melting and the differences are well within the typical error of EDAX analysis

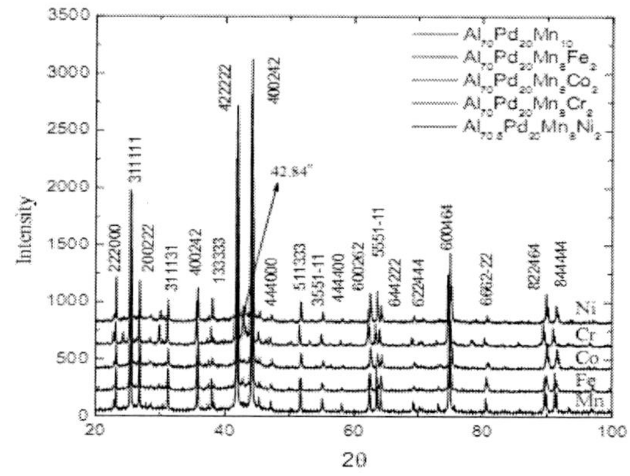

Figure 13. Powder X-ray diffraction patterns of all the as prepared Al70Pd20Mn8TM2 (TM=Fe, Cr, Co and Ni) samples. The indices have been given based on Elser's 6-Dimensional indexing scheme.

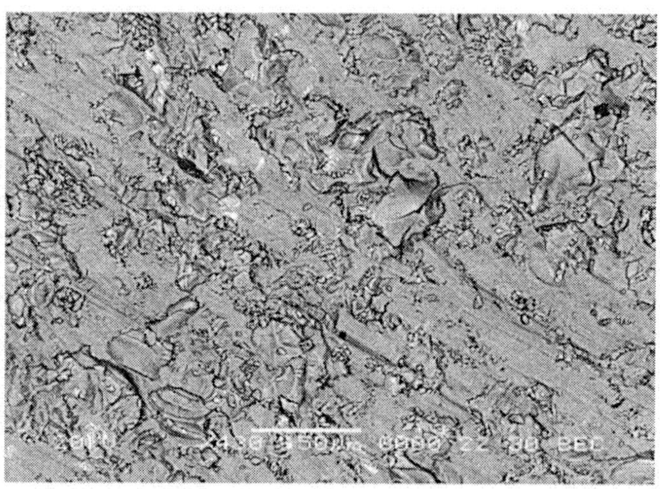

Figure 14. Backscattered SEM micrograph of Al70Pd20Mn8Cr2 quasicrystalline sample. Encircled white regions show the discretely distributed Al3Pd2 phase, which are widely apart.

4.1.2. Magnetic Characterization

Figures 15(a) and (c) represents the results of χ-T carried out at 1000 and 50 Oe, and M-H measurements of $Al_{70}Pd_{20}Mn_8Ni_2$ sample. These features of χ-T and M-H results observed for the $Al_{70}Pd_{20}Mn_8Ni_2$ sample are typical for all the transition metal substituted samples studied here. They only differ in their magnetization values. As can be seen in the inset of Figure 15(a), the χ-T data of zero-field cooled (ZFC) and field-cooled (FC) measurements, carried out at 50 Oe field, bifurcate at about 180K. But no such bifurcation is observed for the ZFC and FC χ-T data of the same sample while the χ-T is carried out at 1000Oe field. The observed features for 1000Oe χ-T data are identical to that of the $Al_{70}Pd_{20}Mn_{10}$ single crystal data [31]. The match between the present data, on polycrystalline samples, to that of the single grained quasicrystals, approves the data as well as the sample quality. The $(\chi - \chi_0)^{-1}$ vs. T plot of χ-T data done at 1000 Oe, shows two distinct slopes, as indicated by the straight lines in the Figure 15(b) and its inset. In literature such a features have been attributed to the presence of two types of moments in the sample [51,52], the one that are nearest neighbor exchange interacting, and the other, which is paramagnetic type. We fitted the χ-T data for all the studied samples, above 217K, i.e. the region corresponding to the first slope as indicated by straight line in Figure 15(b), by Curie-Weiss law given by the equation 3.1.

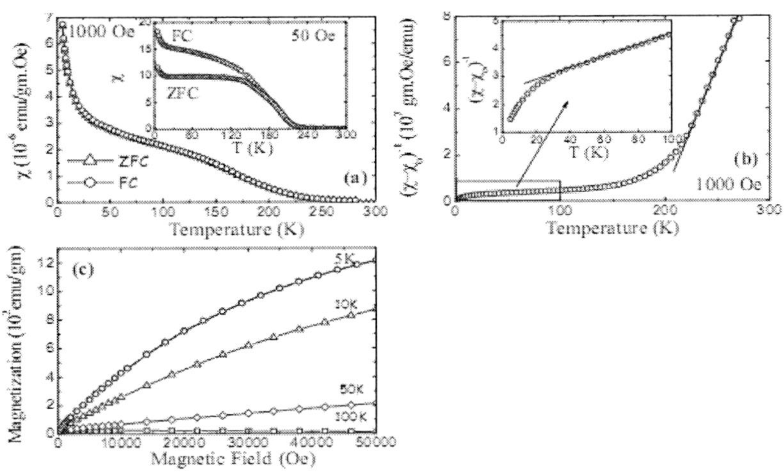

Figure 15. (a) Zero-field cool (ZFC) and field-cool (FC), at 1000 Oe, χ-T data for Al70Pd20Mn8Ni2 sample. The inset shows the ZFC and FC data in the form of χ (107emu/gm.Oe) vs T at 50 Oe. (b) (χ- χ0)-1 vs. T plot 1000Oe χ-T data, insets show the magnified low temperature region of (χ- χ0)-1(105gm.Oe/emu) vs T data. Straight lines indicate the presence of two types of slopes. (c) M-H curves at 5K, 10K, 50K and 100K.

The refined values of parameters C and θ_c, obtained corresponding to all the studied samples are tabulated in Table-6. Positive value of θ_c, for all the studied samples indicates the presence of ferromagnetic interaction between the moments in this range of temperature. From Table-6 it can be noted that the magnetization of the samples decreases as we go through Mn, Fe, Cr, Co, to Ni. Since the M-H curves show non-saturating trend even up to the field of 5-Tesla and a temperature of 5K, the saturation magnetization M_s for each sample was estimated by plotting M vs. 1/H curve, and the extrapolating the data to infinite field [23]. This value of M_s is used to evaluate the spin S ($S=3K_BC/M_sg\mu_B-1$) of the moments, see Table-6. Assuming all the Mn to be magnetic, each with $5\mu_B$, Curie constant ($C= N\mu_{eff}^2 \mu_B^2/3K_B$) for $Al_{70}Pd_{20}Mn_8Ni_2$ was calculated to be $85.5x10^{-4}$emu.K/gm.Oe, where as the experimentally measured value for $Al_{70}Pd_{20}Mn_8Ni_2$ was found to be only $9.6x10^{-6}$emu.K/gm. Oe. Comparing the values of spin and the Curie constant, calculated from the experimental data with that of the theoretically expected ones, it is found that only 0.06% of the all the Mn sites are magnetic in the case of $Al_{70}Pd_{20}Mn_8Ni_2$ sample. The percentages of magnetic moments for other samples are tabulated in Table-6. This estimate is in accordance with the spin-polarized band-structure calculation [37], other bulk magnetization measurements [37,46] and also the results from microscopic magnetic-probes, like neutron scattering [32].

Table 6. Some of the refined and determined values of physical parameters involved in Curie-Weiss law, for Al70Pd20Mn8TM2 samples

TM_2	χ at 5K (FC) (emu /gm. Oe)	M_s (emu /gm)	C (emu.K/g m.Oe)	θ_c (K)	Spin (S)	% Magnetic Mn
Mn	7.9E-4	0.999	1.3E-4	238.9	1.97	2.4
Fe	1.23E-4	0.968	5.0E-5	220.6	0.16	2.25
Cr	4.33E-5	0.568	5.0E-5	140.1	0.98	0.21
Co	3.96E-5	0.288	3.0E-5	181.5	1.33	0.079
Ni	6.68E-6	0.119	9.62E-6	194.1	0.77	0.058

The observed bifurcation of ZFC and FC χ-T data with the application of 50 Oe field, indicates the formation of clusters by a group of moments in this temperature range, which starts freezing below 180K. On lowering the temperature the moments within the cluster, get ordered with an effective finite moment. The occurrence of such a bifurcation in the ZFC and FC χ-T data is a typical character of cluster-glass type magnetic structure [31,54]. Keeping in view the large divergence of the FC χ-T data with respect to ZFC data at 180K and the presence of very dilute magnetic moments, it appears that the moments are existing in the form of clusters (quasicrystals do contain icosahedral clusters of Mn atoms) of just few moments two or three [37]. The enhanced bifurcation between the ZFC and FC χ-T-data, in presence of a field of 50 Oe, and its complete suppression at 1000Oe, indicates that inter-cluster interaction is very weak and can be taken as nearly paramagnetic [31].

Table 7. Table showing the Pre and Post minima slopes of (σ -T) curves of Al70Pd20Mn8TM2 samples and shift in Tmax on application of 8-Tesla field

TM2	χ at 5K (emu /gm. Oe)	ρRT ($\mu\Omega$-cm)	Tmin (K)	Tmax (K)		Pre-minima Slope (Ω-cm.K)-1	Post-minima slope (Ω-cm.K)-1
				Zero Field	In Field		
Mn2	7.9E-4	1126	126.5	9.10	12.91	0.186	-0.63
Fe2	1.23E-4	1144	143.83	9.34	11.10	0.103	-0.66
Cr2	4.33E-5	1044	101.38	2.77	6.75	0.310	-1.72
Co2	3.96E-5	1953	122.34	2.88	6.11	0.114	-0.85
Ni2	1.83E-5	2022	67.12	-	-	0.464	-0.72

The other group of moments contains free moments, which shows its existence in the lower temperature regime, as indicated by the straight line in the inset of Figure 15(b). The linear feature indicates paramagnetic behavior of the moments with weak antiferromagnetic interaction associated with the free moments. The non-saturating trend of the M-H curve also indicates the presence of antiferromagnetic interaction within the sample. Below 26K, $(\chi - \chi_0)^{-1}$-vs-T plots show a clear deviation from the paramagnetic behavior. This feature is common for all the studied samples.

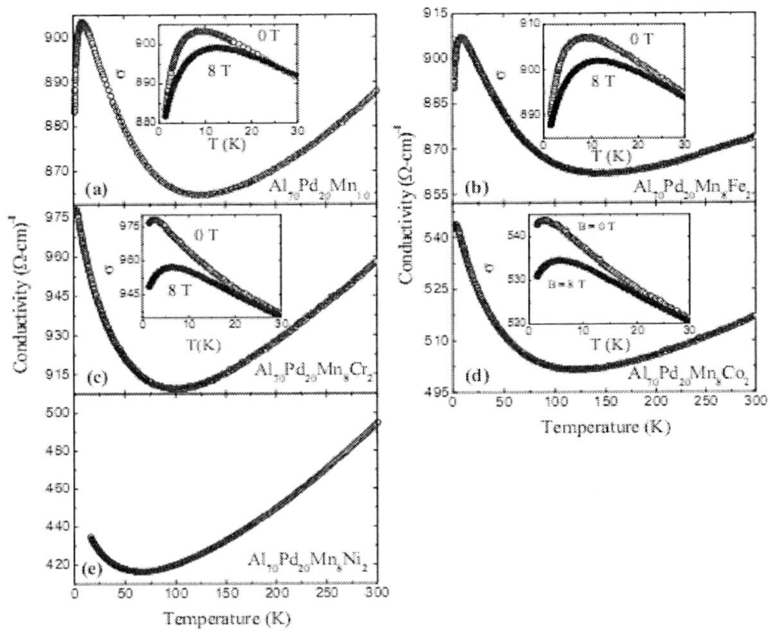

Figure 16. σ -T curves for Al70Pd20Mn8(TM)2 ,for TM = Mn, Fe, Cr, Co and Ni, quasicrystalline samples The circles are the observed data points and the lines are the fit to the data. The insets highlight the low temperature σ -T behavior in the zero-field (open circles) and in the 8-Tesla field (solid circles). High temperature shift of σ-T maxima on application of field can be seen.

4.1.3. Conductivity Vs. Temperature

σ-T curves for $Al_{70}Pd_{20}Mn_{10}$ and $Al_{70}Pd_{20}Mn_8(TM)_2$,for TM = Fe, Cr, Co and Ni, quasicrystalline samples are shown in Figure 16(a), (b), (c), (d) and (e) respectively. The σ-T variation for all the studied samples is qualitatively same. Each shows a pair of minima and maxima except for $Al_{70}Pd_{20}Mn_8Ni_2$, sample for which no maxima is observed till up to 1.4K. In the present investigation, σ-T measurements of all the samples were also carried out in the presence of 8-Tesla magnetic field. This will illuminate the possible origin of the observed transport behavior. It can be seen that corresponding to each composition, the σ-T maxima reduces and shifts to higher temperature with the application of external magnetic field, see the corresponding insets of Figure 16(a-d). The reduction in peak value basically means the occurrence of positive magneto-resistance. A summary of all these results is presented in Table-7, together with the corresponding magnetization values. Table-7 depicts that ρ_{RT} increases in general with decreasing total magnetization.

Another definite correlation is found in the temperature (T_{max}) of σ-T maxima and the magnetization, where, T_{max} increases with increasing total magnetization.

4.1.3.1. (σ -T) Minimum

In this series of samples also, the occurrence of conductivity minima is in accordance with the other observations [22,24,27,40], where it has been attributed to weak anti-localization effect appearing due to strong spin-orbit scattering of conduction electrons by Pd atoms. Nevertheless there are few points that indicate that although the observed σ-T variation is due to weak-localization but the minima is not due to spin-orbit scattering. These points can be illustrated as follows:

1) *First* and the most important point which is unpredicted for this particular series is that even though the Pd concentration is fixed (20%) down the series, the position of minima is getting shifted drastically with different TM doping, as is given in Table-7. Now, if the σ-T minima would have been emerging due to the to weak anti-localization effect appearing due to strong spin-orbit scattering of conduction electrons by Pd atoms, then the position of σ-T minima, should have been fixed if the concentration of Pd, which is known to give rise to strong spin-orbit scattering, remains fixed. This observation clearly indicates that there is no role of Pd in the occurrence of minima. This point very strongly states that although the observed σ-T variation is due to weak-localization but the minima is not due to spin-orbit scattering due to Pd.
2) Secondly, the average post-minima σ-T slope (slope at temperatures below T_{min} but above T_{max}) is in general more than thrice the corresponding pre-minima slope (slope at temperatures above T_{min}), see Table-7. This observation, even qualitatively, is not commensurate with the expectations from weak anti-localization theory [56] (due to spin-orbit scattering, as described in details in section 3.1.3.1 of Part I). The experimentally observed pre-minima slopes (between 240K-300K) are almost linear; if at all curved, it has slight positive curvature. This means that in the present case, from 240K to 300K, weak-localization is being de-phased by inelastic scattering which has a temperature dependence given by $\tau_i = C.T^{-p}$, with $p \cong 2$ (e-ph interaction in dirty-metallic limit).

   This argument also clearly indicates that in the present case, σ-T minima might be originating due to dephasing of weak-localization being caused by an altogether different type of scattering mechanism and not due to spin-orbit scattering.
3) *Thirdly,* in the case of Al-Pd-Mn quasicrystals, Mn is the only content which is expected to carry a localized moment and hence the magnetism in these quasicrystals. Now, in this series the position of σ-T minima is shifting invariably with the replacement of Mn, by other TM, indicates that the Mn and hence magnetic scattering instead of spin-orbit scattering is playing leading role in the origin of minimum in the σ-T variation.
4) Magnetic measurements on all the samples studied in this part , shows that they are magnetic in nature and hence the origin of the appearance of σ-T minima may lie in its magnetic properties. Moreover it is observed experimentally that there is a steep rise in the σ-T variation below minima. In light of these points, along with the detailed description given in 3rd point of section 3.1.3.1, it is obvious that post-minima σ-T variation is being dominated by a random dephasing magnetic (spin-flip) scattering, which is strongly temperature dependent. The dephasing rate

increases continuously with decreasing temperature. Presumably this may be due to Kondo-like spin-flip scattering.

5) As discussed in Part-I , Prejean et al, 2002 [46] and Dolinsek et.al, 2002 [45], have also argued that the σ-T minima (or ρ-T maxima) is probably due to magnetic scattering and not due to spin-orbit scattering effect. The polarization of conduction electrons about the magnetic Mn has been realized in NMR experiments and the magnetic properties are attributed to the Kondo-effect [29,30].

Thus, from the above-described points, we can qualitatively say that the occurrence of σ-T minima cannot be attributed to weak anti-localization effect appearing due to strong spin-orbit scattering of conduction electrons by Pd atoms, but it may be due to the presence of Kondo-type scattering.

4.1.3.2. σ -T Maximum

Akiyama *et al* [40] and C. R. Wang *et al* [24] have shown that there occurs linear region with negative slope in the ρ-ln(T) plot and thus have attributed the occurrence of ρ-T minima (or σ-T maxima) to the presence of Kondo scattering. Conventionally, the occurrence of linear region with negative slope in ρ-ln(T) plot is known to indicate the presence of Kondo scattering, but our in-field σ-T data completely negate this. Since, if the σ-T maxima would have appeared due to conventional type of Kondo effect and then it should have shifted to lower temperature with a negative magneto-resistance. But the present observation of in-field σ-T is just opposite.

4.1.3.3. Possible Origin of σ -T Behavior

As described in section 3.1.3.3 the origin of the observed σ -T behavior for transition metal doped Al-Pd-Mn quasicrystals should also be the same. Being well within the bad metallic limit the σ-T behavior of these quasicrystals also can be well described by the weak-localization effect which is already experimentally well established, even at temperatures as high as 200K [60].

As discussed in section 3.1.3.3, following the equations (3.6), (3.7) and (3.8) we fitted the observed σ-T data by non-linear least square refinement. The fits to the data are shown as continuous curves in Figure 16. The fitting parameters are σ_0, D, τ_i, R_{sf}, S and T_K. In Table-8 we have presented σ_0, Bi, B_{sf}(max), S and T_K. B_{sf} (max) is defined as the maximum scattering field and is given as ($h.R_{sf}/4eD$). The fitted parameters are found to be physical. The first best fitted temperature range i.e. from 40K to 290K, is more or less same for all the compositions. The quality of fits appears to justify the validity of the proposed model. Figure 17 presents the temperature-dependence of spin-flip scattering-field $B_{sf,}$ which has been calculated, using parameters R_{sf}, D, S and T_K, obtained from refinement of the σ-T data. After going through the B_{sf}(max) column, for the temperature range of 40-300K, we see that B_{sf}(max), i.e. the maximum spin-flip rate follow the decreasing trend with decreasing magnetization; but for the 2% Ni doped sample, the behavior is completely different.

The Ni doped sample, although has lowest moment concentration, shows the highest spin-flip rate. This puts the Ni-doped sample in a different category. The moment concentration for Ni-doped sample is $(1/40)^{th}$ of the undoped sample. Since the moment concentration is much lower, the moment-moment distance will be larger and the interaction between the moments will be weak. Hence an individual moment will be less affected by the field of the neighboring moment. Thus, each moment will be free to flip with a faster rate.

Table 8. Refined values of physical parameters involved in Weak-localization including spin-flip scattering for Al70Pd20Mn8TM2 samples

			High Temperature Range (40K-300K)			Low Temperature Range (1.4K-20K)		
TM 2	σ_0 (Ω-cm)$^{-1}$	B_i (h/8πe) (cm^{-2})	S	T_K (K)	B_{sf} (Max) (h/8πe) (cm^{-2})	S	T_K (K)	B_{sf}(Max) (h/8πe) (cm^{-2})
Mn	814.84	2.88E8x T^2	0.095	28.25	1.61E13	0.38	9.09	2.32E13
Fe	823.46	1.38E8x T^2	0.064	32.59	4.73E12	0.39	8.57	8.47E12
Cr	850.84	7.39E8x T^2	0.079	28.32	7.41E12	0.46	2.63	1.84E13
Co	471.62	1.19E8x T^2	0.076	27.91	2.43E12	0.55	2.82	5.69E13
Ni	330.00	1.49E9x T^2	0.228	0.35	7.62E13	-	-	-

Thus, the Ni-doped samples are close to a Kondo-system of nearly isolated moments and therefore it does not show a maxima in its σ-T variation [65,66]. The other samples, which show a σ-T maxima together with the σ-T minima belong to the category of a Kondo-system of interacting moments.

The refined parameters do not fit to the data below 40K. According to the refined values of T_K there should have been maxima at around 30K, but the experimentally observed maxima occur at lower temperature, i.e. effective T_K has shifted to lower value. This shift may be attributed to the changing magnetic state of the sample as indicated by the feature in $(\chi- \chi_0)^{-1}$-vs-T plot below ~26K. Such a feature may appear due to emergence of a ferromagnetic interaction below ~26K, which would push T_K to lower temperatures [66,67]. Due to change of the magnetic state, the parameters Rsf, S and TK are expected to change and hence the low temperature region of the σ-T data were fitted with the equation 3.8, but keeping the parameters σ0, D, and τi to same values as obtained by the refinement of the σ-T data in high temperature range. The parameters Rsf, S and TK were let free to get refined to new values.

The fitting was good only in the temperature range from 1.5K-20K. The fitting of σ-T data in the range 1.5K-20K can be seen in Figure 18. As is obvious from Figure 18, these fits very well account for the observed σ-T maxima and hence we attribute them to Kondo-maxima.

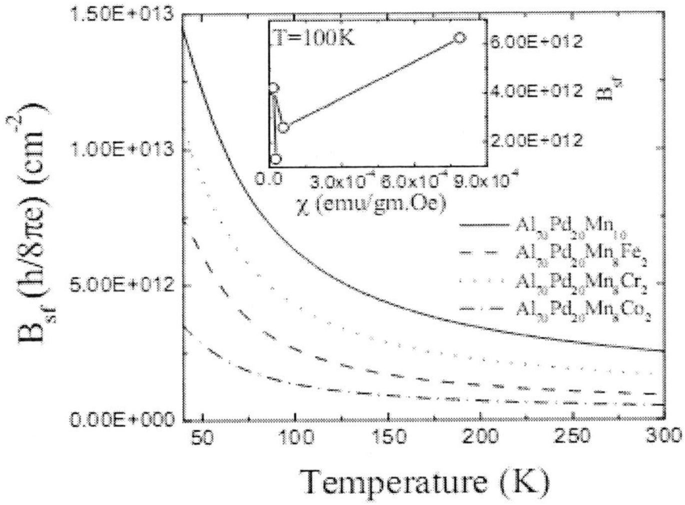

Figure 17. Temperature dependence of spin-flip scattering field for four quasicrystalline samples, which have been derived after refinement of the σ-T data. Inset shows an increasing trend of spin-flip scattering field with the total magnetization of the sample.

The temperature range from 20-40K may correspond to a region of changing magnetic state, where perhaps no unique value of parameters Rsf, S and TK are defined. It should be noted that the temperature range of 20-40K is coincident with the features below ~26K in the $(\chi- \chi_0)^{-1}$-vs- T plot. The values of R_{sf}, S and T_K obtained from the refinement of low-temperature data are also given in Table-8. It can be seen that at low-temperatures, values of S corresponding to each composition has increased and the Kondo temperature T_K has decreased. The increase in the spin value effectively means the enhancement of interaction between the moments. Increased interaction between the moments will slow down the spin-fluctuations and hence spin-flip scattering will decrease. Effectively this means that the T_K would decrease [66,67]. That is what has actually been observed. It can be noted that spin values obtained from refinement of the σ-T data is much lower as compared to the spin values obtained from magnetization measurements.

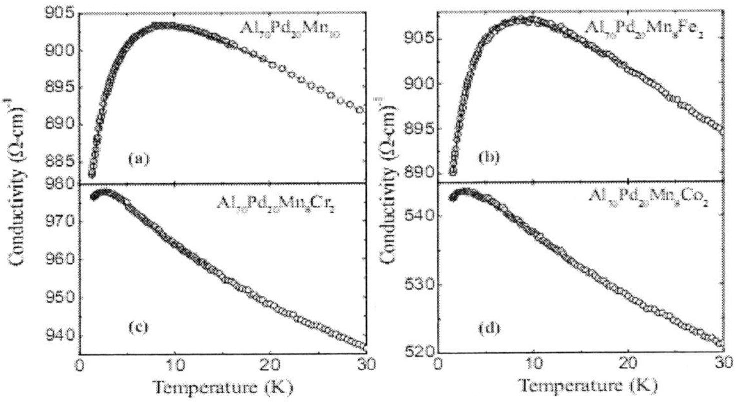

Figure 18. σ-T curves in the temperature range of 1.4-40K showing the occurrence of σ-T maxima for (a) Al70Pd20Mn10, (b) Al70Pd20Mn8Fe2, (c) Al70Pd20Mn8Cr2 and (d) Al70Pd20Mn8Co2 quasicrystalline samples. This also highlights the quality of the fits to the data.

This may be due to the fact that the spin value obtained from σ-T data, is the net spin of a cluster as seen by the conduction electrons during its spin-flip scattering from the cluster in zero-field. Since in the zero-field, order of moments within a cluster appears to be a result of frustrated interaction, the net spin will be the result of nearly random arrangement of moments hence low net spin value would result. Whereas, the spin value as obtained from the magnetization measurements is the spin value of almost one single moment because it has been obtained using saturation magnetization value measured after applying high fields, at which inter moment interactions gets suppressed and only single moment character remains.

4.1.4. Magneto-Resistance

The magneto-resistance ($\Delta\rho(B)/\rho$) data for all the samples is presented in Figure 19. It is clear from the figure that magneto-resistance at 6K are positive for all the samples and follows an H^2 behavior, as expected [61,62,68,69,72,73] from weak-localization theory in high temperature limit. This is so because, at high temperatures, as the field is increased, spin-flip gets suppressed due to which in a weak-localized system (as in the present case), dephasing is suppressed. Decrease in dephasing rate causes increase of resistance and hence a positive magneto-resistance.

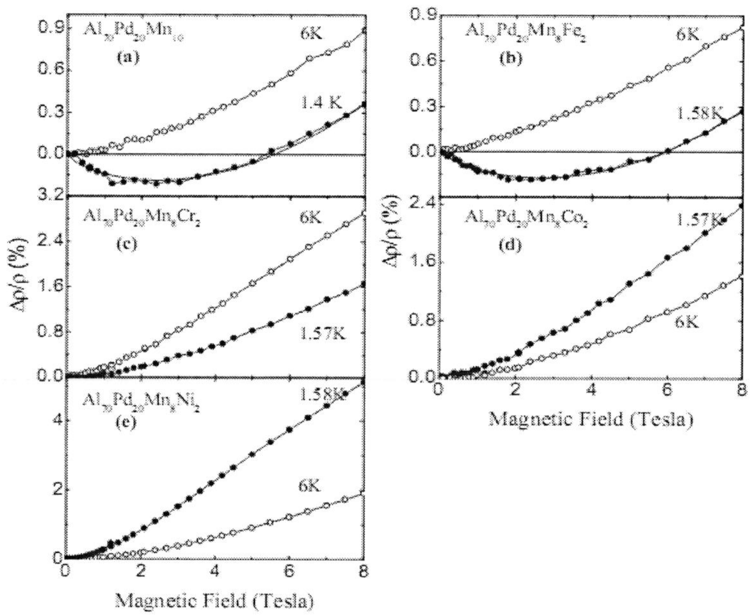

Figure 19. Magneto-resistance curves of (a) Al70Pd20Mn10, (b) Al70Pd20Mn8Fe2, (c) Al70Pd20Mn8Cr2, (d) Al70Pd20Mn8Co2 and (e) Al70Pd20Mn8Ni2 quasicrystalline samples. Continuous curves with 1.4K data in (a) and (b) are fits to the data based on Equation 3.11.

It can be noticed that for $Al_{70}Pd_{20}Mn_{10}$ and $Al_{70}Pd_{20}Mn_8Fe_2$ samples, magneto-resistance becomes negative at lower temperatures. It remains negative for relatively lower magnetic fields, shows a shallow minimum around 2.4 Tesla and finally becomes positive at around 4.5-Tesla for $Al_{70}Pd_{20}Mn_{10}$ sample and at 6 Tesla in the case of $Al_{70}Pd_{20}Mn_8Fe_2$ sample. Now, it can be observed that the magnetization values corresponding to these two samples, which shows negative component of magneto-resistance, are highest, see Table-6.This means that at low enough temperatures, the samples with large magnetization values have negative magneto-resistance, which implies that correlated (or interacting) moments gives negative magneto-resistance. As the temperature is increased, magneto-resistance becomes positive, i.e. when moments are no longer correlated, positive magneto-resistance occurs. Thus, as explained in *Part-I*, when magnetization is low or temperature is high, correlation between the moments is weak and positive magneto-resistance dominates. Thus, in this too the magneto-resistance appears to have two components one positive, with $+H^2$ dependence and the other negative, with $-H^p$ dependence and hence the total magneto-resistance can be expressed as in the equation 3.9. The magneto-resistance curves with both the components are very well fitted with the equation 3.9 with values of $p = 0.65$ and 0.73 for $Al_{70}Pd_{20}Mn_{10}$ and $Al_{70}Pd_{20}Mn_8Fe_2$ sample respectively.

5. Part III

In Part I and Part II, we have tried to provide enough proof that the transport in Al-Pd-Mn quasicrystalline alloys can be explained by quantum interference effect, in the whole studied temperature range, where spin-flip scattering by magnetic Mn is playing the leading role in the origin of the minima as well as the maxima in the $\sigma(T)$ curve instead of spin-orbit scattering due to Pd. Further studies on boron substituted Al-Pd-Mn quasicrystalline provide some more strong evidences, which will certify the dominant role played by the magnetic Mn-atoms in its transport as well as magnetic properties.

Magnetic measurements on Al-Pd-Mn quasicrystalline system [22,23,27] have demonstrated that it exhibits magnetism in the whole icosahedral-phase concentration range i.e. roughly between 7-10% Mn and the number of magnetic moments increases strongly with increasing Mn concentration. But, it is found that an excess Mn concentration usually makes the icosahedral structure unstable. A decrease in average outer electron concentration (e/a) from an optimum value of e/a=1.75 and the strain field (phason strain) due to atomic size difference between Mn and other constituent elements leads to the transformation from quasicrystalline to various types of approximant phases [74]. It was shown by Yokoyama et al [75] that, addition of boron, maintains the quasicrystalline structures in a higher Mn concentration range. Boron is found to relax the strain field around an excess Mn element. Studies have shown that the magnetization value increases with the increase of Boron content in the Mn rich Al-Pd-Mn quasicrystalline alloys. The relation between magnetization and the strain field implies that magnetic ordering is caused by imperfection in the quasicrystallinity, such as large amount of phason strain fields [74]. NMR [76] and FMR [51] studies of this system have indicated that although the x-ray diffraction shows a single phase in these icosahedral alloys, these are magnetically heterogeneous systems, with co-existence of ferromagnetic and non-magnetic Mn atoms and the large magnetization corresponds to the polarization of the conduction electrons through RKKY interactraction, with the ferromagnetic Mn atoms. Dong-Liang Peng et al [74] have also done a systematic magnetic study of this system and have found that this system shows a variety of magnetic behavior with different boron concentration, for example, $Al_{70}Pd_{15}Mn_{15}$ shows a spin-glass behavior, $Al_{70-x}Pd_{15}Mn_{15}B_x$ ($x = 1,1.5$) shows cluster glass and / or super paramagnetic behavior and $Al_{70-x}Pd_{15}Mn_{15}B_x$ ($x = 4.5$) is ferromagnetic at all temperatures and do not completely saturates under a magnetic field of 5T even at 5K. G. de Laissardiere et al [35] have reported that with increasing defect magnetic moment increases. Despite these extensive studies on Al-B-Pd-Mn systems, the simple σ-T behavior has been missing in almost all these reports.

Looking into the definite role of boron substitution in relaxing the phason strain fields about Mn site [74] and also the correlation between the defect/strain and the magnetic Mn concentration [35], it appears that the substitution of boron (B) for aluminum (Al) in the lower Mn containing Al-Pd-Mn alloys, like $Al_{70}Pd_{20}Mn_{10}$, in which the phason strain is much lower than the one with higher, ~15% [74], Mn containing will further relax the strain and change the moment concentration. This indicates that boron substitution may act as a probe for confirming and investigating, the role of magnetic scattering in the electronic transport of Al-Pd-Mn quasicrystals.

Therefore in this Part III we have carried out conductivity vs. temperature (σ-T) measurements in the temperature range of 4-600K and magnetization (M-T) measurements down to 2K at 1000Oe field for $Al_{70-x}B_xPd_{20}Mn_{10}$ (x=0,0.5,1,2and4) quasicrystals. Magnetization vs field (M-H), and field dependent transport measurements like in-field σ-T and magneto-resistance, have also been carried out.

5.1. Results and Discussion

5.1.1. Structural Characterization

Figure 20 shows XRD patterns of all the $Al_{70-x}B_xPd_{20}Mn_{10}$ (x=0,0.5,1,2and4) samples. The diffraction peaks are indexed based on Elser's 6-index system [50]. Powder x-ray diffraction (XRD) characterization of the samples revealed that these samples are single phase and no contaminating phases of Mn-B or Al-Pd are present. The XRD peak widths were found in general to reduce on boron doping, indicating that the structural order of samples improve with substitution of boron in place of aluminum. We have monitored this by estimating the structure correlation-lengths $L = (0.9\lambda/B.\cos\theta)$, shown in Table-9. Here L is the correlation length and B is the full width at half maxima of the diffraction peak. Figure 21(a) and (b) shows the representative selected area electron diffraction (SAD) patterns for these boron doped samples. It reveals the fivefold (a) and twofold (b) symmetries of a face centered icosahedral phase.

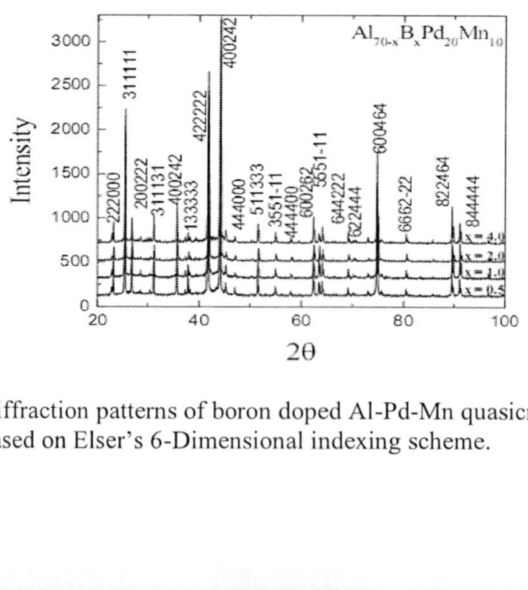

Figure 20. Powder x-ray diffraction patterns of boron doped Al-Pd-Mn quasicrystalline samples. The indices have been given based on Elser's 6-Dimensional indexing scheme.

Backscattering image of SEM, as shown in Figure 21(c), also confirms that the samples are single phase.

The EDAX results clearly shows that the compositions of the alloys are very close to the compositions we tried to make and the differences are well within the typical error of EDAX analysis. Since boron is a light element and its quantity is very less, hence it could not be detected. But its presence is confirmed by the occurrence of systematic in the transport measurements.

Figure 21. (a) and (b): Typical selected area electron diffraction (SAD) patterns of Boron doped sample (B=2%), showing (a) fivefold and (b) Twofold symmetries of face centered icosahedral phase.

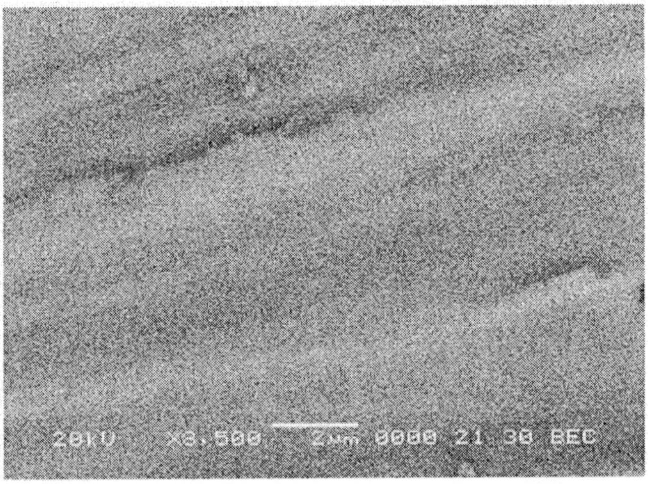

Figure 21.(c): A typical back-scattering SEM micrograph taken from boron doped Al-Pd-Mn quasicrystalline samples.

5.1.2. Magnetic Characterization

Magnetization measurements were carried out on all the samples studied. Figure 22(a) shows the χ-T data at 1000 Oe for the temperature range 0-300K and Figure 22(b) shows the M-H curve at 5K in the field range of 0-5Tesla. This representative χ-T and M-H behavior is similar for all the studied samples.

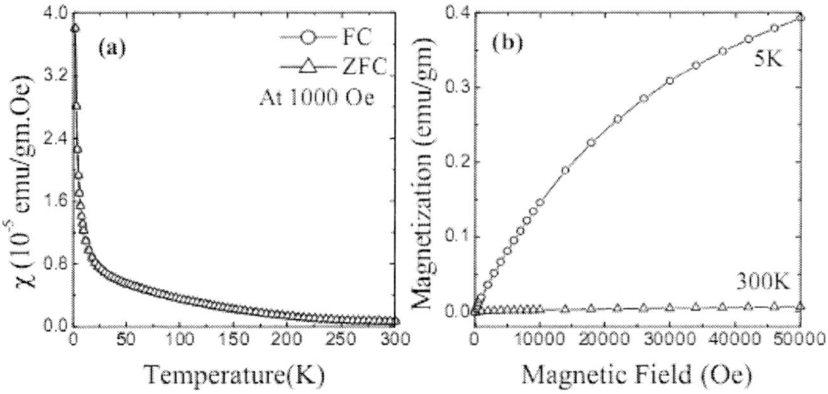

Figure 22. (a) Zero field cool and field-cool χ-T data at 1000Oe and (b) M-H data at 5K and 200K for Al66B4Pd20Mn10.

Table 9. Values of various physical quantities observed for Al-B-Pd-Mn quasicrystals

At.% of Boron	L (Å)	χ at 5K (emu/gm. Oe)	σ_{RT} $(\Omega\text{-cm})^{-1}$	σ_4 $(\Omega\text{-cm})^{-1}$	σ_{min} $(\Omega\text{-cm})^{-1}$	T_{min} (K)	R (σ_4/σ_{min})
0.0	1121	7.9E-4	887.82	-	864.75	126.5	-
0.5	1340	3.03E-5	590.85	913.74	587.87	244.5	1.55
1.0	1354	2.36E-5	686.04	1296.93	683.08	251.6	1.90
2.0	1306	2.07E-5	649.57	1729.56	645.56	253.9	2.68
4.0	1357	1.93E-5	746.98	4000.19	734.16	394.2	5.45

The value of χ at 5K is given in Table 9. It is evident from Table 9 that there is a systematic decrease in the magnetization value, as the boron concentration is increased in the sample. This result is in contradiction with previous reports [74,77]. From figure 22(a) it is clear that there is no divergence between Field cooled (FC) and zero field cooled (ZFC) curves, indicates that the samples are paramagnetic in nature, but the χ-T curve could not be accounted by Curie-Weiss law, in the entire studied temperature range. This deviation from Curie-Weiss law can be attributed to the presence of two types of moments in the sample [51,52]. The first type of moments are coupled Mn-B-Mn moments whereas the other ones are paramagnetic Mn moments, which appears to interact anti-ferromagnetically, as indicated by the non-saturating trend of the M-H curve even up to a field of 5T and at the temperature of 5K see Figure 22(b). We have fitted high temperature χ-T data for all the studied samples, by Curie-Weiss law, using equation 3.1.

We estimated a representative value of moment concentration following the method, as described in section 3.1.2 of Part I. It was found that out of all the Mn present in the sample only ~0.4% is magnetic in nature, which corresponds to 400ppm. This figure is comparable to the moment concentration in conventional Kondo-alloys. Since the concentration of magnetic moment is very dilute, it appears that these moments might be interacting indirectly via RKKY type of interaction.

5.1.3. Conductivity Vs. Temperature

Figure 23 (a), (b), (c) and (d) show the σ-T curves for $Al_{70-x}B_xPd_{20}Mn_{10}$ (for x=0.5, 1, 2 and 4) quasicrystals respectively, in the temperature range of 4-600K. It can be clearly seen that even a slight (0.5%) doping of boron in place of aluminum changes the σ-T behavior of Al-Pd-Mn quasicrystalline samples, drastically. For the sake of comparison σ-T data for undoped $Al_{70}Pd_{20}Mn_{10}$ sample, is shown in the inset in Figure 23 (a). It can be seen that in comparison to the undoped $Al_{70}Pd_{20}Mn_{10}$, the substitution of boron not only decreases the room temperature conductivity of the samples but at the same time it also improves the metallic character of the samples. It is interesting to note that with 4% substitution of boron the negative temperature coefficient of conductivity (metal like character) extends up to ~395K. These σ-T results are the first of its kind for Al-transition metal-based icosahedral quasicrystals. Another interesting point to be noted is that for all the studied compositions there appears a σ-T minimum, as in the case of undoped $Al_{70}Pd_{20}Mn_{10}$ sample, but unlike the case of undoped $Al_{70}Pd_{20}Mn_{10}$, no σ-T maximum is observed in the studied temperature range, see inset of Figure 23 (a).

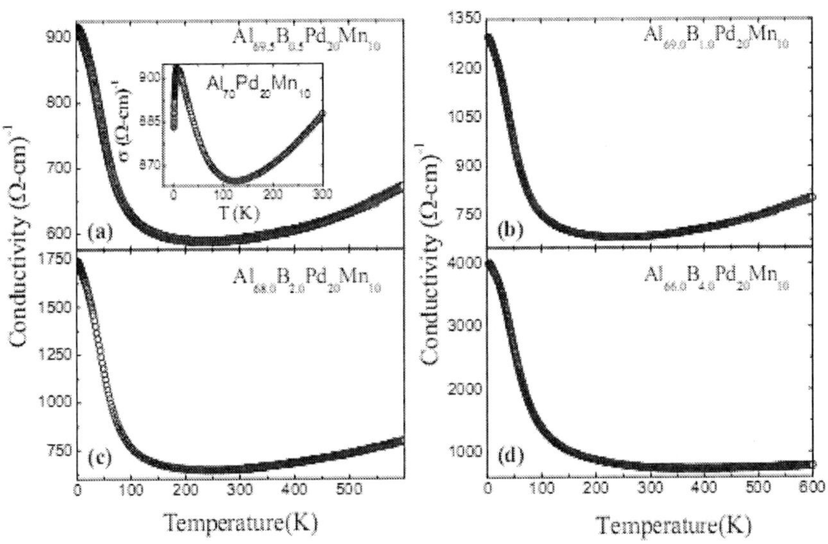

Figure 23. σ -T curves of Al70-xBxPd20Mn10 quasicrystals. The solid lines are the fit to the data based on Eqn.2. The inset in (a) shows the σ-T curve for Al70Pd20Mn10 sample.

The details of the origin of σ-T maximum have been discussed in Part-I. From Figure 23, it should be noted that the ratio R (σ_4/σ_{min}, where σ_4 is the conductivity value at 4K and σ_{min} is the minimum conductivity), the temperature T_{min} and σ_{min} increases with boron concentration, see Table-9. These features indicate that the origin of the observed σ-T behavior is the same for all the boron-doped samples. It also indicates towards the appearance or enhancement of a scattering mechanism whose rate of scattering scales with boron concentration.

While looking at the whole curve from 4K-300K, it appears that the nature of σ-T variation of boron doped and undoped samples are quite different but in actual it is not so. This is evident from the Figure 24, presenting comparison between the σ-T data for $Al_{70}Pd_{20}Mn_{10}$ and $Al_{69}B_1Pd_{20}Mn_{10}$ quasicrystals. It can be seen clearly that the nature of the post and pre minima slopes of the two curves are the same. Thus, the origin of σ-T variation in boron-doped samples appears to be the same as that for the undoped $Al_{70}Pd_{20}Mn_{10}$.

Since the room temperature resistivity of all these samples are well above the Mooij-criteria hence, as discussed in Part-I and like previous other reports [27,38,39,40], the σ-T behavior can be understood based on weak-localization theory.

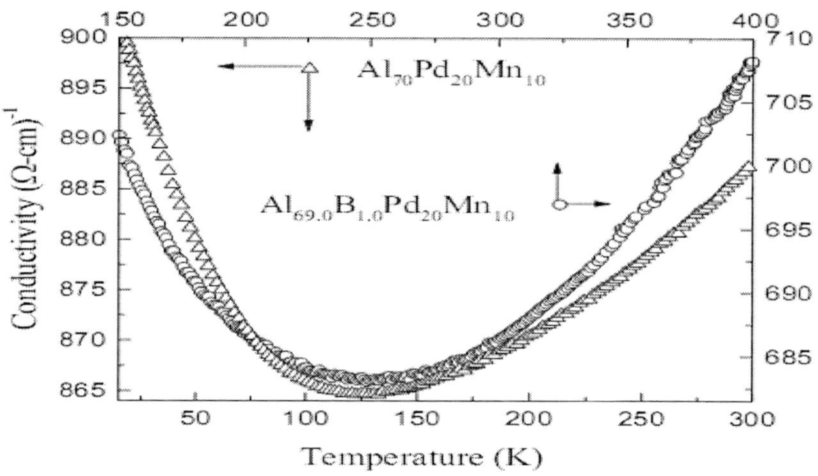

Figure 24. Comparison of σ-T slopes about Tmin of Al70Pd20Mn10 and Al69B1Pd20Mn10 samples.

As discussed earlier in Part-I, unlike in some earlier reports [27,38,39,40] in this series of samples also there are similar features that indicate that although the observed σ-T variation is due to weak-localization but the appearance of minima is not due to spin-orbit scattering. In fact the present data on boron substituted Al(B)-Pd-Mn quasicrystals exhibit it much more strongly. Note worthy points are as follows:

1. The observed drastically high post (below T_{min}) minima slope as compared to the pre minima slope (i.e. above T_{min}) and the dramatic shift of its minima temperatures as high as ~395K, clearly indicates that spin-orbit (anti weal-localization) effect is certainly not the origin of the observed minima for the σ-T variation for these quasicrystalline alloys.
2. The steep rise of the post-minima slope (below T_{min}) in σ-T indicates that it is being dominated by a random dephasing mechanism, which increases continuously and much more strongly with decreasing temperature than that of the previously described alloys in Part I and Part II. As can be seen in the discussions extended in section 3.1.3.3 the spin-flip scattering rate appears to be drastically high in this case.

Following equations 3.6, 3.7, 3.8 and the Annexure-I , we plotted $\Delta\sigma(T)$ –vs- $\pm \ln^{-1}$ (T). We find that for $T \gg T_K$ and $S \ll 1$ $\Delta\sigma(T)\ \alpha \pm \ln^{-1}$. For details see Annexure-I. Such a behavior is clearly seen in the σ vs. $\ln^{-1}(T)$ plots, shown in Figure 25 (a), (b), (c) and (d), where linear regions are indicated by straight lines. Figure 26 gives a plot of $\{[\sigma(T)-\sigma_{min}]/[\sigma_4-\sigma_{min}]\}$ vs.

($ln^{-1}T/ln^{-1}T_{min}$). It shows that plots corresponding to each composition have got nearly scaled. Thus, the occurrence of linearity in σ vs. $ln^{-1}(T)$ plots, as indicated by thick lines in Figure 25, and the perfect scaling, as observed in Figure 26, strongly indicate the dominance of Kondo-type spin-flip scattering in these quasicrystalline alloys. We have obtained the spin-flip scattering field by least square fitting of the data, shown as continuous curves in Figure 23. The refined parameters B_i, B_{sf}(max) (defined as R_{sf}/D), S and T_K are summarized in Table-10. The scattering field parameters are expressed in units of $(h/8\pi e)cm^{-2}$ [70,71] and D is the electronic diffusivity. All the fitted parameters are found physical. Now, if we examine Table-9 minutely, we will find that, the observed values of σ_{RT} and σ_{min} appear to be resultant of the interplay between sample's structural order and its magnetization.

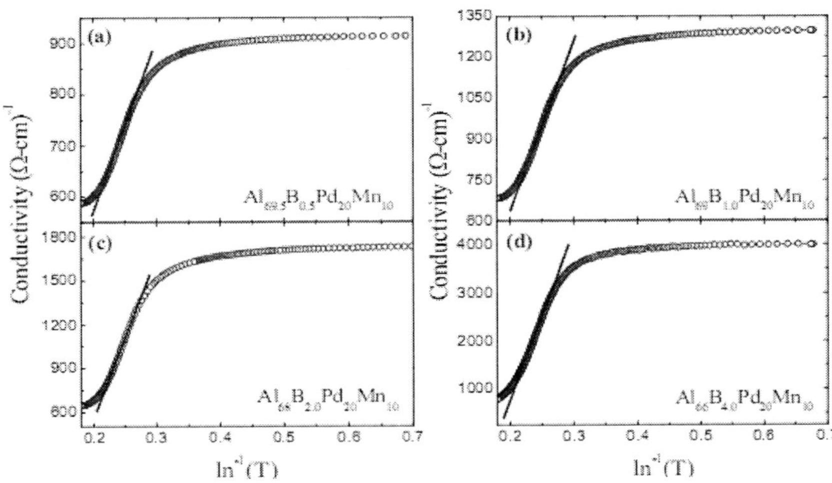

Figure 25. σ-ln-1(T) plots of the σ-T data from Al70-xBxPd20Mn10 quasicrystals. Straight lines indicate linear regions.

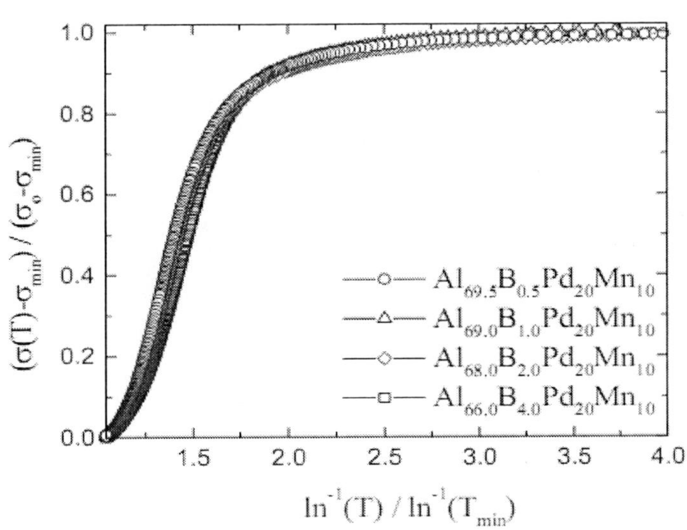

Figure 26. {[σ(T)-σmin]/[σ4-σmin]} vs. (ln-1T/ ln-1Tmin) plots of the σ-T data. A perfect scaling of σ-T data for all Al70-xBxPd20Mn10 samples can be seen.

Table 10. Refined values of parameters involved in Weak-localization including spin-flip scattering

At. % of Boron	S	Bi xT2 (h/4πe) (cm-2)	Bsf (h /4πe) (cm-2)	TK (K) (Fitted)
0.0	0.095	2.88E8	1.61E13	28.25
0.5	0.019	2.13E8	1.54E14	47.32
1.0	0.016	5.12E8	4.38E14	48.68
2.0	0.014	1.04E9	1.02E15	52.60
4.0	0.021	4.70E8	1.86E16	46.79

The overall decrease of conductivities on boron substitution may due to enhancement of structural order on boron doping, as evident from the values of structural correlation length L, in Table-9, but the increasing trend of σ_{RT} and T_{min} within the boron-doped series itself, appears to be due to increase in spin-flip rate with increasing boron concentration see Table-10. It can be observed form Table-10 that value of spin S is very small. This may be due to the reason that the Mn-moments are anti-ferromagnetically interacting through RKKY type interaction, as indicated by magnetic measurements. The coexistence of spin-flip scattering and RKKY is well known [67,78]. Hence during spin-flip interaction, which is of the order of 10^{-13} to 10^{-14} secs only, the interacting conduction electron will basically see a snap-shot of the arrangement of moment, which will most likely have nearly anti-parallel alignment, canceling each others fields. In such situation the interacting electrons will see effectively much lower spin value. Unlike the case of σ-T variation in $Al_{70}Pd_{20-x}Mn_{10+x}$ and $Al_{70}Pd_{20}Mn_{10-x}(TM)_x$ samples the absence of clear σ-T maxima for boron doped samples in the studied temperature range clearly indicates that the moments are not interacting ferromagnetically.

5.1.4. Magneto-Resistance Measurement

The representative magneto-resistance data taken for $Al_{68}B_2Pd_{20}Mn_{10}$ shows $+H^2$ dependence at low-fields and high temperatures. This is obvious from the parabolic nature of the curves at low fields and high temperatures; see Figure 27(a).

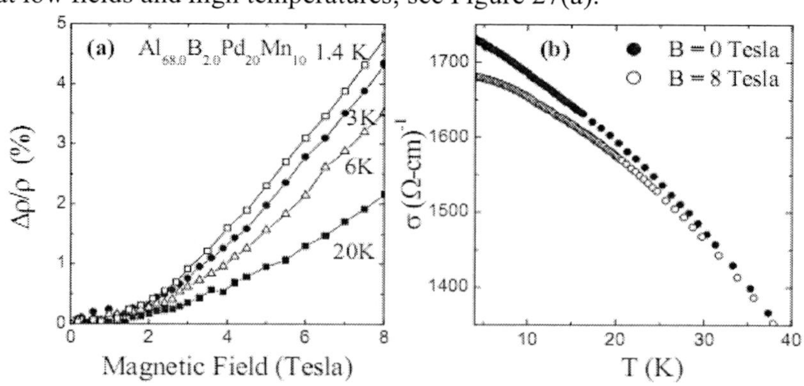

Figure 27. (a) Occurrence of nearly +H2 dependent magneto-resistance of Al68B2Pd20Mn10 quasicrystals. (b) Shows infield σ -T data at 8T for Al68B2Pd20Mn10.

Baxter et al [61] have shown the occurrence of enhanced positive magneto-resistance effect in the case of spin-fluctuating system. The reason for the occurrence of positive magneto-resistance is that with the application of external field the flipping of spins will get suppressed and hence the dephasing rate will also get suppressed. This will in turn cause increase in resistance and hence a positive magneto-resistance. Figure 27 (b) shows zero-field and in-field σ-T data up to 40K, where it is seen that with the application of magnetic field, conductivity decreases. A decrease in conductivity means an increase in resistivity and hence a positive magneto-resistance is seen, as demanded in the case of spin-fluctuating system. It can be seen that at temperatures above ~30K magneto-resistance effect disappears.

CONCLUSIONS

Based on the above described observations, analysis and discussions of zero-field and in-field σ-T, magneto resistance and related magnetic properties measurements, for the samples of series $Al_{70}Pd_{20+x}Mn_{10-x}$ (x=0,1and 2) , $Al_{70}Pd_{20}Mn_{10-x}(TM)_x$ (x=0 and 2 for TM = Fe, Co, Cr and Ni) and $Al_{70-x}B_xPd_{20}Mn_{10-x}$ of a stable quasicrystalline alloys, it can be concluded that the electronic transport in Al-Pd-Mn quasicrystals is completely dominated by quantum interference effects including electron-phonon and the Kondo-type magnetic scattering as the random dephasing processes. The minimum in the σ-T data is not due to anti-weak localization effect, caused by spin-orbit scattering from Pd, rather it is a result of competing scattering processes of electron-phonon and Kondo-type spin-flip scattering. The spin-flip scattering increases with decreases temperature and dominate the electron-phonon inelastic scattering giving rise to $\ln^{-1}$ (T) dependence of conductivity below Kondo temperature. The observed σ-T maxima for $Al_{70}Pd_{20+x}Mn_{10-x}$ and $Al_{70}Pd_{20}Mn_{10-x}(TM)_x$ alloys is due to maxima in the spin-flip scattering rate which is expected while spin-flip scattering of electrons is from a system of interacting moments rather than from isolated moments. The electronic transport in Al-B-Pd-Mn quasicrystals is also fully dominated by quantum interference effects, including spin-flip scattering by moments which appear to have anti-ferromagnetic interaction. The occurrence of $\ln^{-1}(T)$ and the observed scaling of σ-T data in $\{[\sigma(T)-\sigma_{min}]/[\sigma_4-\sigma_{min}]\}$ vs. ($\ln^{-1}T/\ln^{-1}T_{min}$) plots, strongly indicate the dominating role of Kondo type scattering in Al-Pd-Mn quasicrystals.

Observed negative magneto-resistance shows a correlation with the magnetization. The samples with large magnetization have large negative component of magneto-resistance. The presence of H^p (p<1) dependent negative magneto-resistance, in the weakly localized system of Al-Pd-Mn quasicrystals, appears to be a consequence of Kondo-scattering of electrons by a system of interacting moments only.

ANNEXURE I

The spin-flip scattering rate of conduction electrons in weak-localized systems is given [70,71] as

$$\tau_{Sf}^{-1} = R_{sf} \frac{\pi^2 S(S+1)}{\pi^2 S(S+1) + \ell n^2 (T/T_K)}$$

Here S is the spin of the flipping magnetic moment, T_K is the Kondo temperature and R_{sf} is the maximum scattering rate, which is realized at $T=T_K$. In the conditions when either S is very small or when T>> T_K or T << T_K, the terms in the denominator will be related as

$$\ln^2\left(\frac{T}{T_k}\right) >> \pi^2 S(S+1)$$

Hence,

$$\tau_{Sf}^{-1} \approx \frac{C_1}{\ln^2\left(\frac{T}{T_k}\right)}$$ C_1 is a constant.

or

$$\tau_{sf} = C_2 \ln^2\left(\frac{T}{T_k}\right) \qquad C_2 = 1/C_1$$

where C_2 will finally be given as ,

$$C_2 = \frac{1}{R_{sf} \pi^2 S(S+1)}$$

In the case of weak-localization the temperature dependent part of the conductivity is given as

$$\Delta\sigma(T) = \frac{e^2}{2\pi^2 \hbar}\left(\frac{1}{\sqrt{D\tau_{in}}}\right)$$

If the inelastic scattering events are dominated by the spin-flip scatterings then the temperature dependent part will be given as

$$\Delta\sigma(T) = \frac{e^2}{2\pi^2 \hbar}\left(\frac{1}{\sqrt{D\tau_{sf}}}\right)$$

$$= C_3\left(\frac{1}{\sqrt{\tau_{sf}}}\right);$$

with

$$C_3 = \frac{e^2}{2\pi^2 \hbar} \frac{1}{\sqrt{D}}$$

or

$$\Delta\sigma(T) = C_3 \frac{1}{\sqrt{C_2 \ln^2\left(\frac{T}{T_k}\right)}}$$

$$\Delta\sigma(T) = \pm C_4 \ln^{-1}\left(\frac{T}{T_k}\right)$$

with

$$C_4 = \frac{C_3}{\sqrt{C_2}}$$

or,

$$\Delta\sigma(T) = \pm C_4 \frac{1}{(\ln T - \ln T_k)}$$

or,

$$\Delta\sigma(T) = \pm C_4 (\ln T - \ln T_k)^{-1}$$

or ,

$$\Delta\sigma(T) = \pm C_4 \ln^{-1} T_k \left(\frac{\ln(T)}{\ln T_k} - 1\right)^{-1}$$

or,

$$\Delta\sigma(T) = \mp C_4 \ln^{-1} T_k \left(1 - \frac{\ln(T)}{\ln T_k}\right)^{-1}$$

Notice the change in sign of the right hand side term.

The factor in the right hand side bracket can be expanded using the following expansion for the condition of $T<T_K$ only because the following expansion is true only for $x<1$

$(1-x)^{-1}=1+x+x^2+x^3+x^4$------

Thus,

$$\Delta\sigma(T) = \mp C_4 \ln^{-1} T_k \left\{1 + \frac{\ln(T)}{\ln T_k} + \left(\frac{\ln(T)}{\ln T_k}\right)^2 + \left(\frac{\ln(T)}{\ln T_k}\right)^3 + ---\right\}$$

For $T<<T_K$ higher order terms can be neglected and $\Delta\sigma(T)$ can be approximated as

$$\Delta\sigma(T) \approx \mp C_4 \ln^{-1} T_k \left\{1 + \frac{\ln(T)}{\ln T_k}\right\}$$

$$\Delta\sigma(T) \approx \mp C_4 \ln^{-1} T_k \mp C_4 \ln^{-2} T_k (\ln T)$$

In the above expression only the second term is temperature dependent. Thus at temperatures much below T_K the temperature dependent part of the conductivity of a spin-flip dominated weak-localized system can be given as

$$\Delta\sigma(T) \propto \mp(\ln T)$$

Again taking the above expression

$$\Delta\sigma(T) = \pm C_4 (\ln T - \ln T_k)^{-1}$$

$$\Delta\sigma(T) = \pm C_4 \ln^{-1} T \left(1 - \frac{\ln T_k}{\ln T}\right)^{-1}$$

In the condition of $T \gg T_K$ the above expression can be approximated to

$$\Delta\sigma(T) \approx \pm C_4 \ln^{-1} T \left(1 + \frac{\ln T_k}{\ln T}\right)$$

$$\Delta\sigma(T) \approx \pm C_4 \ln^{-1} T \pm C_4 \frac{\ln T_k}{\ln^2 T}$$

At sufficiently high enough temperatures, the second term will be much smaller due to $\ln^2 T$ term in its denominator and hence the temperature variation will effectively follow

$$\Delta\sigma(T) \approx \pm C_4 \ln^{-1} T$$

Thus we can deduce that presence of Kondo type spin-flip scattering of conduction electrons in a weak-localized systems will lead to a

$$\Delta\sigma(T) \propto \mp(\ln T) \text{ when } T \ll T_K$$

and

$$\Delta\sigma(T) \approx \pm C_4 \ln^{-1} T \text{ when } T \gg T_K$$

The above two features of the $\Delta\sigma(T)$ variation may be used as tests for investigating the presence of the Kondo type spin-flip scattering in weak-localized systems.

References

[1] M. Kohomoto and B. Sutherland, *Phys. Rev.* B 34, 3849 (1986).

[2] T. Odagaki and D. Nguyen, *Phys. Rev.* B 33, 2184 (1986).
[3] S. J. Poon, Adv. in Phys. 41, 303 (1992).
[4] F. S. Pierce, S. J. Poon and Q. Guo, *Science* 261, 737 (1992).
[5] a. N. P. Lalla, R. S. Tiwari and O. N. Srivastava, J.Phys.: Condens. Matter 7, 2409 (1995); N. P. Lalla, R. S. Tiwari and O. N. Srivastava, *J. Phys. : Condens. Matter* 1, 7603 (1989).
[6] Ralph Rosenbaum, Tim Murphy, Bruce Brandt, Chang-Ren Wang, Yuan-Liang, Shr-Wen Wu, Shui - Tien Lin and Juhn - Jong Lin, *J. Phys.: Condens. Matter* 821-831, (2004).
[7] Z. D. Huseyin and Oner Yildirhan, *Phys. Rev.* B 42, 6831 (1990).
[8] M. Krajci, M. Windish, J. Hafner, G. Kresse and M. Mihakovic, *Phys. Rev.* B 51, 17355 (1995).
[9] M. Krajci, J. Hafner, M. Mihalkovic, *Europhys. Lett.* 34, 207 (1996).
[10] M. Krajci, J. Hafner, M. Mihalkovic, *Phys. Rev.* B, 55, 843 (1997).
[11] C. J. Jenks, S. L. Chang, J. W. Anderegg, P. A. Thiel and D. W. Lynch, *Phys. Rev.* B 54, 6301 (1996).
[12] A. P. Tsai, A. Inoue and T. Masumoto, Japan, *J. Appl. Phys.* 26, L1505 (1987):27, L1587 (1988).
[13] A. P. Tsai, A. Inoue, Y. Yokoyama and T. Masumoto, Mater. Trans. Japan Inst. Metals 31, 98 (1990); *Philos. Mag. Lett.* 61, 9 (1990).
[14] A. P. Tsai, Y. Yokoyama, A. Inoue and T. Masumoto, Japan. *J. Appl. Phys.* 29, L1161 (1990).
[15] A. P. Tsai, A. Inoue and T. Masumoto, *Philos. Mag. Lett.* 62, 95 (1990).
[16] C. Dong, J. M. Dubois, M. de Boissiue, M. Boudard and C. Janot, *J. Mater. Res.* 6, 2637 (1991).
[17] S. W. Kycia, A. I. Goldman, T. A. Lograso, D. W. Delaney, M. Sutton, E. Dufresne, R. Bruning and B. Rodricks, *Phys. Rev.* B 48, 3544 (1993).
[18] M. de Boissieu, M. Durand-Charre, P. Bastie, A. Carabelli, M. Boudard, M. Bessiere, S. Lefebvre, C. Janot and M. Audier, *Phil. Mag. Lett.,* 65, 147 (1992).
[19] Y. Yokoyama, T. Miura, A. P. Tsai, A. Inoue and T. Masumoto, *Mater. Trans.* JIM, 22, 97 (1992).
[20] A. P. Tsai, Y. Yokoyama, A. Inoue and T. Masumoto, *J. Mater. Res.* 12, 2646 (1991).
[21] S. Matsuo, H. Nakano, T. Ishimasa and Y. Fukano, *J. Phys.: Condens. Matter*, 1, 6893 (1989); C. R. Lin, S. T. Lin, C. R. Wang, S. L. Chou, H. E. Horng, J. M. Cheng, Y. D. Yao ans S. C. Lai, J. *Phys.: Condens. Matter,* 9, 1509 (1997); Y. Yamada, Y. Yokoyama, K. I. Matono, K. Fukura and H. Sunada, *Jpn. J. Appl. Phys.* 38, 52 (1999).
[22] P. Lanco, T. Klein, C. Berger, F. Cyrot-Lackmann, G. Fourcaudot and A. Sulpice, *Europhys. Lett.* 18, 227 (1992).
[23] M. A. Chernikov, A. Bernasconi, C. Beeli, A. Schilling and H. R. Ott, *Phys. Rev.* B 48, 3058 (1993).
[24] C. R. Wang, S. T. Lin and Y. C. Chen, *J. Phys.: Condens. Matter,* 6, 10747 (1994).
[25] S. Nimori, A. P. Tsai and G. Kido, *Physica* B, 237-238, 567-568 (1997).
[26] S. Matsuo, H. Nakano, T. Ishimasa and M. Mori, *J. Phys. Soc. Jpn.* 62, 4044 (1993).
[27] K. Saito, S. Matsuo, H. Nakano, T. Ishimasa and M. Mori, *J. Phys. Soc. Jpn.* 63, 1940 (1994).

[28] J. C. Lasjaunias, A. Sulpice, N. Keller, J. J. Prejean and M de Boissieu, *Phys. Rev.* B 52, 886 (1995).
[29] V. Simonet, F. Hippert, M. Audier and G. Trambly de Laissardiere, *Phys. Rev.* B 58, R8865 (1998).
[30] F. Hippert, V. Simonet, M. Audier, Y. Calvayrac, R. Bellissent, G. Trambly de Laissardiere and D. Mayou, *Mat. Res. Soc. Symp. Proc.* 643, K14.2 (2001).
[31] S. Nimori and A. P. Tsai , *J. Magn. Magn. Mater.* 241,11 (2002).
[32] F. Hippert, M. Audier, H. Klein, R. Bellissent and D. Boursier, *Phys. Rev. Lett.* 76,54 (1996).
[33] G. Trambly de Laissardiere, D. Mayou and D. Nguyen Manh, *Europhys. Lett.* 21, 25 (1993).
[34] G. Trambly de Laissardiere, D. Nguyen Manh, L. Magaud, J. P. Julien, F. Cyrot-Lackamnn and D. Mayou, *Phys. Rev.* B 52, 7920 (1995).
[35] G. Trambly de Laissardiere and D. Mayou , *Mater. Sci. and Eng.* A 294-296, 621 (2000).
[36] J. Hafner and M. Krajci, *Phys. Rev.* B 57, 2849 (1998).
[37] M. Krajci and J. Hafner, *Phys. Rev.* B 58, 14110 (1998).
[38] M. A. Chernikov, A. Bernasconi, C. Beeli and H. R. Ott, *Europhys. Lett.* 21, 767 (1993).
[39] S. Takeuchi, H. Akiyama, N. Naito, T. Shibuya, T. Hashimoto, K. Edagawa and K. Kimura, *J. Non-Cryst. Solids,* 153&154, 353 (1993).
[40] H. Akiyama, T. Hashimoto, T. Shibuya, K. Edagawa and S. Takeuchi: *J. Phys. Soc. Jpn.* 62, 639 (1993).
[41] S. Matsua, H. Nakano, T. Ishimasa and M. Mori, *Solid State Communications,* 92, 811 (1994).
[42] M. Rodmar, B. Grushko, N. Tamura, K. Urban, O. Rapp, *Phys. Rev.* B 60, 7208 (1999).
[43] R. Escudero. J. C. Lasjaunias, Y. Calvayrac and M. Boudard, *J. Phys.: Condens. Matter*, 11, 383 (1999).
[44] Enrique Macia, *Phys. Rev.* B 66, 174203 (2002).
[45] J. Dolinsek, M. Klanjsek, Z. Jaglicic, A. Bilusic and A. Smontara, *J. Phys.: Condens. Matter,* 14, 6975 (2002).
[46] J. J. Prejean, C. Berger, A. Sulpice and Y. Calvayrac, *Phys. Rev.* B 65, R140203 (2002).
[47] de Laissardiere G T and Mayou D 2000 *Material Science and Engineering*, A 294.
[48] S. Banerjee, R. Goswami, K. Chattopadhyay and A. K. Raychaudhury, *Phys. Rev.* B 52, 3220 (1995).
[49] G. N. Banerjee, S. Banerjee and R. Goswami, *J. Phys.: Condens. Matter,* 15, 2317 (2003).
[50] V. Elser, *Phys. Rev.* B 32, 4892 (1985).
[51] D. Bahadur, C. M. Srivastava, M. H. Yewondwossen and R. A. Dunlap, *J. Phys.: Condens. Matter* 7, 9883 (1995).
[52] M. E. Mc Henry, V. Srinivasan, D. Bahadur, R. C. O'Handley, D. J. Lloyd and R. A. Dunlap, *Phys. Rev.* B 39 3611 (1989).
[53] M . Scheffer and J. B. Suck, *Material Science and Engineering,* A 294 629 (2000).
[54] R. Mathieu, P. Nordblad, D. N. H. Nam, N. X. Phuc and N. V. Khiem, *Phys. Rev.* B 63, 174405 (2001).

[55] Ryuji Tamura, Takayuki Asao, Mutsuhiro Tamura and Shin Takeuchi, *J. Phys.: Condens. Matter* 11 10343 (1999).

[56] J. S. Dugdale, The Electrical Properties of Disordered Metals (Cambridge Univ. Press) (1995).

[57] N. Rivier and M. J. Zuckermann, *Phys. Rev.Lett.*21, 904 (1968).

[58] J. Delahaye and C. Berger, *Phys. Rev.* B 64, 094203 (2001).

[59] J. Delahaye, C. berger and G. Fourcaudot, *J. Phys.: Condens. Matter* 15, 8753 (2003).

[60] M. Ahlgren, P. Lindqvist, M. Rodmar and O. Rapp, *Phys. Rev.* B 55, 14 847 (1997).

[61] D. V. Baxter, R. Richter, M. L. Trudeau, R. W. Cochrane and J. O. Strom-Olsen, *J. Phys. France* 50, 1673 (1989).

[62] J. J. Lin and J. P. Bird, *J. Phys.: Condens. Matter* 14, R501 (2002).

[63] D. Jha and M. H. Jericho, *Phys. Rev.* B 3, 14 (1971).

[64] J. W. Loram, T. E. Whall and P. J. Ford, *Phys. Rev.* B 3, 953 (1971).

[65] M. T. Beal-Monod, *Phys. Rev.* 178, 874 (1969).

[66] K. Matho and M. T. Beal-Monod, *Phys. Rev* B 5, 1899 (1972).

[67] K. H. Fischer, Z. *Physik* B 42, 27 (1981).

[68] B. L. Altshuler and A. G. Aronov, Electron-Electron Interactions in Disordered Systems, edited by M. Pollak and A. L. Efros, (North-Holland, Amsterdam) 4 (1985).

[69] H. Fukuyama, Electron-Electron Interactions in Disordered Systems, edited by M. Pollak and A. L. Efros, (North-Holland, Amsterdam) 153 (1985).

[70] C. V. Haesendonek, J. Vranken and Y. Bruynseraede, *Phys. Rev* B 58, 1968 (1987).

[71] R. P. Peters, G. Bergmann and R. M. Mueller, *Phys. Rev* B 58, 1964 (1987).

[72] P. A. Lee and T. V. Ramakrishna, *Rev. Mod. Phys*. 57, 287 (1985).

[73] H. Suhl, Phys. *Rev.Lett.,* 20, 656 (1968).

[74] Dong-Liang Peng, K. Sumiyama, K. Suzuki, A. Inoue, Y. Yokoyama, K. Fukura, H. Sunada, *J. Magnetism and magnetic materials,* 184, 319 (1998).

[75] Y. Yokoyama, A. Inoue and T. Masumoto, *Materials Transactions,* JIM, 33, 1012 (1992).

[76] T. Shinohara, Y. Yokayama, M. Sato, A. Inoue and T. Masumoto, *J. Phys.: Condens. Matter* 5, 3673, (1993).

[77] Y. Yokoyama, A. Inoue, H. Yamauchi, M. Kusuyama and T. Masumoto, *Jpn. J. Appl. Phys.* 35, 3533 (1996).

[78] V. Simonet, et al., *J. Non. Crys. Solids* 334&335,408 (2004).

In: Quasicrystals: Types, Systems, and Techniques
Editor: B.E. Puckermann, pp. 47-76
ISBN 978-1-61761-123-0

Chapter 2

LOGARITHMIC PERIODICITY – PROPERTIES, TESTS AND UNCERTAINTIES

Antony J. Bourdillon
UHRL, P.O.Box 700001, San Jose, CA 95170

ABSTRACT

'Logarithmic periodicity' refers to three features in quasicrystals: firstly the ideal structure is uniquely icosahedral and infinitely extensive; secondly, the diffraction patterns contain corresponding orders that are geometrically spaced; and thirdly the mathematical description of electronic states is by special Fourier transforms in logarithmic order. This periodicity is driven by the low enthalpy in the subcluster. The model differs from most mathematical models because the three dimensional tiles share edges not faces. Experimental evidence of several types supports the model, beyond its conceptual simplicity. The principal three sources are: electron diffraction; electron microscopy; and diffraction simulations. The variety of properties and predictions are consistent with available experimental data in the binary quasicrystals such as Al_6Mn. Though logarithmic periodicity describes ideal solids with perfect icosahedral symmetry, the structure is defective in realization. While the defects should be expected in rapidly cooled and metastable solids, they imply uncertainties that require further refinement. If dendritic crystal growth depends on deposition of supercluster planar quads, the higher the order, the more nearly icosahedral.

1. INTRODUCTION

Is there a model that is perfectly icosahedral and infinitely extensive? We seek approximants consistent firstly with structural data, secondly with physical principles including stoichiometry and enthalpy, and thirdly with crystal growth.

Quasicrystals were discovered twenty eight years ago. The diffraction patterns from Al_6Mn were shown, roughly, to have icosahedral symmetry [1,2]. The symmetry is inconsistent with the fourteen Bravais lattices that completely describe crystals. Meanwhile, electron microscope images suggested that the newly discovered materials were 'aperiodic', and so they have often been called. However the diffraction patterns are as sharp as from regular crystals, so an extensive structure is sought that has the required angular symmetries. Cuts into multidimensional space [3] were supposed to provide an explanation, but the explanation was incomplete while tiles were undecorated and fudge factors were incorporated into simulations [4,5,6][1]. The first problem for physics is whether the data represent Bragg diffraction. They don't, because the orders, *n*, are generally, though not uniquely, logarithmic instead of linear. The second problem is structural. It is not necessary to model with more than one unit cell. The diffraction patterns can be indexed, tiled and simulated using a single unit [2] as is the norm in crystallography and consistent with the driving force. A third problem is dimensional: because each diffracted beam arises from a combination of multiple interplanar spacings, the quasi-lattice parameter is a compromise. The concept of logarithmic periodicity overcomes these problems at the same time as providing a comparatively conventional solution for the structure and diffraction patterns. This structure, consistently with the model, is driven by a unique subcluster. By comparison, the most common alternative approach that is used to explain the properties of quasicrystals, by mathematical modeling with dual or multiple tiles [7], combine Ptolemaic complication with uncertain decoration. The variants of such approaches assume face sharing which is not necessary, and they display a pre-Newtonian lack of physicality. The scope of the present paper does not extend to what are called the axial quasicrystals [7]. The subcluster can be used in an obvious way to model axial structures, since opposing faces are mutually parallel. However, icosahedral symmetry is simpler, and that is why it presents the greater challenge to insight and the greater promise of understanding. In this paper we review the consequences for a structure that is driven by the icosahedral subcluster.

The diffraction patterns of quasicrystals have unique features. Though they do not generally follow Bragg's law, there are partial exceptions [2,8,9]. The order in Bragg diffraction is described by the positive integer *n.* In high energy electron diffraction, the wavelength is much smaller than the interplanar spacing, $\lambda \ll d$, and the scattering angle is approximately twice the Bragg angle $\Theta \approx 2\theta$ so that the law can be written $n\lambda \approx d\Theta$. The corresponding formula for quasicrystals has a logarithmic order, $n\lambda \approx d'\tau^{-m}\Theta$, where m is a positive or negative integer, and the golden ratio is $\tau \equiv \left(1+\sqrt{5}\right)/2$ (appendix 1). In this case the quasi-lattice parameter and interplanar spacing must be corrected [8, appendix A] for the Compromise Spacing effect (CSE) to be described below. The correction is indicated by the prime on *d*'. Notice that the (uncorrected) parameter, q, is a basis for the diffraction patterns. With the indexation adopted, the subcluster cell length is $2q/0.947$, where the divisor corrects for the Compromise in spacing. Divide by τ to find the size of the subcluster edge. The logarithmically periodic solid and its calculated 'structure factors' fully explain the geometric series, including the exceptions, and likewise the absence of higher Bragg orders, illustrated in appendix 1. The concept of structure factors was originally defined for periodic solids where the factors are the scattering powers of a comparatively small unit cell for various indexed beams. The concept is usually used to identify forbidden lines. In the present case what are forbidden are high order Bragg diffraction and high order Laue zones. Now the

1 The effort is doubtful because the authors failed to notice, even after calculations, that their original data is not icosahedral (section 5.2.4).

'structure factor' is the scattering power of all the atoms in a very large cluster, calculated by the same formalism as used traditionally in finite cells. In principle, our cluster is infinite, but finite clusters are used for computational reasons by approximation. The formalism is thus applied independently of periodicity. Beside demonstrated consistency with data, the novel method uncovered new effects that are not accessible from small clusters (cf. [10]: from small clusters the diffraction peak is broad, so that inaccurate structure factors have been calculated by standard methods; but in larger superclusters all values are zero at the uncompromised Bragg angle.)

Any physical geometric series becomes Fibonacci when the ratio is scaled to τ , the golden ratio. Owing to the prevalence of τ in their geometries [11], 5-fold symmetries in two dimensions and icosahedral symmetry in three dimensions are natural associates to the geometric series. The icosahedron is to τ as the sphere is to π , so also is the Pythagorean star [11] to the circle. In21tuitively, it should not be surprising that the icosahedral subcluster is the most beautiful way of including geometric diffraction patterns with 5-fold symmetries. Less surprising is the evidence in favor.

In this review, the discussion is limited to evidence in binary quasicrystals of the type Al_6Mn. The structure examined has remarkable explanatory power for the wide range of characteristic phenomena that are exceptional in the context of crystallography. Ample experimental evidence is described to support the model. The core proof is collected into a series of lemmas (appendix 2) reproduced from *Quasicrystals and quasi drivers* [8]. The argument is further elucidated in the compilation *Quasicrystals' 2D tiles in 3D superclusters* [9]. Those books are more comprehensive than this summary review, which focuses on evidence by assuming some prior explanatory concepts and experimental detail.

2. MODEL

The concept of logarithmic periodicity begins with the icosahedral subcluster. This contains a Mn atom at its center, with 12 Al atoms coordinated icosahedrally about it (figure 1a). 12 subclusters are joined by edge-vertex-edge sharing to form an icosahedral cluster. 12 clusters share edges to form a supercluster order 1 and the series is extended to infinity. All subclusters, clusters and superclusters are uniquely oriented. The unique orientation would be violated by face sharing, and this fact separates quasicrystals from crystals.

Figure 1b shows icosahedra represented as triadic golden rectangles [2]. This representation, though hardly used in quasicrystal studies, is well known in mathematics [11]. The representation is particularly well suited to unit cells that share edges and has many advantages described below. The dimensions of each rectangle are $1 \times \tau$. The edge length we take as the icosahedral unit. In each triad, the rectangles intersect at right angles. When the triad represents the subcluster, the unit is the diameter of the Al atom. Twelve triads form an icosahedral cluster. The structure is repeated for superclusters order 1,2,3…∞. Each level of subcluster, cluster and supercluster, with increasing order, scales in length and breadth by the multiple τ^2 compared with its preceding component. This value is the 'stretching factor' in three dimensional tiling. The tiling 'forces the border' [12]. It turns out that the subcluster is dense, so that its low enthalpy can be seen as the driving force for the structure. Consider next the tiling and supertiling; defects at cluster and supercluster centers will be considered in a later section.

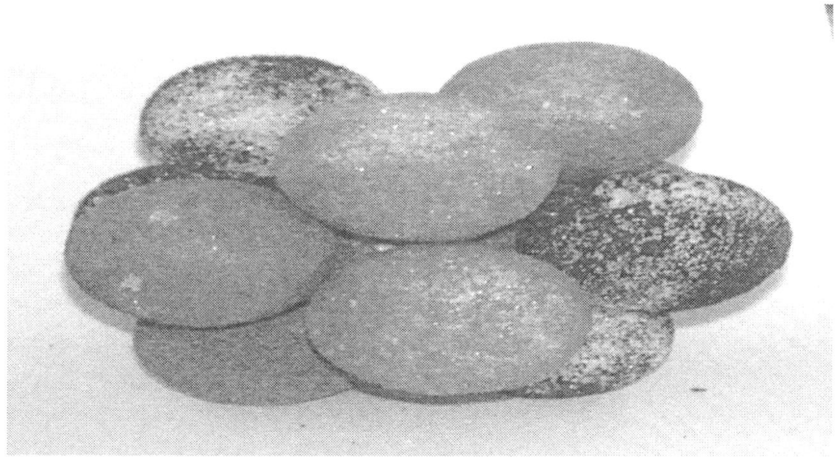

Figure 1a. An icosahedral subcluster contains a central Mn atom surrounded by twelve Al atoms. Their locations can be represented by the corners on triadic golden rectangles [2] (reprinted from Bourdillon [8]).

Figure 1b. Twelve golden triads can be formed into an icosahedral cluster when the thirteen spherical atoms in each subcluster are replaced by edge sharing icosahedra. The structure can be extended indefinitely with unique symmetry. A triad of golden rectangles, each having dimensions $1 \times \tau$, can be used to represent the icosahedral subcluster (figure 1a). With dimensions scaled by τ^2, the triad can also represent a cluster, or when scaled by τ^{2+2n} a supercluster of any order n=1… ∞. The view is from close to the [001] axis.

According to the model, constituent subclusters share edges which tile a dodecahedral surface inside the cluster [9]. This surface is illustrated in figure 2 and is sandwiched between two icosahedral surfaces to make a three dimensional tile. The closed dodecahedral surface forms between 'triple points' at the meeting of three adjacent subclusters. Though the dodecahedral tiling is a closed surface, the sandwich structure is pseudo space filling. The pseudo space filling operates around 'hopping sites':

Definitions:

- A Triple point is a point where vertices of three adjacent icosahedra meet. In the model for quasicrystalline Al_6Mn, this site is occupied by an Al atom. The cluster is held together by a network of 20 triple points.
- A 'hopping' site is a pair of sites on two adjacent icosahedra such that the sites are separated by a distance (typically τ^{-1}) less than the diameter of Al (~1). The sites are accommodated either by one of them being vacant, or by hopping of the Al between sites.

Thus the triple points form a highly coordinated and strongly bound closed network on the corners of a regular dodecahedral surface.

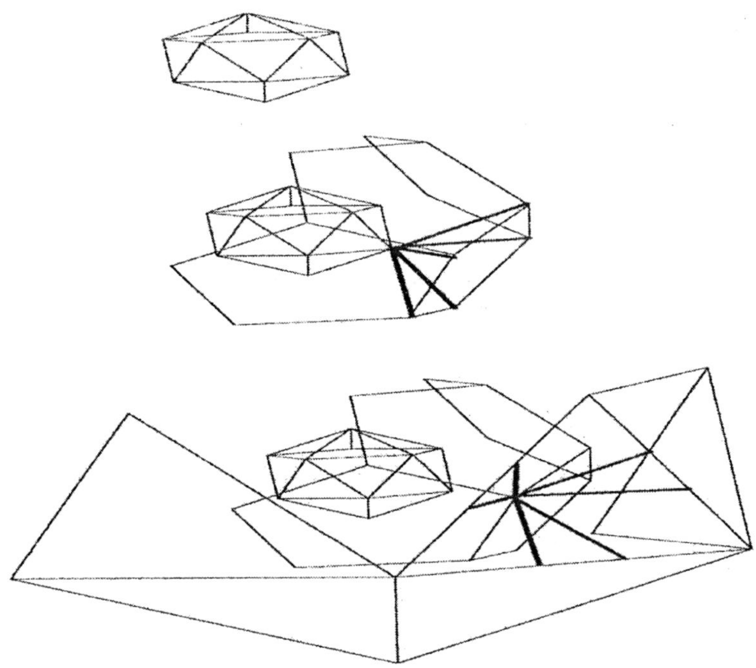

Figure 2. (Center), the shared edges of a cluster of icosahedral subclusters (figure 1) map onto the surface of a regular dodecahedron, shown cut away. Here, the sides are pentagonal with unit length. The dodecahedron is sandwiched between (top) a regular icosahedral 'hole', of (triangular) side length $1/\tau$, and a regular icosahedral shell (bottom) of side length τ^2 Thick lines represent subcluster edges, of unit length, that join the 'hole' to triple points at dodecahedral corners (center). The triple points connect, by subcluster edges, to shell sides (bottom). The sandwich is a 3-dimensional tile that 'stretches' with factor . The group theoretical symmetry is identical in the three concentric Platonic solids (reprinted from Bourdillon [9]).

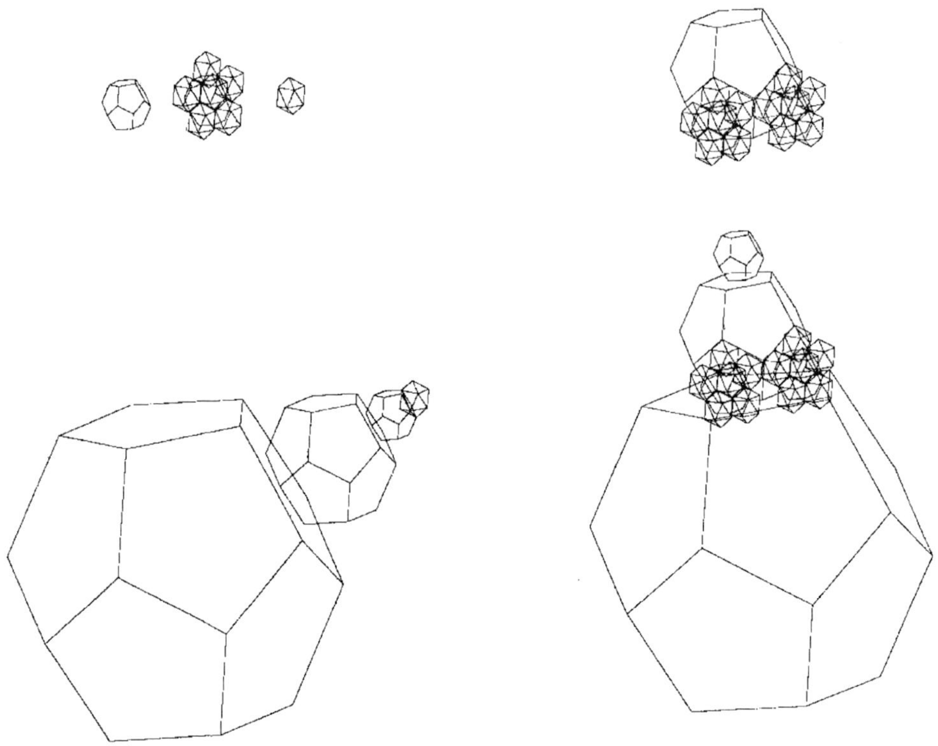

Figure 3 shows how the tiles extend in space. At lower left, Dodecahedral tiles for a supercluster order 2, for a constituent order 1 and a further constituent cluster are shown in relative positions. The cluster dodecahedron shares edges with a corresponding icosahedral subcluster. At upper left, a cluster made from icosahedral subclusters on a dodecahedron skeleton. The cluster is itself icosahedral with side τ^2, and with central icosahedral 'hole' of side $1/\tau$. At upper right cluster and subclusters on a supercluster order 1. Lower right, clusters share edges with dodecahedra on superclusters orders 1 and 2.

The figure shows, in broad lines, connections between the three surfaces. These are icosahedral edges of unit length. The dodecahedron is the reciprocal Platonic solid to the icosahedron. The two have the same point group symmetry and combine concentrically in the figure so as to produce the icosahedral diffraction pattern.

A concept can be so economical that it contains special interest even without supporting evidence. So it was that several years passed before experimental evidence confirmed general relativity. Special relativity was equally economical, but in this case the Michelson Morley experiment had already demarcated the solution. Evidence available for logarithmic periodicity in quasicrystals is wider but less conclusive. We understand this is because the structure is, on the one hand, defective so that diffraction is inconclusive; while on the other, high resolution electron microsocopy (HREM), which is notoriously ambiguous, becomes most convincing when it is pinned to a model.

3. Properties

3.1. Observations

The concept of the logarithmically periodic solid has many explanatory advantages. These are made obvious by the representation by golden triads.

3.1.1

It is easy to see in figure 1b that the golden triads are aligned. The subclusters are therefore all uniquely aligned with icosahedral symmetry, and so also are the clusters and superclusters.

3.1.2.

It is easy to see that the diffraction pattern is discrete. The subclusters are centered on virtual lattice points that can be described by $a+b\tau$ where a and b are positive or negative integers.

3.1.3.

It is easy to see how the structure can be extended infinitely with unique symmetry.

3.1.4.

It is easy to demonstrate the match between optimum defocus HREM [13,8] and the clusters and supercluster orders 1 and 2.

3.1.5.

It is not hard to simulate diffraction patterns from models, which confirm the section for supercluster order 2 [9].

3.1.6.

It is easy to calculate 'structure factors' for the ideal solid to compare with electron and X-ray diffraction. This ease depends on two convenient features. Indexation can be performed by simple inspection [8], after noticing the effects of double diffraction [9,8]. Secondly, all atoms in the regular and extensive structure can be located easily on Cartesian coordinates. It turns out that the calculated beam intensities, make an excellent match with ranked intensities derived from the data of Shechtman et al [1,2]. The result is significant evidence for the model and will be discussed in further detail below.

Many types of data have been shown to correlate with the structure and this is partly because of the convenience of the structural representation by triadic golden rectangles. The advantages described above are part of the evidence, i.e. in support of the conceptual economy of the theory.

---

2 after taking into account the presumed transcription error mentioned earlier and discussed below.

3.2. Consequences

3.2.1. Indexation

Given a diffraction pattern from a three dimensional solid that does not follow Bragg's law, how are the dimensions of the diffracting elements to be defined? A significant advantage of the logarithmically periodic solid is that it is infinitely extensive, properly decorated, and defined. How then are those dimensions to be attached to the individual diffraction beams? The first step is indexation. Icosahedral symmetry is most easily indexed on a cubic reciprocal cell [8,4,5,6][cf. 3] and this has the added advantage that Cartesian coordinates can be used to locate features in real space. Knowing the atomic radii of Al and Mn in metals, the next step is to assign indices to the individual diffracted beams. Using Bragg's law as an approximation, it is possible to try to fit the data to simulated diffraction beams. In this process there is good news and bad news. Even the bad news is good and bad. The first lucky strike is that there is only one more or less obvious attribution that can be made to fit. It is the one that indexes the third decagon (or ring), the bright one in the 5-fold diffraction pattern, according to the lattice spacing $\tau/2$, i.e. half the length of a golden rectangle. Intuitively, it is not surprising that this is the interplanar space that produces the brightest diffraction and this was subsequently confirmed by calculation. The attribution is analyzed in appendix A of ref [8]. Unfortunately, the fit is unacceptably bad as the atomic radii are too large by 10%. This is the first bad news but we proceeded nevertheless. The first indexed beam is labeled $(2/\tau,0,0)$ and the rest of the indexation follows by inspection [9], after special account is taken for the double diffraction in the 2-fold axial pattern described below. The reciprocal unit cell is the unit cube in icosahedral units, based on the edge length in the icosahedral subcluster. Methods of indexation are not generally unique, but the one adopted is unique in having a unit cubic cell in reciprocal space. The method is again unique in having the cell specifically decorated. The unit will be scaled to experimental values in the context of modeling. Atoms must fit allotted spaces. With the resulting indexation, 'structure factors' for the logarithmically periodic structure can be approximately calculated using conventional formulae but applied to a very large cluster. The method calculates the scattering due to each atom in the cluster by adding, in each diffracted bream, the scattered amplitudes and squaring the final addition for intensities.

Indexation of the 5-fold axial pattern and of the 3-fold axial pattern in Al_6Mn is more straight forward and logical than is the case for the denser 2-fold diffraction pattern. Here the pattern is found to be constituted from two parts, a Fibonacci part in geometric series and a Bragg part in regular linear series [8,9]. The two interact through double diffraction and this appears as the explanation for the high density of the pattern. The two parts have simple patterns: a vertical-horizontal cross and a diagonal X-cross. When the patterns are recognized, indexation proceeds in the obvious way by vector additions [9]. Having the unit cube as the unit cell in reciprocal space facilitates the ensuing calculations.

3.2.2. The Compromise Spacing Effect

Using the derived indexation, the 'structure factors' for the model were calculated in the conventional way. Again the calculation is preliminary and the second level of bad news appears. The preliminary approximation involves ignoring less certain parts of the structure at the 'holes' [8] of superclusters and limiting the calculation to a finite order of supercluster. A third order supercluster contains 250,000 atom sites and provides stability in calculated values. First results showed very weak diffraction with scattered intensities. They conflicted with experimental data. However this finding was followed by double good news. When the scattering angle was scanned, the diffraction peak was found at a smaller angle than would be predicted by Bragg diffraction appropriate to applied indices. Scattering factors maximized, consistently, at angles 0.947 smaller than the prediction based on Bragg's law. Analytically, the difference correlates with a compromise in the multiple interplanar spacings [8] that contribute to each diffracted beam in the geometric series.

3.2.3 Dimensions

There is a most important consequence: the quasi-lattice parameter is 5% larger than previously proposed. The corrected value is 0.218 nm. This parameter corresponds to the value of 0.412 nm for the length of the subcluster. The atomic radii are given extra space in the modeling and this is critical. Supposing that some Al electrons are hybridized onto the central Mn atom in each dense subcluster, the space allotted by the model is in the acceptable range. The CSE that has just been described is a new physical effect that is both derived from, and gives support to, the concept of the logarithmically periodic solid. The bad news turned out good and there are further consequences.

3.2.4. Enthalpy, the Driving Force

Figure 4 represents two Mn atoms: one rattles in the face-centered cubic matrix; the other is tightly centered in a subcluster. Each is twelve fold coordinated. In metals. Mn has a diameter 12% smaller than Al. Knowing the true quasi-lattice parameter, the two volumes can be compared. Based on the experimental value for d_0, the subcluster is 17% smaller. The corresponding advantage in enthalpy appears as the driving force for the logarithmically periodic solid. Having this explanation for the structure, it is natural to progress to further evidence for this type of solid.

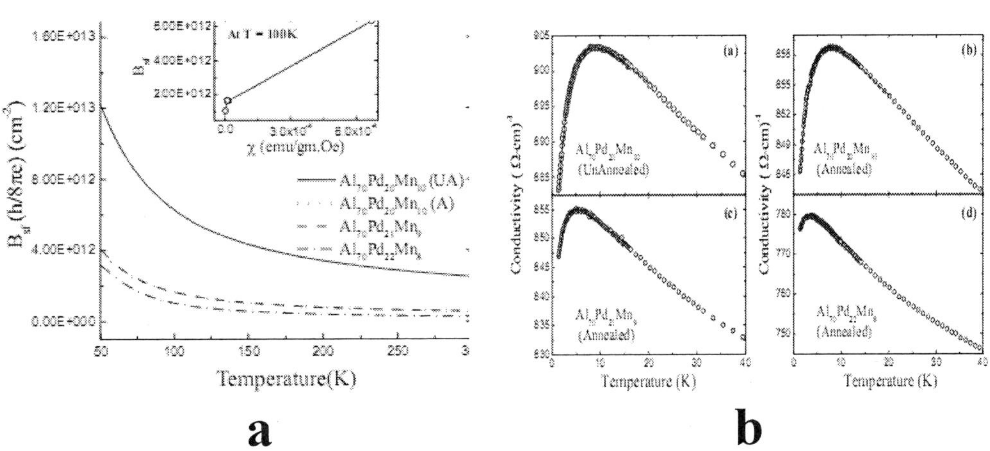

Figure 4. (a) Illustration of (111) planes of cross-sectioned atoms in matrix fcc Al, containing dissolved Mn, in solid fill at the center. The smaller size of the solute in the close packed structure gives room for "rattle".(b) View of corresponding atoms in the icosahedral subcluster oriented with the three-fold (111) normal close to vertical. The atoms in the central plane through Mn are offset, as indicated by arrows, so that all adjacent Al atoms are separated by the same unit icosahedral side length as in figure 1a. The central Mn atom is tightly bound with no rattle. The volume occupied by the structure (b) is 17% smaller than (a) (reprinted from Bourdillon [8]).

3.2.5. Angular Filtering

Angular filtering explains the sharp diffraction patterns that are produced by material that is structurally defective. As part of checks for the consistency of simulations, the 'structure factor' program was run on some known solids and some speculative idealizations for the structure. When 'structure factors' were calculated for matrix fcc Al, using a sample with a similar number of atoms as a supercluster order 3, the lines were found broader than for the logarithmically periodic solid. It became clear that the Fibonacci series interplanar spacings work as a narrow bandpass angular filter for the diffraction [9]. This result extends to rapidly solidified material where defect densities are high.

3.2.6. Double Diffraction

The Fibonacci sequence is conducive to double diffraction. Each member of the series is the sum of its two preceding members. Each member is also τ times the preceding member, i.e. in geometric series. Two members in sequence can double diffract to produce a third member. This can occur as easily as double diffraction between orders in Bragg diffraction from crystals.

The 2-fold axial pattern was originally indexed using double diffraction as a hypothesis. It explained the high density of the pattern and its derivation from two different types of pattern. The hypothesis is confirmed by evidence for this type of diffraction in convergent beam electron diffraction (CBED) [14]. Though double diffraction is mentioned here as a property, the evidence that it provides for the whole structure will be further discussed in section 4.2.2. The physical origin is described in ref [9].

3.2.7. Electronic States

The logarithmic periodicity in the diffraction patterns in quasicrystals suggests that their electronic band structures are likewise logarithmic. We consider two regimes: the low energy states -10eV<e<10eV that occupy the valence and lower conduction bands; and the high energy states e~100,000 eV, typical in transmission electron microscopy. The Bloch wave formalism applies to both regimes.

Quasi Bloch waves are defined by the following procedure. Solving the Hamiltonian for a free electron, the operator equation can be written simply for the one-dimensional case:

$$\left(\frac{1}{2m}p^2 + U(x)\right)\psi(x) = \left(\frac{1}{2m}p^2 + \sum_m U_m \exp(iG_0\tau^m x)\right)\psi(x) = \varepsilon\psi(x) \qquad (1)$$

i.e. similar to the crystalline case [15] where the potential is now summed over modified reciprocal lattice vectors:

$$\sum_G U_G \exp(iG.x) = \sum_m U_m \exp(iG_0\tau^m.x) \qquad (2)$$

with $G_o=2\pi/d_o$, i.e. 2π times the inverse of the quasi-lattice parameter [ref. 8 appendix A]. Notice that the reciprocal lattice vectors, G, have become logarithmic. There is however a special complication because the zero order beam coincides with the lattice vector $G_o\tau^{-\infty}$, and $U_{-\infty}$ is therefore finite. Meanwhile, since superclusters are centro-symmetric:

$$U_g = U_{-g} = U_g^* \tag{3}$$

so that U_g is real. Allowing for finite $U_{-\infty}$, we then find that a quasi-Bloch theorem can be applied, with the bases:

$$\psi_m(x) = u_m(x)\exp(iG_0\tau^m x) \tag{4}$$

where

$$\psi(x) = \psi(k.x) = \sum_m C_m(k)\exp(iG_0\tau^m.x) \tag{5}$$

This equation is similar to ref [16] equation 9.7 with the differences that the vector $(\boldsymbol{k}+\boldsymbol{g}).\boldsymbol{r}$ is altered and simplified to $G_0\tau^m x$ while the summation is taken over m instead of $\boldsymbol{g}$. The solution follows as a set of equations:

$$\{K^2 - (G_0\tau^m)^2\}C_m(x) + \sum_h U_m C_{m-h}(x) = 0 \tag{6}$$

where h is finite, and the magnitude of the electron wave vector

$$K = \sqrt{\frac{2m\varepsilon}{h^2} + U_{-\infty}} \tag{7}$$

The last term under the square root sign being the mean crystal potential. With these adaptations, the low energy and high energy regimes can be analysed. The two beam case and multiple beam cases in high energy diffraction can be analyzed in the normal way [16] and graphically represented as below. This analysis applies to radial systematic rows, but with a little vector adaptation, intermediate beams in the diffraction patterns can be included.

The band structures in figure 5 allow for reflections at quasi Brillouin zone boundaries. The scales are logarithmic in fig 5a, and linear in figure 5b. The latter is useful for comparing with the conventional description of crystal band structures. The bands are quasi reduced by including only those above the primary free electron band, dependent on k^2. The bands in the figures correlate changes in wave vector, whether increasing or decreasing, with changes in wave energy. Thus the slope for the primary is 2, with increasing k, and the slope is -1 when k is decreasing: the same change in wave vector, whether increasing or decreasing, corresponds to the same change in energy. The nearly free electron band structures are consistent with the logarithmic diffraction patterns and with the general concept of logarithmic periodicity, but they are less dependent on details of the structure, e.g. decoration of cells.

These low energy electron bands describe valence states close to the Fermi level and also low energy excited states, $-10\text{ev}<e<10\text{eV}$. The Fermi level should occur at the filling of about three bands. The lowest energy bands have very long wavelength. Their relatively high density tends to stabilize the logarithmic structure.

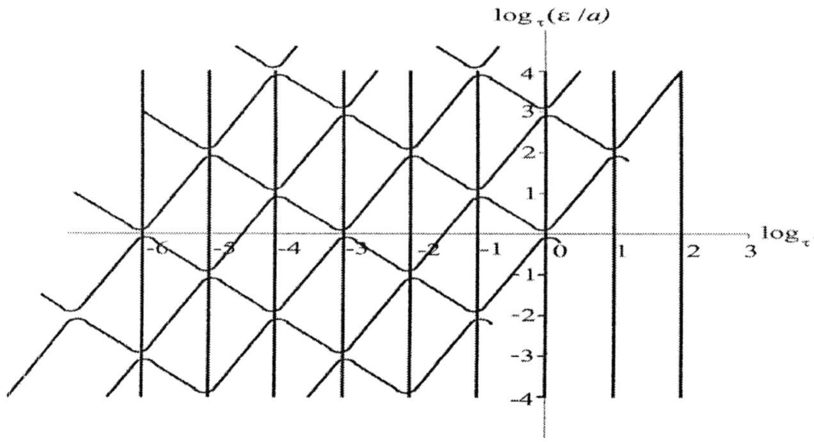

a

Figure 5a. Logarithmic, nearly free-electron band-structures for an ideal, logarithmically periodic, solid. Vertical lines represent quasi Brillouin zone boundaries. The constant, $a = \hbar^2 / 2m$. Units are icosahedral, with unit side length on the subcluster (reprinted with permission from Elsevier [2]).

b

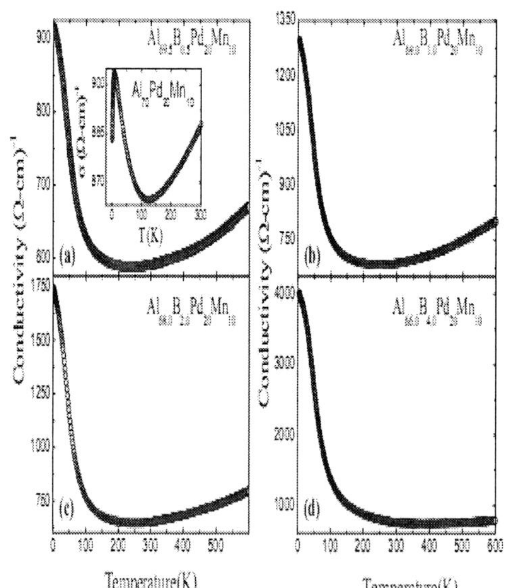

Figure 5b. The same band-structures plotted on linear scales (reprinted with permission from Elsevier [2]).

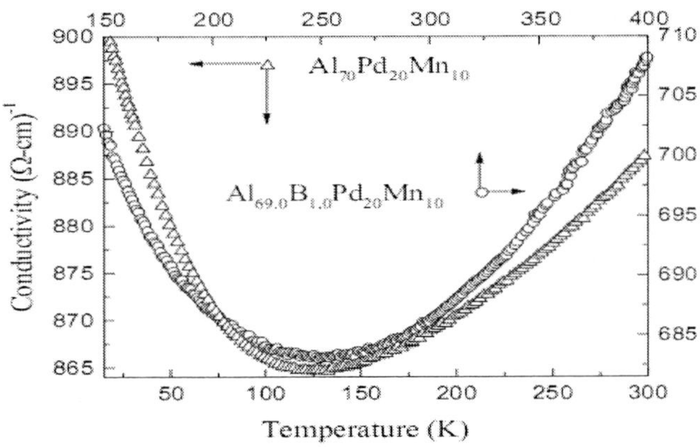

Figure 6a. Dispersion curves (hyperbolic thick lines) spanning three reciprocal quasi-lattice vectors (on vertical thin lines 1,2,3). The prime on the horizontal component of the wave vector kx indicates it is normalized to the reciprocal lattice vector between beams 1 and 2. Brillouin zones are bisecting planes on the vertical dark lines and do not coincide with diffracted beams at the thin verticals (cf. crystals) (reprinted from Bourdillon [9]).

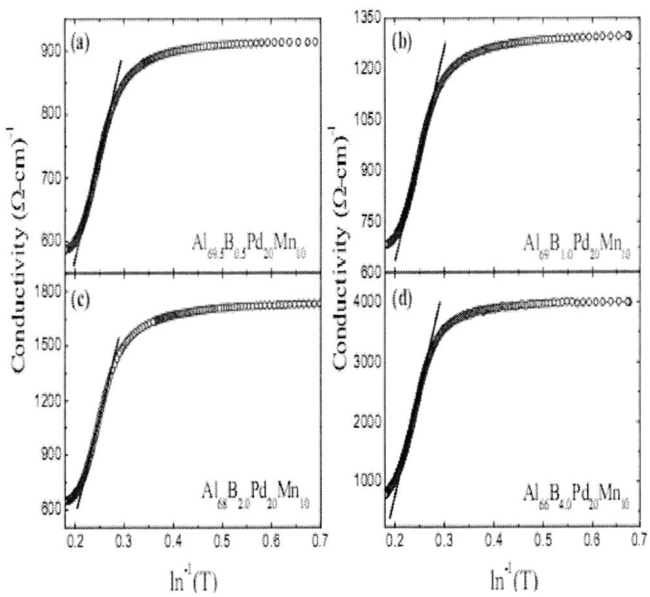

Figure 6b. Lin-log wave vector dispersion curves (dashed lines) on a systematic row of 6. Arcs (full lines) are of the same type as in figure 4a but plotted on a different scale. The scale is broken near zero order. Dotted curves indicate how the bands may be expanded for longer rows (reprinted from Bourdillon [9]).

At much higher energy, dispersion curves describe the motion of highly excited states, for example 100 kV electron beams used in imaging. A discontinuity in the abscissae (figure 6b) allows a representation that runs approximately from positive logarithmic values to negative.

HREM simulation programs [17] require adaptation to the peculiar periodicity inherent in quasicrystals. The logarithmic periodicities in the quasi-Bloch waves correspond to translational properties in Fourier transforms (FT). Where the transform,

$$FT\{f(x)\} = g(k), \quad (7)$$

in one dimension for simplicity. It follows, in translation [18],

$$FT\{f(x+a)\} = \exp(iak)\, g(k). \quad (8)$$

Then

$$h(a) = FT\{g(k)\} \quad (9)$$

where equation (8) is a mathematical relation and (9) provides a distribution of discrete values for the locations of quasi Bragg planes, h(a), in logarithmic periodicity. Conversely, with exponential multiplication,

$$FT\{\exp(i\alpha)g(k)\} = f(x-\alpha). \quad (10)$$

By writing τ^{-m} for a or α, the transference to quasicrystals is obvious. Notice a significant difference between crystals and quasicrystals. In both cases solutions to the Hamiltonian equation produce wave responses. In the case of crystals, the linearly periodic waves superpose, whereas in quasicrystals, translations occur by the product of wave functions and corresponding summations of indices. An illustration of these phase relations is given in appendix 1. Geometric phase shifts are transferred from the quasi lattice to the wave functions in a way that corresponds to the arithmetic phase shifts in Bragg diffraction. The phase shifts carry corresponding momentum transfers to diffracted waves in high energy beams.

The transforms described by equations 7-10 may be extended to three dimensions. This extension is conveniently implied in the 'structure factor' calculations to be described below. The equations describe mathematical effects of translation within logarithmic periodicity. There is an outstanding need for novel HREM simulations of quasicrystal defects and of 'holes'. For example, an explanation for the 'hull shaped' inserts at the supercluster center has previously been proposed [8] and so has the 3-fold structure at cluster centers in the 5-fold axial pattern [9]. It would be interesting to correlate these after development of the appropriate software. In QC diffraction, the quasi-lattice shifts the phase of the scattered beams. After lens aberrations and defocusing of diffracted beams, the HREM image maps the foil potentials.

4. Evidence

When the logarithmically periodic model was first proposed, it was as an ideal model that explained many of the most important features in the data from quasicrystals. These features included the symmetry and the sharpness of the diffraction due to both the virtual grid and to the inifinite extensibility of the structure. The model is also consistent with stretching in tiling geometry, except that our tiles are edge sharing within 3-dimensional space. The structure allowed for simulations of the diffraction patterns which were facilitated by a convenient system of indexation. The simulations were consistent with logarithmic periodicity and implied fundamental modifications to Bragg diffraction, including the absence of high orders and the new CSE. With this effect, dimensions could be attributed that fitted the model. From this point the correspondence between model and data became uncannily close. Diffraction pattern intensities were correctly simulated. An error was observed in the original data [1,2]. When the model was compared with HREM, a remarkable likeness was again found. The model proposed for an observed specimen was used to simulate icosahedral diffraction at a section of supercluster order 2. This information is further confirmation for the efficacy of the model. Other effects were also consistent with data so that the evidence for logarithmic periodicity proved more incisive than for any other model.

4.1. Simplicity, Symmetry, and Sharpness

Several types of evidence for logarithmic periodicity have been described in previous work. The purpose of this section is to return to the core evidence with both brevity and clarity. The prelude to the evidence has already been described in forgoing sections, namely the *simplicity and comprehensive application of the basic concept*, including its consequences. *The basic concept is the driving force due to the dense subcluster which shares edges to propagate icosahedral symmetry.* The extension is infinite. A concept is not normally regarded as evidence but remarkable consistency in its application is a special type of evidence that is sometimes described as beauty in a theory. In the following sections the description that is needed is for independent experimental tests. Following this outline of evidence, it will prove useful to discuss the extension of the model. Ideally it extends to infinity, but in realization there are limitations, and this is where evidence is likely to become not universal, but specimen dependent.

Above the range of supporting evidence for logarithmic periodicity, three types stand out: high energy electron diffraction; simulations of the diffraction; and HREM from thin foils. models. Meanwhile, double diffraction is observed in CBED and a test of detailed symmetry is described.

4.2. Ranking of Beam Intensities and Calculated 'Structure Factors'

Initially, 'structure factors' for the ideal model were calculated as an exercise. Surprisingly they made an excellent match with ranked experimental intensities (table I) from electron beam diffraction patterns. The calculations were performed without fudge factors and were made possible because the model is properly decorated with each atomic site identified. The confidence level is therefore high, but is reduced before defects at 'holes' and elsewhere are fully accounted for. The defects contribute minority effects, and so diminish the confidence only a little. Their insertion, including further definition of the defects, remains a refinement for the future. The result that compares calculated intensities with data is given first because of its significance. Subsidiary steps to this result are further evidence and are indented in sub sections below.

The method used in the 'structure factor' calculations is described elsewhere [8,9]. A few features will be mentioned here. The formula used is standard but the supercluster cell is large. All Mn atoms are counted once, since they are central in subclusters. Al atoms are of four types. At triple points the applied atomic scattering amplitude of Al is divided by 3 and the scattering power is counted three times. At 'hopping sites' the atomic scattering amplitude is divided by two to account for 50% occupation. At cluster 'holes' the scattering amplitude is divided by 4 to account for occupancy. At sites on edges shared between clusters, the scattering amplitude is halved. This becomes the default value for the Al scattering in superclusters. The value conveniently reduces the effect of outer shells where the atoms are numerous. The effect contributes to observed convergence in values with increasing supercluster size. The calculated intensities depend on the size of supercluster evaluated, and are relative. Notice that the cluster contains 12 subclusters. Each subcluster has 5 'triple points' and 6 'hopping sites'. The occurrence of 'hopping sites' does not affect expected stoichiometry, because the scattering power there is the same as from Al atoms shared on edges between cells.

While the 5-fold pattern and 3-fold diffraction patterns are relatively uncomplicated and correlate well with calculated 'structure factors' and intensities [8], the 2-fold pattern provides the best test of the method. This is a consequence of both the high relative density of the diffracted beams in the last pattern, but also because of the complexity of the pattern. Figure 7a shows how the two component patterns – one Fibonacci; the other Bragg – superpose by double diffraction. The indexation convention shown in figure 7b was adapted to the calculations based on the cubic unit cell in reciprocal space.

The overall computed result, while convincing, is unique since no fudge factors were employed [cf.3,4,5]. The 'structure factors' that have been calculated for the well defined and properly decorated ideal model have been useful in explaining a large number of other features in quasicrystal data. Some of those features are briefly described in the following sub-paragraphs. We shall see that when the diffraction data is confirmed by HREM, the evidence becomes doubly convincing.

Meanwhile a critical comment should be made about the use of structure factors in structure determination by electron or X-ray diffraction. The comment is necessary because quasicrystals pose a special case.

In high energy electron diffraction, pattern symmetry is a more common structural determinant than beam intensities. This is because, in crystalline specimens, many details contribute to the generated intensities besides structure factors [16]. Some of these details are specific to a measurement including precise specimen orientation, foil thickness, deviation parameter, foil bending, extinction distance, local specimen temperature, diffuse scattering including Kikuchi lines, etc. Moreover, at highly diffracting conditions, intensities oscillate with foil thickness by the pendulum effect. In axial patterns from crystalline specimens, the reflection sphere determines intensities in zero order and higher order Laue zones; but in the dense, 2-fold pattern from quasicrystalline Al_6Mn that sphere does not dominate the intensity distributions. It becomes feasible then to consider the 'structure factors' for discovering the atomic structure, as is done by X-ray diffraction. In crystalline specimens, such as silicon, there are only two values of structure factor and these make reflections allowed or forbidden. The patterns produced by quasicrystals are of a different type. The method we adopted was therefore to compare calculated structure factors with measured values. The thinner the specimen, the more accurately will our 'scattering factors', in the kinematic approximation [16], simulate diffracted intensities.

X-ray diffraction patterns could, in principle, provide similar data. The absorption rate is much lower, and so therefore is the atomic sensitivity. Specimens are therefore larger. However, the binary quasicrystals are microscopic and therefore inaccessible. Ternary systems add another degree of freedom for their structures and can be grown to larger dimensions. They also give rise to further complications so we are not concerned with them here.

Table I shows an excellent match between the calculated intensities and rankings that can be observed in the data of Shechtman et al. [1]. The indices correspond, in symbolic form, with those shown on figure 7b and with quasi-Miller indices based on the cubic unit cell in reciprocal space. In quasicrystals the diffraction peak is shifted from the Bragg angle expected from any particular quasi-Miller index owing to a compromise arising from multiple contributing spacings. This is the CSE described above and illustrated more comprehensively elsewhere [8,9]. The intensity is the sum of squared scattering factors calculated from peak height times full width half maximum. On the bottom line of the figure is the calculated intensity, within noise, for a line which is forbidden but appears in one of Shechtman's diffraction patterns. This line should be compared with E1, ranked 9! The calculation enables us to predict the reorientation and this has been supported by quasi Bragg planes drawn on our model [ref. 8, appendix D].

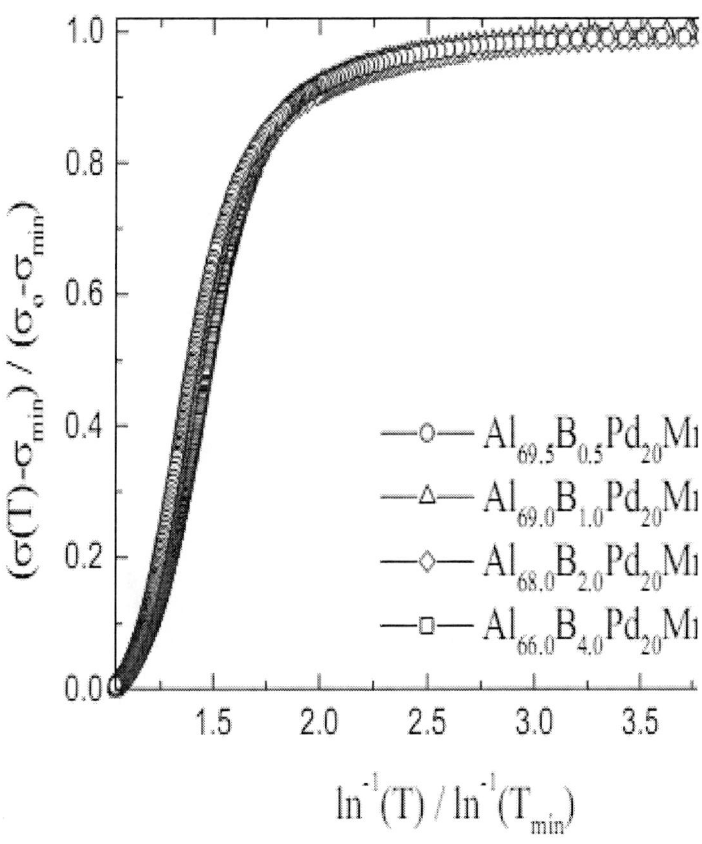

Figure 7a. Composition and indexation of two-fold [001] axial diffraction pattern (bottom) for Al6Mn, based on a Fibonacci cross and Bragg diagonals (top). The two systems double diffract (center) and superpose (reprinted from Bourdillon [9]).

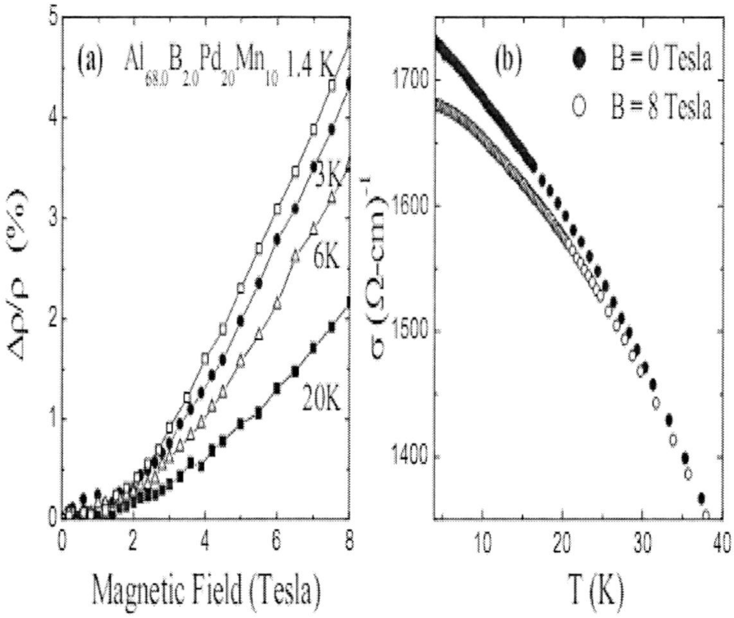

Figure 7b. Indexation convention derived from the double diffraction illustrated in figure 7a and used in structured factor calculations used in table I. Brighter diffracted beams are filled. They include the cross and the Bragg diagonals. (reprinted from of Bourdillon [8,9,]).

4.2.1. Logarithmic Periodicity

Logarithmic periodicity is a characteristic feature in quasicrystals. Not only are the higher orders logarithmic, but higher order Bragg diffraction, with exceptions as in the 2-fold axial pattern, are forbidden. Logarithmically periodic diffraction patterns are supported both by the evidence of 'structure factor' model calculations and by illustration (appendix 1).

The unusual periodicity is a consequence of Fibonacci variations in interplanar spacings. They occur naturally in the logarithmically periodic model. Maps have been drawn elsewhere [8] that project quasi Bragg planes in a supercluster onto two dimensions. The variations in spacings filter scattered waves into logarithmic series. The filtering depends on the properties of the Fibonacci sequence operating on the exponents of wave formulae, and they are simulated through 'structure factors'..

4.2.2. Double Diffraction in CBED

Double diffraction was used to explain the dense diffraction pattern on the 2-fold axis. It was implied in the indexation and calculations. There is also strong evidence for this interpretation in fortuitous convergent beam electron diffraction (CBED) [14]. The data was fortuitous because it was observed in a thin foil with a horizontal planar defect. This caused interference fringes to occur in the convergent beam pattern. The interference fringes are typically tangential to the diffracted orders. However, the interference lines were frequently oblique, and this was understood as a consequence of double diffraction.

In the 5-fold axial pattern, Bragg higher orders are forbidden both in primary diffraction and in double diffraction. The double diffraction occurred only at large angles to the primary, i.e. in the second dimension of the 2-dimensional pattern. The physical origin of double diffraction is discussed in detail elsewhere [9].

Table I. Excellent match between calculated intensities with ranked experimental values for the beams indexed on the positive quadrant in figure 7b

Index symbol	h,k,l,	Intensity	Experimental rank
	0,0,0		10
A1	$2\tau^{-3},0,0$	20k	6
A2	$0,2\tau^{-3},0$	20k	6
A3	$-2\tau^{-3},0,0$	20k	6
A4	$0,-2\tau^{-3},0$	20k	6
B1	$2\tau^{-2},0,0$	300k	7
C1	$2\tau^{-1},0,0$	570k	9
D1	$2,0,0$	540k	8
E1	$\tau^{-1},1,0$	960k	9
F1	$2\tau^{-1},2,0$	230k	5
F5	$3\tau^{-1},1,0$	132k	5
E1+B4+B1		89	3
E!+B4+A1		57	3
E1+B4		545k	8
E1+B4+A3		37	close to 0-order
A2+C1		4.3k	4
A2+B1		2.6k	4
A2+A1		230	3
E1+A4		230	3
E1+A4+A3		1.4k	3
E1+B4+A2		680	1
B2+C1		75k	5
B2+B1		25k	4
B2+A1		250	3
E1+B1		330k	5
E1+A1		1.2k	3
E1+A3		780	5
C2+C1		200k	5
C2+B1		70k	4
C2+A1		1.9k	2
E1+A2		15k	5
Rotated E1	Forbidden	17	0

4.2.3. Bragg Anomaly in the 2-Fold Pattern

The logarithmically periodic solid also provides the explanation for the Bragg anomaly in the diffraction pattern on the 2-fold axis. The pattern is anomalous in the quasicrystal because the beam spacings in the diagonal X-cross are not in the geometric series that is most usual in quasicrystals, but in linear series. The atomic maps, that are produced by projections of a supercluster onto a plane [8], showed planes of atoms that are the scattering agents of diffraction. Normal to the diagonal direction, planes can be seen some of which have regular spacing, while others follow the more common – in these solids – Fibonacci spacings. The latter do not diffract, but the former do. Why does this happen? The logarithmically periodic solid reveals the reason and it is supported by 'structure factor' calculations. The reason is that while the linearly spaced planes are aligned, the Fibonacci spaced planes are not [8]. All other planes that scatter in logarithmic order – from all three major axes – are aligned. The scattering from the aligned planes is coherent; from misaligned planes it is incoherent. We know of no other model that can provide this type of explanation.

4.2.4. 2-Fold Pattern Orientation Anomaly

Likewise, the 'structure factor' calculations support the supposition of the transcription error in the data of Shechtman et al. [1]. Their data are not in fact icosahedral [2] and it is easy for anyone to verify this. The calculations demonstrate what the pattern should be for icosahedral symmetry. No other model had previously demonstrated the fact over a long period, and the fact had not even been noticed at least at the review level. Parity violation in the weak nuclear force was not discovered by such inattention to detail. When detail is not acknowledged, there is little chance for an agreed solution. A theory that discovers and explains the detail has something to commend it. This observation has further importance. In logarithmic periodicity, defects constitute an important but secondary feature. It is not sensible to search for them until the pattern is correctly established. The importance is equally great for other models [e.g.19].

4.3. Diffraction Due to Clusters

If we take as an example the 5-fold axial pattern $[01\tau]$, and isolate the third bright decagon including the indexed beam $(2/\tau,0,0)$ [8,9] then it can be shown, with few exceptions, that the remainder of the pattern can be constructed in principle by double diffraction. There is an important exception, namely the inner decagon that includes $(2/\tau^3,0,0)$. This corresponds to the interplanar spacing characteristic of a cluster. It is the diffraction equivalent of the cluster images to be described in the following section.

4.4. HREM Images of Clusters and Superclusters

After diffraction and its simulation, the third most important evidence comes from the optimum defocus HREM of very thin films [13]. This shows repeated evidence of clusters that match in geometry and dimension the logarithmically periodic clusters. The signature for the cluster shows as a decagon of circles that surround a 3-fold pattern at its center (section

4.4.2). Sectional patterns have been demonstrated that match exactly, in both aspect and dimension, the experimental data, and images have been modeled using triadic golden rectangles. Figure 8a shows a simulation of a section of supercluster order 1 viewed from the 5-fold axis. This section is outlined in figure 8b with a regular pentagon which is in turn copied, in ref. 8, onto the image reprint.

The three patterns are not only strikingly similar but have matching scales. There is a further coincidence. The model is consistent with cluster 'holes' that produce 3-fold patterns in the 5-fold axial pattern (section 4.4.2) and with the 'hole' in supercluster order 1 where ingrowths are simulated as hull structures [8], rounded at one end and pointed at the other. Thus figure 8b models the HREM data of Bursill and Peng. Simulated diffraction [9] from the model in figure 8b, taken in conjunction with incoherent scattering from alternative models, implies that figure 8b is a section of a supercluster order 2. The 'holes' at the centers of both clusters and of the supercluster order 1 are signatures of the logarithmic periodicity. Notice however, that when a specimen foil is more than one cluster thick, ~1.4 nm, these signatures will be masked by overlapping layers. By contrast, equivalent signatures of higher order superclusters are less likely in HREM because of the combined effect of complicating 'hole' inserts and of the thinner sampling sections, when compared with cluster size. The images will then seem more like the 'aperiodic' structures often discussed [3-7], with the logarithmic periodicity hidden.

a

Figure 8a. Signatures for clusters and for a supercluster order 1. Simulated 5-fold axial view of a section of a supercluster order 1, made up of a pentagonal plane of clusters while omitting one (lower) half of the supercluster and opposite apical cluster. Each filled circle represents a subcluster containing one Mn + 12 Al atoms. The subclusters are decagonally arranged in clusters. The cluster centers show 3-fold symmetry due to 3-fold occupation of Al sites near each cluster center. With variations, this 3-fold center is evident in all clusters. The figure identically matches HREM data [13] that is modeled in figure 8b. The contrast corresponds [8] to the photographic negatives.

b

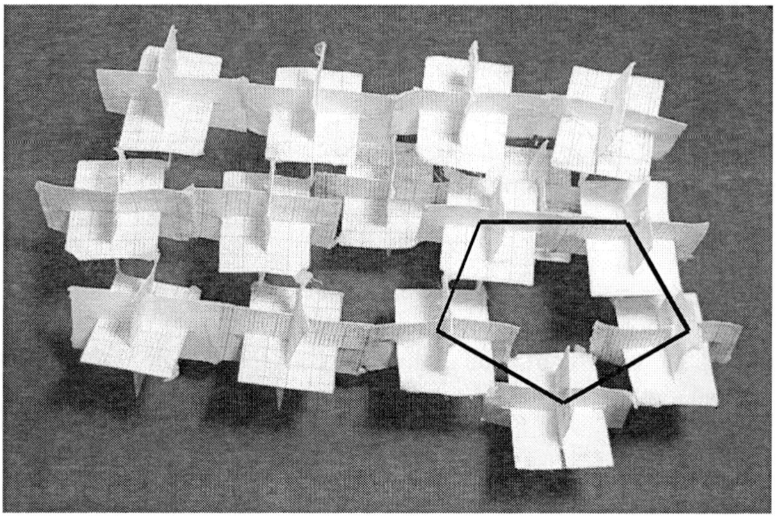

Figure 8b. Golden triad structures representing icosahedral clusters, exactly model the HREM [13]. Each corner of the golden rectangular triads locates an icosahedral subcluster consisting of a central Mn + 12 Al atoms. In HREM, 3-fold centers of regular decagons mark the cluster centers [9]. Note the section of supercluster order 1 outlined with a pentagon that connects cluster centers as in figure 8a. At the center of this supercluster order 1 are, in HREM, 5 hull shapes that are interpreted as ingrowths at the center. Simulated diffraction shows the model is a section of a supercluster order 2. (Reprinted from Bourdillon [8].).

Is there an untutored eye that cannot match [8] the patterns that are common in figures 8a and 8b with the HREM in ref. [13]? Let its owner know their scales also match.

Not all HREM can be modeled in the same way. Similar obstacles apply to imaging of thick specimens as of high orders. In very thin foils, sections of superclusters can sometimes be identified. However, the thicker a foil, the more difficult is it to identify the section being imaged. For this reason, many images acquired with the highest resolution [20] are difficult to interpret in terms of logarithmic periodicity, and loosely appear 'aperiodic'.

4.4.1. 'Structure Factor' For The HREM Model Structure

Corresponding 'structure factors' were consistently calculated [9] for the $[01\tau]$ 5-fold diffraction plane modeled (figure 8b) from the optimum defocus image [13]. Alternative and hypothetical types of model were proved to produce incoherent diffraction. The incoherence in scattering from other models is confirmatory evidence for the supercluster order 2. Even if not unique, the confidence level due to these combined simulations is reasonably high.

4.4.2. The 3-Fold Cluster Center in the 5-Fold Pattern

The 5-fold pattern of Bursill and Peng [13] has a special feature with positive interpretative value and problematic explanation. Identifying the cluster of subclusters by their dimensions and symmetry, the cluster centers are distinguished by 3-fold patterns that vary around a limited set of orientations. Why are they 3-fold and what is their explanation?

Logarithmic periodicity provides an obvious explanation. The 3-fold symmetry correlates with the occupancy of 'hole' sites in the cluster. These sites are arranged on a regular icosahedron of side $1/\tau$, i.e. shorter than the diameter of Al. Most of the sites are therefore 'hopping' sites. The available space allows the occupancy of 3/12 atoms. The image due to the cluster center has been previously modeled in a rough way [9]. The orientation of the 3-fold pattern is not uniform. This corresponds with the fact that there are 20 permutations for the occupation of the three sites. It thus appears that the 3-fold occupation of the 12 icosahedral sites is the reason for the symmetry. The variable 3-fold pattern, that had originally appeared as an anomaly in the 5-fold symmetry, turns out to be confirmatory evidence for the partial occupation of sites in the cluster 'hole'.

5. Uncertainties

5.1. Extension

In concept, logarithmic periodicity extends to infinity. The sharpness of the experimental diffraction patterns implies considerable extension, though whether this extension is a rigid supercluster structure or a more loosely decorated grid is open to simulation. HREM shows structures at the cluster and supercluster levels. The dimension of the cluster is evident in both electron and X-ray diffraction and this is the feature that appears most consistently in optimum defocus HREM. The model shown in figure 8 produces an icosahedral diffraction pattern, apparently because it is a section of supercluster order 2. Beyond this level, evidence becomes weaker. HREM also suggests there are planar defects like grain boundaries [13,8], and these might limit the extension of the logarithmic periodicity. Another planar defect in an electro-polished specimen produced a fortuitous CBED pattern with significant interference fringes (section 4.2.2). What is the path towards establishing the extension of this type of periodicity in binary specimens that cannot be grown into large single crystals?

One path is to understand the growth mechanism and morphology. Another is to understand equivalent structures in ternaries, quaternaries etc. A third path is to understand the defects inherent in logarithmic periodicity, so that X-ray diffraction and other methods can be more accurately applied. These ways are for the future; for the present consider inherent defects in logarithmic periodicity.

5.2. Defects

Pauling observed a long time ago that while the icosahedral structure occurs in crystallography within larger unit cells, the cluster would contain a central hole [21]. Actually, the cluster center is so dense that it has vacancies, and evidence for these has just been described (4.4.2). At the high order supercluster levels the holes become less certain and more problematic.

5.2.1. The Aperiodic Cluster 'Hole'

As mentioned earlier in section 4.4.2, the 'holes' that are imaged in thin foils, with optimum defocus, are signatures to the cluster structure. The centers of the subclusters are unproblematic. The subcluster has a central Mn atom. The cluster centers are less simple. There are 20 permutations for the 9 vacancies on 12 sites. In consequence, two parts to the HREM optimum dcfocus image [13] are a logarithmically periodic part and an aperiodic part (figure 8a). Supposedly, the scatter due to the cluster centres is only weakly diffractive. It seems likely therefore, that the various images arise above destructive interference produced, under defocus, by the zero order beam with the diffracted beams that are due to the logarithmically periodic parts.

5.2.2. The 'Hole' in Supercluster Order 1

Supertiles in logarithmic periodicity form closed dodecahedral surfaces in two dimensions and these surfaces are sandwiched between pseudo space filling icosahedral surfaces (figure 2). In the supercluster order 1, the side on the dodecahedron has dimension τ^2. This lies between the side lengths τ on the 'hole' and τ^4 on the outer shell. Consider the 'hole'. An icosahedron with side τ has length τ^2. Thus the side of the 'hole' is equal to the length of the subcluster. The volume of the 'hole' is greater than the volume of a subcluster and less than the volume of two. How can the space be filled?

Consider first a structure that might be based on the subcluster and then refer to HREM to search for a match. Suppose an Al atom is added to every side of the icosahedral subcluster. The twenty faces translate to twenty corners of a dodecahedron. The diameter is shorter than the right magnitude to fit the 'hole', and the shape does not match. It should be supposed therefore that the defect is more complex. Try continuing along the path taken. Loosely add another atom to the center of each 5-sided dodecahedral face. The resulting icosahedron, loosely packed, has side 6% smaller than τ. The figure is therefore a possible filling for the supercluster 'hole'. It represents a grouping of 45 atoms arranged in the concentric series, central atom, icosahedron, dodecahedron and icosahedron. The group symmetry is constant and one with the supercluster. However it is not clear how such a structure should appear in HREM, so an alternative has also been suggested.

The most common defect structure in the images of Bursill and Peng have been called hull structures. Each is pointed at one end and rounded at the opposite end. In the 5-fold section of the supercluster, five hull structures point towards its center. They are represented in figure 8a and seem to be ingrowths. We have guessed [8] that the hull structures might be semi-subclusters grown on substitutional Mn, but the detail is far from clear. A similar structure is repeated elsewhere in their micrograph, especially along an apparent planar defect that runs parallel to one of the 5-fold lines of symmetry [13].

5.2.3. The 'Hole' in Superclusters of Higher Order

Similar considerations apply to 'holes' of order higher than 1. As there exists no HREM data, any proposed solutions are speculative. At order 2, the side of the 'hole' has length τ^3, where the unit is, as before, the side length of the subcluster. At these dimensions, crystal growth becomes an increasing determinant of defects in the structure.

5.2.4. Glassy Structures

The uncertain structures of supercluster 'holes' reopens the possibility of glassy structures. Logarithmic periodicity supposes an infinite extension of superclusters with increasing order. The problematic 'holes' also increase. Given the binding of subclusters, there is an alternative solidification route that bypasses clusters. To understand this, consider three subclusters agglomerating from the melt. Let them join by edge sharing with a 'triple' point where the three subclusters meet on a single Al atom. The subclusters are identically oriented. The centers of the three subclusters inhabit a plane. Now add a fourth subcluster and suppose that it is energetically favorable for it to join also at a triple point. There are only two ways in which this can be done. They result in either the concave quad and the planar quad (figure 9). The evidence of clusters supports the predominant occurrence of concave quads. Each concave quad has four 'hopping' sites separated by less than one Al diameter. While the concave quad is the unique conceptual basis of logarithmic periodicity, the alternative planar quad facilitates defective structures. Such seems to be consistent with the data of Bursill and Peng and our interpretation of ingrowths at cluster centers and outgrowths at planar defects.

5.3. Limitation to Binary Systems

Logarithmic periodicity has explanatory power in the description of binary quasicrystal systems. Al_6Mn is like other binaries [19] in satisfying the following conditions that are evident for the formations of icosahedral subclusters:

– -a chemical ratio of 6:1 between solvent and solute atoms, and
– -a ratio of their radii Rsolute/Rsolvent= $\sqrt{\tau^2+1}-1$.

Figure 9. Unique set of quads having two triple points: planar quad (left) and concave quad (right). When the triadic golden rectangles represent a subcluster, each rectangle corner locates an atom. The cluster is made from four overlapping concave quads.

Binaries are simpler than ternaries and more complex systems. Scientific method chooses simple systems to build models for others that are less amenable to calculation. Extension of the method would require adaptation, but many of the lessons learned, such as the CSE, will transfer, to enhance and give credence to the more complex and speculative developments.

5.4. Quasicrystal Growth Mechanisms

Crystals grow by attraction of melt atoms to the intersection of a growing plane on the liquid bulk [15]. Typically, the plane rotates about a screw axis as the crystal grows in the direction of the screw. It is difficult to think of a screw axis in a supercluster, so growth is problematic. Screw dislocations are also a product of deformation. Whereas they are commonly found and easily identified in electron microscopy, they are not so evident in quasicrystals. A particular problem that is attached to the growth of the binary Al_6Mn is rapid solidification in the quasicrystal. Observation of growth is therefore extremely difficult. We begin by supposing that the subcluster agglomerates in the melt phase before sticking to a cluster. If the growth of superclusters proceeds by a similar sticking, first at triple points and secondly into icosahedra, then the precipitates would facet as icosahedra. However, to the extent that the growth depends on temperature gradients a directional texture should be anticipated. Since the quasicrystal is a product of segregation [14], the texture could be dendritic when nucleation is slow, or it could be approximately spherical when nucleation is more rapid in larger thermal gradients. It may well be difficult to differentiate the two types owing to the large number of symmetry axes in the icosahedron.

With some assumptions, these suppositions are supported by video observations of growth in a ternary quasicrystal $Al_6Pd._{0.43}Mn_{0.29}$ [22]. Growth was observed in real time by use of synchrotron radiation illuminating a growth cell under Bridgman directional solidification. Depending on pulling rate, planar facets were observed growing on quasicrystals, or, with larger gradients, precipitates were observed to nucleate and grow, developing facets as they grew larger. The images suggest that the growth is by acquisition of planes of superclusters. This would be facilitated by the planar quad configuration rather than concave superclusters, with greater order of supercluster allowing the greater icosahedral symmetry in the bulk.

Pd, with an outer 4d shell, is like Mn, with an outer 3d shell, but the metallic radius of the former is 7% smaller. Both have a radius significantly smaller than the Al atom. A comparison with Al_6Mn would be predicated on an assumption that the stable ternary is structured like the metastable binary Al_6Mn. Our diffraction simulations have shown that planes of the binary clusters are not sufficient to produce icosahedral diffraction patterns in samples smaller than about 100 nm [9]. Measurements of icosahedral symmetry in the bulk therefore imply that growth in Al_6Mn occurs by deposition of preformed superclusters and not by sticking of individual atoms or clusters.

Conclusion

The literature on quasicrystals is massive, but it is also unsatisfactory because it has failed to prove their structure [23,24]. "What is most intriguing, of course, is whether we are concerned with a material having singular structural properties because of the chemistry of *Al* and *Mn*, or whether the principles suggested by the quasi-crystal concept will find more widespread application [13]." The intrigue has lasted one quarter of a century. The motivation for the present work began with an effort to find both the explanation for and meaning of the fortuitous convergent beam electron diffraction [14] that had remained unexplained for many years. The present exploration began with a model that is unequivocally icosahedral and infinitely extensible. On this model it was possible to calculate anew the X-ray and electron responses. Surprisingly, it was found that 'structure factors', experimental diffraction intensities, and optimum defocus HREM images matched the model and did so precisely. The proof depended on some new physical effects that included details of the diffraction and the CSE.

The subcluster is the driving force in the model. The solution is beautiful in both mathematics [11] and in physics. Whatever the extension, the model is so productive it must remain a useful description for the local level. Though there is sufficient evidence for clusters and superclusters up to the second order, the detail of how the structure extends further in reality is not so far verified experimentally. The extension may vary from specimen to specimen depending on formation processes. Ternary quasicrystals and systems that do not follow the binary stoichiometry of 6:1, with appropriate atomic radii, may require different and less physical interpretations for the moment.

APPENDIX 1. QUASI BRAGG DIFFRACTION

The varied interplanar spacings in quasicrystals have many effects that differ from those due to regular spacings in Bragg diffraction from crystals. One feature we called the CSE (section 3.2). Quasicrystal diffraction, like Bragg diffraction, is dictated by interference between waves reflected at Bragg planes. Crystals filter in arithmetic orders, quasicrystals, having spacings in Fibonacci series f^n [11], filter in logarithmic orders. When wave functions multiply, phases add.

$$e^{2\upsilon ir.g\tau/\lambda} = e^{2\pi ir.g/\lambda}.e^{2\pi ir.g/(\lambda\tau)} \qquad \text{(A1)}$$

where

$$f^{n+1} = f^n + f^{n-1} \qquad \text{(A2)}$$

and

$$f^n = \tau .f^{n-1}. \qquad \text{(A3)}$$

Here, the reciprocal lattice vector is represented by g, varying between adjacent planes by factors τ^m, with the logarithmic order m a positive or negative integer. λ represents the wavelength, assumed monochromatic in this treatment. The position coordinate is $\boldsymbol{r}$ and is vectorial, like $\boldsymbol{g}$ in three dimensions. Equations A2 and A3 are equally useful as definitions of the series. The relationships, A1-A3, are illustrated in the figure A1. Notice that the three reflections on the right have the same local condition, $\lambda/2 = d\sin(\theta)$, for respective interplanar spacings *d,* after correction for the CSE. By assumption, the effects are generally the same for X-radiation as for electrons. However, for the latter example, the figure is compressed in the vertical direction because in high energy electron diffraction the reflections are glancing. Then a continuity condition for probability amplitudes of an electron incident on a Bragg plane follows from equation A2. There are three branches into three different scattering angles. A typical quasi Bragg plane bounds several interplanar spacings and causes multiple branches into a geometric series of scattering angles. (The figure suggests that 2nd order Bragg diffraction is theoretically possible in quasicrystals. Compare the left and right sides of the figure. But the high orders are made improbable by the very large scattering angles and deviation parameters on the concomitant high logarithmic order deflections. 'Structure factor' calculations do not support the suggestion, but rather forbid these deflections.) Further detail, including indexed scattering powers from very large superclusters, is given in ref. [25].

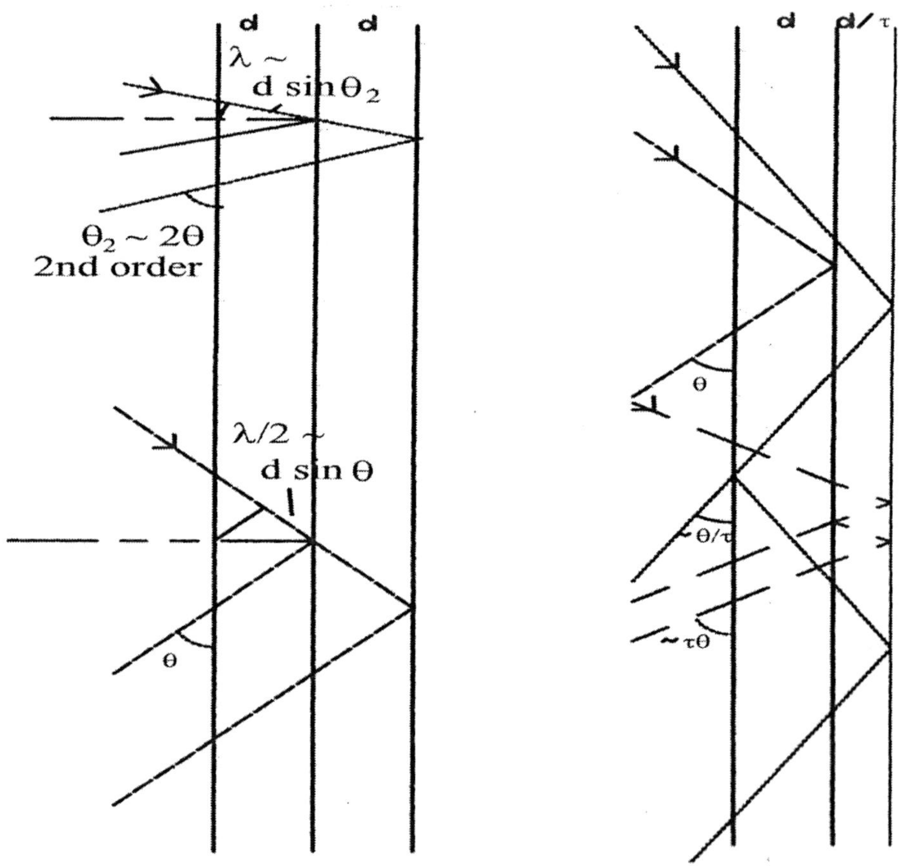

Figure A1. Quasicrystal diffraction (right) compared with Bragg diffraction (left) in regular crystals. Rays represent beams that interfere constructively after reflections from vertical Bragg planes. Approximating for small angles in electron diffraction, scattering angle $\Delta\sigma(T) \approx \pm C_4 \ln^{-1} T \pm C_4 \frac{\ln T_k}{\ln^2 T}$ in quasicrystals or $\Delta\sigma(T) \approx \pm C_4 \ln^{-1} T$ in crystals. In high energy electron diffraction, the reflections are glancing. (reprinted from Bourdillon [8]).

Appendix 2. Lemmas, Proofs and Corollaries

Construction of lemmas and proofs in *Quasicrystals and quasi drivers* [8]. Where prior explanation was needed, lemmas were sometimes placed after proofs in the original argument. Here, proof 3 and corollary 1 have been rearranged in logical order.

Lemma 1.The diffraction patterns can be completely indexed using simple inspection.
Lemma 2. The diffraction has two parts: Fibonacci and Bragg.
Proof 1. The intensities of diffracted beams are correctly calculated.
Lemma 3. The icosahedral subcluster is dense and therefore chemically stable.
Lemma 4. Space is filled by icosahedral substructures plus defects.
Proof 2. The supercluster order 1, clusters and subclusters are observed.
Corollary 1. The driving force is the low enthalpy of the subcluster.
Lemma 6. The Fibonacci series is filtered by alternating Fibonacci interplanar spacings.
Lemma 7. Double diffraction does occur, but only at near normal angles.
Lemma 8 and corollary 2. The measured interplanar spacings are larger in Fibonacci series diffraction than in corresponding Bragg diffraction.
Proof 3. Corresponding defects are observed.
Corollary 3. Atomic planes that are misaligned in the direction of their normals do not diffract.

Reference

[1] Shechtman, D.; Blech, I.; Gratias, D; Cahn, J.W. *Phys. Rev. Lett.*, 53, 1951 (1984).

[2] Bourdillon, A. J. (2009) *Sol. State Comm.* 149 1221-5 (2009)

[3] Elser, V. *Phys. Rev. B,* 32 4892-4898 (1985)

[4] Duneau, M.; Katz, A. *Phys. Rev. Lett.* 54 3688-2691

[5] Levine, D.; Steinhardt, P. J. *Phys Rev B*, 34 596-616 (1986)

[6] Cahn, J. W. ; Shechtman, D.; Gratias, D. *J. Mat. Res.* 1 13-26 (1986).

[7] e.g. Steurer, W. *Z. Kristallogr.* 219 (2004) 391-446

[8] Bourdillon, A.J. *Quasicrystals and quasi drivers,* (2009) UHRL ISBN 978-0-9789839-1-8.

[9] Bourdillon, A.J. *Quasicrystals' 2D tiles in 3D superclusters* (2010) UHRL ISBN 978-0-9789839-2-5. Yamaguchi, T.; Fujima, N. J. *Phys. Soc. Jap.* 57 4206-4218 (1988)

[10] Yamaguchi, T.; Fujima, N.J. *Phys. Soc. Jap.,* 57 4206-4218 (1988)

[11] Huntley, H.E. *The divine proportion, a study in mathematical beauty,* 1970 Dover ISBN 0-468-22254-3

[12] Sadun, L. *Topology of tiling spaces,* American Mathematical Soc. 2008 ISBN 978-0-8218-4727-5

[13] Bursill, L.A.; Peng, J.L. *Nature,* 316, 50-51 (1985). Notice that the authors' suggested model is not properly icosahedral.

[14] Bourdillon, A.J. *Phil. Mag. Lett.* 55, 21-26 (1987)

[15] Kittel, C. *Introduction to solid state physics*, Wiley, 1976.

[16] Hirsch, P.; Howie, A.; Nicholson, R.B.; Pashley, D.W.; Whelan, M.J. *Electron Microscopy of thin crystals,* Kreiger 1977

[17] www.hremresearch.com/Eng/download/documents/HREMcatE.html and www.hremresearch.com/Eng/simulation.html

[18] e.g. Riley, K.F., *Mathematical methods for the physical sciences,* Cambridge University Press, 1974

[19] Takakura, H.; Gomez, C.P.; Yamamoto, A.; De Bressieu, M.; Tsai, P. *Nature materials* 6 58-63 (2007)

[20] In addition to ref 20, see Portier, R.; Shechtman, D.; Gratias, D.; Cahn, J.W. *J. Elect. Microsc.,* 10, 107 (1985); Hiraga, K.; Hirabayashi, M.; Inoue, A.; Matsumoto, T.; *Sci. Rep.Inst. Tohoku Univ. Ser. A*, 32, 309 (1985), Chattopadhyav, K.; Ranganathan, S.; Subanna, G.N.; Thangaraj, N. *Scr. Met.* 19, 767 (1985), Knowles, K.M.; Greer, A.L.; Saxton, W.O.; Stobbs, W.M. *Philos. Mag B,* 52, L31, (1985).

[21] Pauling, L. *Letters to Nature,* 317, 512-514 (1985). Unfortunately the author chose not to evaluate electron diffraction data.

[22] Nguyen-Thi, H.; Gastaldi, J.; Schenk, T.; Reinhart, G.; Mangelinck-Noel, N.; Cristiglio, V.; Billia, B.; Grushko, B.; Hartwig, J.; Klein, H.; Baruchel, J. ArXiv 040910 spotlight quasicrystals (2008)

[23] Senechal, M. What is a Quasicrystal? *Notices to the American Mathematical Society*, 53 886-7 (2006).

[24] Proceedings of ICQC9, Zurich, July 6-11, 2008.

[25] Bourdillon, A.J., Indexed scattering powers in a logarithmically periodic solid, to be published in *International J. of Condensed Matter, Advanced Materials and Superconductivity* (2010).

In: Quasicrystals: Types, Systems, and Techniques
Editor: B. E. Puckermann, pp. 77-105
ISBN 978-1-61761-123-0

Chapter 3

VACANCIES IN QUASICRYSTALS

***Kiminori Sato*[1]**
Department of Environmental Sciences,
Tokyo Gakugei University, 4-1-1 Koganei,
Tokyo 184-8501, Japan

ABSTRACT

Positron annihilation studies of quasicrystals (QC's) and their related materials (crystalline approximants) are reviewed. We describe why a positron, anti-particle of electron, is suitable for probing vacancies locally in aperiodic QC's. A series of positron annihilation spectroscopy is then briefly outlined. Positron lifetime spectroscopy reveals high concentration of structural vacancies more than 10^{-4} in atomic concentration for QC's and crystalline approximants studied. Chemical environments around the structural vacancies are investigated by coincident Doppler broadening spectroscopy. In addition, the concentration of structural vacancies is discussed based on the positron diffusion data obtained by a variable-energy slow positron beam. Besides the structural vacancy, we refer to other two kinds of vacancies: thermally formed high-temperature vacancy and electron-irradiation induced vacancy. Finally, the structural phase transition in QC's probed through the local atomic and electronic structures around structural vacancies is presented.

1. INTRODUCTION

Quasicrystals (QC's) are aperiodically-ordered intermetallic systems with long-range order but without atomic lattiçe periodicity [1]. Due to this unique so-called quasi-periodic structure, diffraction and scattering techniques employing electron, x-ray, and neutron cannot be easily applied for the structural analysis of QC's. Detailed local atomic structure as, e.g., vacancy is thus poorly discussed, though they are expected to be associated with the

1 e-mail: sato-k@u-gakugei.ac.jp.

mechanism of stabilization, high-temperature processes of diffusion, plastic deformation, and phase transition.

We employ a positron, anti-particle of electron, which provides information on vacancy-type defects in materials without being interfered by the structural aperiodicity unlike the diffraction and scattering techniques. Since the early works on QC's by positron lifetime spectroscopy [2], positron trapping sites higher than 10-4 in atomic concentration have been consistently clarified for stable QC's [2-20]. These trapping sites cannot be removed by long-term annealing more than 3 months at the temperature of 1073 K close to the melting point [21]. In the case of plastically deformed [9] or electron irradiated QC's [14,21], no additional lifetime component is detected, because the structural vacancy concentration is so high that the detection of an effect of dislocations on the positron lifetime is hindered.

Here, we review the application of positron annihilation spectroscopy to the study of QC's and their related materials (crystalline approximants). In the following section, a series of positron annihilation spectroscopy is briefly outlined. We first characterize structural vacancies in QC's by making full use of positron lifetime and coincident Doppler broadening spectroscopy as well as the local atomic structures of crystalline approximants. The concentration of structural vacancies is discussed based on the positron diffusion data obtained by a variable-energy slow positron beam. Besides the structural vacancy, we refer to other two kinds of vacancies: thermally formed high-temperature vacancy and electron-irradiation induced vacancy. Finally, the structural phase transition in QC's probed through the local atomic and electronic structures around structural vacancies is presented.

2. Positron Annihilation Spectroscopy

2.1. Positons in Materials

When energetic positrons are implanted into a condensed matter, they rapidly lose their energy and reach thermal equilibrium within a few ps. After thermalization, positrons begin to diffuse in materials. The average positron diffusion length L_+ is typically of the order of 100 nm in metals [22]. During diffusion positrons interact with surrounding atoms and finally annihilate with their electrons. Positrons are Bloch delocalized in the perfectly ordered matrix in crystalline materials, characterizing the annihilation rate λ_b of given materials. At vacancy-type defects such as monovacancies, the potential felt by positrons is lowered due to reduction in the Coulomb repulsion, which results in a lower energy level than for delocalized-free state. The transition from the delocalized state to the localized one is called as positron trapping. Since local electron density at the defect site is lower than in defect-free region, positron lifetime τ_d (the reciprocal of annihilation rate λ_d) of the trapped positrons becomes longer than in defect-free region τ_b (the reciprocal of λ_b).

Annihilation of a positron-electron pair produces γ-ray photons, which have a total energy of $2m_0c^2$. Here, m_0 and c are the electron rest mass and the velocity of light, respectively. Energy-momentum conservation requires the release of two or more photons. In the center-of-mass frame of the positron-electron pair, two annihilation γ-rays are emitted in the opposite directions to each others. In the laboratory frame, in which positron is considered to be in the rest, γ-ray energies are Doppler shifted and deviate from 180°, depending on the momentum P_L of the annihilating pair. This is called as Doppler broadening spectroscopy.

The experimental techniques utilizing the positron lifetime and momentum distribution of annihilation γ-rays enable us to probe the vacancy-type defects and their chemical environment, respectively. In addition, the positron diffusion experiment with a variable-energy slow positron beam provides us the information on the concentration of vacancy-type defect.

2.2. Positron Lifetime Spectroscopy

^{22}Na positron source is used in the present positron lifetime spectroscopy. The positron source, sealed in a thin foil of Kapton, is mounted in a sample-source-sample sandwich. ^{22}Na emits a positron together with a 1.27 MeV birth γ-ray. The positron implanted into the sample annihilates with the electron, providing 511 keV annihilation γ-rays. The average lifetimes at individual positron states are measured as the time difference between the birth and 511 keV annihilation γ-rays. The positron lifetime spectrum is recorded with a fast-fast coincidence system employing a photomultiplier tube with a scintillator. The scintillators detect the birth and annihilation γ-rays. The 1.27 MeV birth and 511 keV annihilation γ rays are energy-selected by constant-fraction-differential discriminators (CFDDs), and the timing pulse from one of CFDDs is delayed by a time-delay module (DELAY). Subsequently, a time-to-amplitude converter (TAC) produces an analog output, whose height is proportional to the time interval between 1.27 MeV and 511 keV γ rays. Analog signals from TAC are transferred to a multi-channel analyzer (MCA) with an analog-to-digital-convertor (ADC). The time resolution of the system is ~ 220 ps full width at the half maximum (FWHM). For each spectrum at least 1.0×10^6 coincident counts are corrected. After subtracting the background, positron lifetime spectra are numerically analyzed by POSITRONFIT code [23].

2.3. Coincident Doppler Broadening Spectroscopy

In the Doppler broadening spectrum, the positron annihilation events with valence electrons contribute to the low-momentum part of the momentum distribution, whereas tightly bound core electrons contribute to the high momentum part. One-detector Doppler broadening measurements have been used to study line-shape variations in the low-momentum region. However, line-shape variations due to the core electrons in the high-momentum region cannot be measured easily by one-detector Doppler broadening spectroscopy because of the poor signal-to-noise (S/N) ratio. Coincident measurements of the energies of two positron-electron annihilation photons gives the possibility to increase the S/N ratio up to 10^5 and extends the range of measurements of the Doppler broadened line up to $50 \times 10^{-3}\ m_0c$, which can be used for a local chemical analysis of elements [24-26].

The coincident measurements are performed by measuring the energies of the two annihilation quanta E_1 and E_2 with a collinear set-up of two high-purity Ge detectors. The Doppler broadening spectra are obtained by cutting the E_1, E_2 spectra along the energy conservation line $E_1+E_2 = (1022 \pm 1)$ keV, taking into account the annihilation events within a strip of ± 1.6 keV. Low- and high-momentum parts of Doppler broadening spectrum are analyzed by taking the S and W parameters covering the central area $(-2.5\text{-}+2.5) \times 10^{-3}\ m_0c$

and the wing area $\pm(25\text{-}38)\times10^{-3}$ m_0c with normalization to the total area of the spectrum, respectively. Besides the *W* parameter, ratio representation normalized to that of the reference sample is often employed in order to highlight the difference of spectral shape at the high momentum region.

2.4. Positron Diffusion Experiment

Positron diffusion experiments are performed by Doppler broadening measurements with a magnetically guided slow positron beam, detailed elsewhere [27]. The Doppler broadening spectra obtained at each positron incident energies are analyzed by taking the *S* parameter. The measurement is repeated several times so as to ensure good statistical precision.

The average positron-diffusion length L_+ can be generally estimated from the analysis of the *S*-*E* plot. After implantation at a positron incident energy *E*, positrons rapidly thermalize. The spatial distribution of the thermalized positrons *P*(*E*, z) along the incident direction (*z*) can be described by a scaling function [28-29]

$$P(E,z)=N_{lm}\left(\frac{u}{C_{lm}}\right)\exp\left[-\left(\frac{u}{C_{lm}}\right)^{m}\right], \tag{1}$$

where N_{lm} is a normalization constant, and C_{lm}, *l*, and m are parameters which depend on material in consideration [29-30], and *u* is defined as,

$$u=\frac{z}{\langle z(E)\rangle}, \tag{2}$$

where $\langle z(E)\rangle$ is the mean implantation depth. $\langle z(E)\rangle$ is assumed to be given by

$$\langle z(E)\rangle=\frac{A_i}{\rho_i}E^{\alpha_i+\beta_i\ln E}, \tag{3}$$

where *E* and ρ_i are the incident energy of the positron beam and the density of the sample, respectively. The parameters A_i, α_i, and β_i depend on the atomic number of the constituent elements as confirmed by Monte Carlo simulation [31].

After thermalization, positrons begin to diffuse in the sample. The one-dimensional positron diffusion is described by the following equation

$$D_+\frac{\partial^2 N(z,t)}{\partial z^2}-\frac{\partial N(z,t)}{\partial t}-\Gamma N(z,t)=0, \tag{4}$$

where D_+ is the positron diffusion coefficient, *N*(*z*, *t*) is the positron density as a function of both time and position, and Γ is the effective annihilation rate of the positron. The observed *S* parameter is given by a linear combination of contributions from different annihilation sites

$$S(E) = \sum_i S_i F_i(E), \tag{5}$$

where $F_i(E)$ is the fraction of positrons annihilating in the i-th state characterized by the S_i parameter. The fraction $F_i(E)$ can be obtained by solving the diffusion equation, subjected to the positron implantation profile and boundary conditions. By fitting the measured S parameters to Eq. (5), one can obtain the average positron-diffusion length L_+.

3. Structural Vacancy

3.1. Al-Based QC

Positron lifetime spectroscopy consistently yields one dominant component with lifetimes of 210-250 ps for QC's studied [2-20]. As demonstrated in Figure 1, the free positron lifetime in defect-free solids characteristically decreases with increasing valence electron density (dotted line). The positron lifetimes measured in Al-based QC's are much longer than the free positron lifetimes in pure metals [32], intermetallic compounds TiAl, Ti_3Al, and FeAl [33], and in semiconductors Si [34], SiC [35], and diamond [36] (dashed line). The positron lifetimes in Al-based QC's are rather close to the experimental values for single vacancies in SiC [35] and FeAl [33]. This behavior is practically independent of the crystallographic structure or of the chemical composition of QC's. Furthermore, in the case of plastically deformed [9] and electron irradiated AlPdMn alloys [14,21], no additional lifetime component is detected. This is due to the high concentrations of structural vacancies in Al-based QC's, causing saturation trapping of positrons. This can be concluded irrespective of the low conduction electron density of QC's because the positron lifetimes are still longer than those in the covalently bonded semiconductors (see Figure 1).

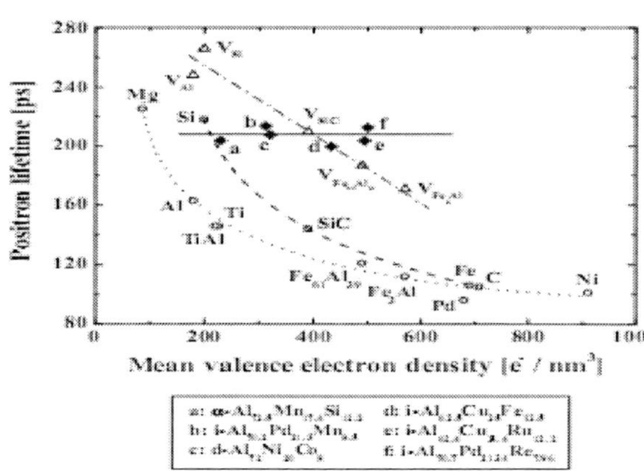

Figure 1. Positron lifetimes measured for a: crystalline approximants α-AlMnSi [10], QC's (b: AlPdMn [9-10,13-14,16], c: AlNiCo [4,7,18], d: AlCuFe [6,8], e: AlCuRu [8,17], f: AlPdRe [10] (closed diamonds and solid line), pure metals [32] and intermetallic compounds [33] (open circles and dotted line), semiconductors [34-36] (open squares and dashed line) plotted over the mean densities of valence electrons (outer s, p, d electrons). For comparison, positron lifetimes in vacancy-trapped states in pure

metals, in intermetallic compounds [32], in Si [34], in SiC [35], and in diamond [36] are added (open triangles and dash-dot-dash line).

It is worth comparing the positron lifetimes in QC's to those in crystalline approximants, since QC's are believed to possess similar basic clusters to those in the approximants. α-AlMnSi 1/1-phase is considered to be a low-order crystalline approximant similar to Al-based QC's in its local atomic structure. The structural analysis for α-AlMnSi has been conducted by conventional diffraction methods [37].

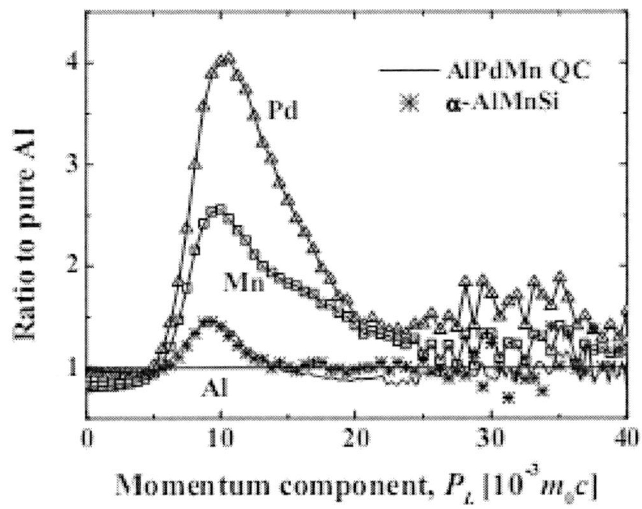

Figure 2. Coincident Doppler broadening spectra of the QC AlPdMn (solid line) and crystalline 1/1-approximant α-AlMnSi (stars) together with those of pure Mn (open squares) and pure Pd (open triangles). Each spectrum is normalized to the Doppler broadening spectrum of pure Al (horizontal line).

This atomic structure is composed of b.c.c. packing of Mackay icosahedron (MI) clusters, with a lattice constant of 12.68 Å. The MI cluster contains an Al icosahedron in the first shell and its central site is completely vacant. The central site of the MI cluster located in the b.c.c. lattice sites is thus the most likely candidate for positron trapping center. It should be noted here that positron lifetimes observed for Al-based QC's are similar to that for the α-AlMnSi approximant (see Figure 1). This suggests the existence of the same type of the structural vacancies in the Al-based QC's.

Coincident Doppler broadening ratio spectrum of AlPdMn QC normalized to pure Al is shown in Figure 2 together with those of pure metals and □-$Al_{68.31}Mn_{21.21}Si_{10.48}$ crystalline approximant. In the core electron region higher than 20×10^{-3} m_0c, the spectra of QC's are essentially identical to that of pure Al, indicating Al chemical environment around the positron annihilation sites. In addition, the high momentum spectrum of QC is identical to that of α-$Al_{68.31}Mn_{21.21}Si_{10.48}$ approximant, which contains the MI clusters with Al_{12} icosahedron. This together with similar positron lifetimes clearly demonstrates that there exist structural vacancies surrounded by Al atoms in AlPdMn QC's as the center sites of Al_{12} icosahedra in α-AlMnSi approximant.

3.2. CdYb Binary QC

The cubic crystal Cd_6Yb can be considered as the crystalline approximant of the $Cd_{5.7}Yb$ QC [38-39]. The observed positron lifetime spectra for both the QC and cubic crystal are 230 ps and 234 ps, respectively. Considering that the atomic densities of $Cd_{5.7}Yb$ QC (4.29×10^{28} m^{-3}) and cubic crystal Cd_6Yb (4.31×10^{28} m^{-3}), which are estimated from the composition and density [40], are close to that of pure Cd (4.63×10^{28} m^{-3}), significantly high values were obtained. Here we have estimated the positron lifetimes in the free state for $Cd_{5.7}Yb$ QC and cubic crystal Cd_6Yb approximating those to be the compositionally weighted average of the constituent element values (τ_{Cd}=190 ps [41] ps and τ_{Yb}=260 ps [42]). The estimated values are 201 ps and 200 ps for $Cd_{5.7}Yb$ QC and cubic approximant crystal Cd_6Yb, respectively, which are significantly shorter than the lifetimes observed here. Therefore it is unlikely that the obtained positron lifetimes of 230 ps and 234 ps are due to annihilations in the defect-free region of the specimens.

In the case of crystalline approximant Cd_6Yb, its atomic structure has been determined from a single crystal x-ray diffraction analysis: it can be described as b.c.c. packing of three-layered icosahedral atomic clusters [40]. The first shell surrounded by the dodecahedral second shell of 20 Cd atoms consists of four Cd atoms, which occupy four sites among equivalent eight sites, the other four sites being vacant. These vacant Cd sites are likely to serve as trapping sites for positrons and the observed long positron lifetime in the cubic crystalline approximant Cd_6Yb most probably corresponds to that at these vacancy sites. We obtain essentially the same positron lifetime for the QC $Cd_{84.6}Yb_{15.4}$, suggesting the existence of the same type of the structural vacancies in the QC.

Figure 3 shows the Doppler broadening spectra of the $Cd_{5.7}Yb$ QC and crystalline approximant Cd_6Yb together with those of pure metals. Figure 3 (a) shows the raw spectra whereas the ratio representation normalized to the pure Cd spectrum are shown in Figure 3 (b). The Doppler spectra of the QC and approximant are slightly different from each other in the momentum region around 5×10^{-3} m_0c. This may reflect the different electronic structure in the valence electron region between the two phases. In the core electron region higher than 20×10^{-3} m_0c, the spectra are essentially identical to each other, indicating the same chemical surroundings of the trapping sites. By comparing the raw spectra with the pure Cd spectrum in the inset of Figure 3 (a), we find that the chemical surroundings of the trapping site are dominated by Cd atoms. The ratio spectra in Figure 3 (b) clearly exemplify Cd-rich chemical surroundings of the vacancies in the QC and approximant.

Judging from essentially the same positron lifetimes and Cd-rich chemical environment observed in the $Cd_{5.7}Yb$ QC and crystalline approximant Cd_6Yb, the following picture can be reasonably presented. The QC is composed of the same local cluster units as that in the approximant, where the tetrahedron with four Cd atoms is located at the center. Positrons are localized after thermalization around the 4Cd tetrahedra inside the dodecahedral shell of 20Cd in the two phases and are annihilated therein.

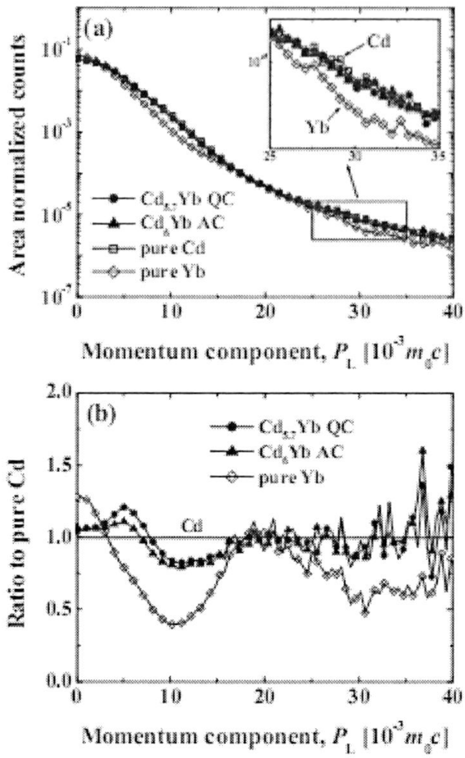

Figure 3. (a) Coincident Doppler broadening spectra of the icosahedral QC $Cd_{5.7}Yb$ (full circles) and cubic crystalline 1/1-approximant Cd_6Yb (full triangles) together with those of pure Cd (open squares) and pure Yb (open diamonds). Each spectrum is normalized by the total number of counts. The inset shows the blown-up section in the high-momentum core electron region. (b) Doppler broadening ratio spectra taken from Figure 1 (a). Each spectrum is normalized to the Doppler broadening spectrum of pure Cd (horizontal line).

3.3. MgZnSc QC

$Zn_{85}Sc_{15}$ is regarded as the crystalline 1/1-approximant of the $Zn_{81}Mg_6Sc_{13}$ QC [43]. Positron lifetime spectroscopy yields a single component for both the QC (207 ps) and approximant (215 ps). The positron lifetimes in the free state for the $Zn_{81}Mg_6Sc_{13}$ QC and cubic $Zn_{85}Sc_{15}$ can be estimated by a compositionally weighted average of the lifetimes for the constituent elements [τ_{Mg} = 237 ps, τ_{Zn} = 134 ps, and τ_{Sc} = 199 ps [44]]. The estimated values are 149 ps and 144 ps for the QC and approximant, respectively, which are significantly shorter than lifetimes observed in the present experiments. The observed lifetimes of the QC and approximant are much longer than the free-state lifetime for Zn (134 ps) that has atomic density of 6.57×10^{28} m^{-3} similar to the QC (6.07×10^{28} m^{-3}) and approximant (6.18×10^{28} m^{-3}). They are rather close to the value for monovacancy in Zn (220 ps) [45]. It is thus most likely that the observed positron lifetimes of 207 and 215 ps are due to annihilations in the structural vacancy-type sites.

Figure 4 shows the Doppler broadening spectra of the $Zn_{81}Mg_6Sc_{13}$ QC and cubic $Zn_{85}Sc_{15}$ together with those of pure elements. Raw spectra are displayed in Figure 4 (a), whereas the ratio representation normalized to the Zn spectrum is presented in Figure 4 (b).

The Doppler broadening spectra of the QC and approximant are similar to each other over the entire momentum range studied. The close similarity in the core electron region above 20×10^{-3} m_0c indicates that the chemical surroundings of the positron trapping sites are essentially the same in the two phases.

The Doppler broadening ratio curves of the QC and approximant are horizontal and parallel to the data of pure Zn (Y=1) in the core electron region. They differ from those of pure Mg and Sc with specific positive slopes, as indicated by straight lines in Figure 4 (b). The horizontal ratio curves indicate that the positron wave functions in the QC and approximant dominantly probe an electron shell in Zn atoms. Therefore, positrons in the QC and approximant annihilate with electrons at the structural vacancy-type sites surrounded by Zn atoms.

According to the structural model of Andrusyak et al. [46], the approximant $Zn_{85}Sc_{15}$ is described as b.c.c. packed three-shell clusters with icosahedral symmetry, consisting of an innermost 20Zn dodecahedral shell with a radius of 3.7 Å, an intermediate 12Sc icosahedral shell with a radius of 4.9 Å, and an outermost 30Zn icosidodecahedral shell with a radius of 5.7 Å.

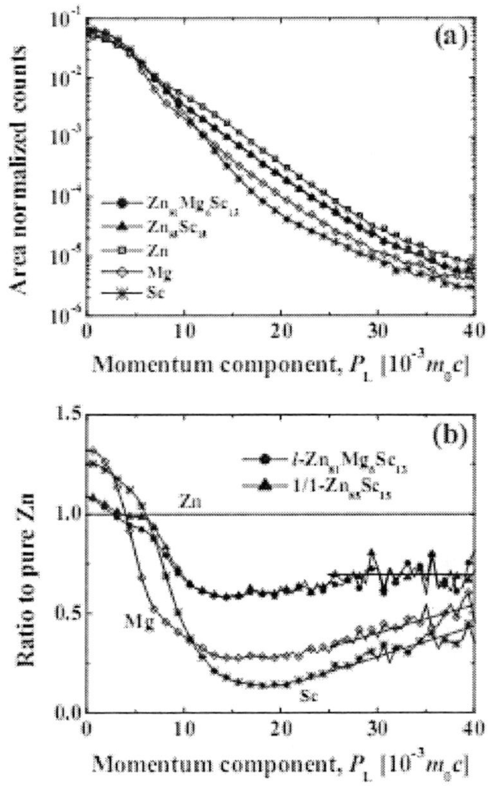

Figure 4. Coincident Doppler broadening spectra of the icosahedral QC $Zn_{81}Mg_6Sc_{13}$ (open circles) and cubic crystalline 1/1-approximant $Zn_{85}Sc_{15}$ (stars) together with those of pure Zn (open triangles), pure Mg (open squares), and pure Sc (open diamonds). Each spectrum is normalized to the total number of counts. (b) Doppler broadening ratio spectra taken from Figure 1 (a). Each spectrum is normalized to the Doppler broadening spectrum of pure Zn (Y=1). The straight lines in the core electron region are drawn for guiding the eye.

The three-shell cluster is very similar to that of the Cd-based binary approximant Cd_6Yb except for the interior of the first shell [40]. The interior of the dodecahedral first shell in $Zn_{85}Sc_{15}$ is empty, whereas it is filled with four Cd atoms forming a tetrahedron in Cd_6Yb. Thus, according to the model a large space of about four vacancies are present in the approximant $Zn_{85}Sc_{15}$.

As mentioned above, positron lifetime spectroscopy revealed the existence of structural vacancy-type sites with a positron lifetime of 234 ps for the approximant Cd_6Yb. The lifetime is slightly shorter than that for monovacancy in pure Cd [41] and ascribed to the annihilation inside the 20Cd dodecahedral cluster containing four Cd atoms inside. The observed positron lifetime in the approximant $Zn_{85}Sc_{15}$ is slightly shorter than that for monovacancy in pure Zn [45] and hence cannot be explained by the annihilation in the completely empty space inside the dodecahedral first shell. Since no other element except for Zn is detected by the coincident Doppler broadening measurement, Zn atoms are presumably placed inside the first shell of the approximant as are 4Cd atoms in the approximant Cd_6Yb. Interestingly, the $Zn_{81}Mg_6Sc_{13}$ QC exhibits the same Zn-rich chemical environment as the approximant in the coincident Doppler broadening spectrum. No sign of substituted Mg atoms is detected from positron annihilation spectroscopy. We believe that positrons annihilate at the same local cluster unit with the 20Zn dodecahedral first shell with several Zn atoms placed in the central site for the QC as in the approximant.

4. Structural Vacancy Concentration

4.1. Al-Based QC

It is generally difficult to derive quantitative information for QC's and crystalline approximants, because they contain high concentration of structural vacancy giving rise to saturation trapping. Here, we employ positron diffusion experiments using a variable-energy slow positron beam in order to estimate the vacancy concentration. First positron diffusion experiments using a slow positron beam have been conducted for an AlPdMn QC [47-48]. Figs. 5 and 6 shows the S parameter data observed for α-AlMnSi crystalline approximant and Al-based QC's (AlPdMn, AlPdRe, and AlCuFe) as a function of positron incident energy (S-E plot). The measured S parameters for α-AlMnSi crystalline approximant and Al-based QC's rapidly increase in the lower energy region, beyond which it is saturated. The rapid increases observed here are typical for QC's and approximants, and are due to high concentration of structural vacancy-type sites, in agreement with other positron annihilation experiments.

As is already demonstrated by positron lifetime and coincident Doppler broadening spectroscopy for α-AlMnSi crystalline approximant, positrons are exclusively trapped by the structural vacancy located at the central sites of MI clusters. Here, we define the average positron-trapping radius (r_d) that a positron begins to localize for trapping [10,12,19-20]. If positrons enter within the radius r_d from the center of MI clusters (b.c.c. lattice site), they are surely trapped by the vacancies.

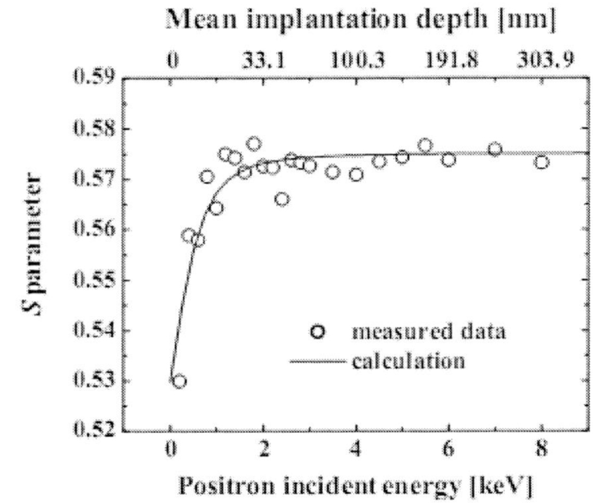

Figure 5. Positron diffusion data of α-AlMnSi crystalline approximant obtained by a slow positron beam. The solid lines are results of fits based on the method described in the text.

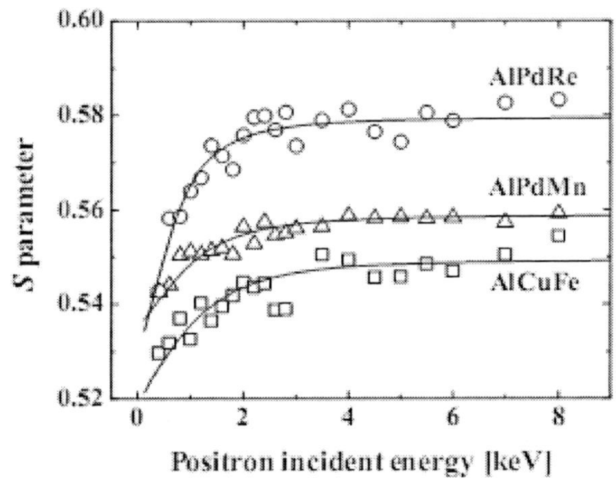

Figure 6. Measured *S* parameter data for icosahedral QC's AlPdRe (open circles), AlPdMn (open triangles), and AlCuFe (open squares) as a function of positron incident energy. The solid lines are results of fits based on the method described in the text.

In the single layer model (surface-bulk structure) as the cases of QC's and crystalline approximants, Eq. (5) can be simply rewritten as $S = S_sF_s + S_bF_b$, where S_s and S_b are the S parameters, and F_s and F_b are the annihilation rates for the surface and bulk, respectively. By calculating the probability of the thermalized positron diffusing without being trapped in the central site of MI clusters, we get the average positron-diffusion length L_+ as a function of the average positron-trapping radius r_d. A solid line in Figure 5 gives the result of a fit performed by the weighted non-linear least-square method, where the trapping radius r_d is determined to be 4 Å.

We assume the same trapping radius r_d of 4 Å for estimating the trapping site density of other Al-based QC's. This assumption is reasonable because nearly the same local atomic structure as that in the approximant is observed for the QC by the positron lifetime and coincident Doppler broadening spectroscopy. For the sake of convenience, b.c.c. cluster packing is assumed, although we know that the QC has the quasiperiodic packing of the icosahedral clusters. In the analysis of the measured *S* parameters, the lattice constant for the hypothetical b.c.c. cluster packing of the QC structure was treated as a free parameter under constraint r_d = 4 Å. The results of the fit are shown by solid lines in Figure 6. The lattice constant for AlPdMn, AlPdRe, and AlCuFe QC's was determined to be 15.85 Å, 13.75 Å, and 16.17 Å, resulting in a structural vacancy density of 5.0×10^{20} cm^{-3}, 7.7×10^{20} cm^{-3}, and 4.7×10^{20} cm^{-3}.

4.2. CdYb Binary QC

Figure 7 shows the *S* parameter for the $Cd_{5.7}Yb$ QC (open squares) and its crystalline approximant Cd_6Yb (full squares) as a function of positron incident energy and mean implantation depth. It is clearly seen from Figure 7 that the measured *S* parameters for the two phases increase in the energy region from 0 to 10 keV, beyond which they are saturated. The rapid increase, typical for the QC's and their related materials, is due to the dense distribution of structural vacancy-type sites giving rise to the saturation positron trapping.

We note that the *S* parameter for the QC increases more slowly than the approximant in the shallow energy range from 0 to 10 keV (see inset in Figure 7). The similar surface *S* parameters at the lowest incident energy for the two samples indicate that no surface effect is involved in the dissimilar variations of the *S* parameters with incident energy. Hence, the data in Figure 7 demonstrates the lower positron diffusivity in the bulk of the approximant than that of the QC. The distinct positron diffusivities arise from the difference of the density of trapping sites between the two phases.

According to the x-ray structural analysis by Larson et al. [49], the Cd_4 tetrahedron in the approximant would be disposed in one of six possible orientations. Furthermore, Tamura et al. [50] observed that the Cd_4 tetrahedra are orientationally ordered at low temperatures, i.e., the central Cd_4 tetrahedra is ordered below 110 K. From the estimated low activation energy, they suggested high frequency rotation of the Cd_4 tetrahedra at room temperature. On the basis of these reports, we expect that the Cd_4 tetrahedra are in highly disordered state in the approximant.

The scaling method [28-30] has been applied to the structure with four Cd atoms placed inside the dodecahedral second shell. We assume that the density of the trapping sites in the approximant is given by $2/a^3$, where a is the lattice constant (a = 15.64 Å). By calculating the probability of the thermalized positron diffusing without trapped around the tetrahedron, we obtain the average positron-diffusion length in the approximant as a function of the trapping radius r_d. A solid line in Figure 7 gives the result of a fit performed by the weighted non-linear least-square method, and the trapping radius r_d is determined to be 3 Å for the approximant.

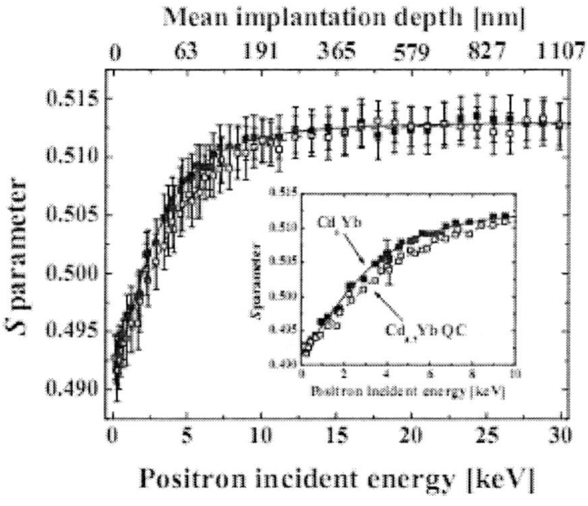

Figure 7. Measured S parameter data of icosahedral QC $Cd_{5.7}Yb$ (open squares) and cubic crystalline 1/1-approximant Cd_6Yb (full squares) as a function of positron incident energy. The solid and dashed lines are results of fits based on the method described in the text. The inset shows the blown-up section in the energy range from 0 keV to 10 keV.

Since the $Cd_{5.7}Yb$ QC and crystalline approximant Cd_6Yb have a similar local atomic structure as demonstrated by the positron lifetime and high momentum Doppler broadening spectroscopy, we employed the same trapping radius of ~ 3 Å in obtaining the density of the trapping sites in the QC $Cd_{5.7}Yb$. The results of the weighted non-linear least-square fit are shown in Figure 7. The density of the trapping sites for the QC is obtained as 4.1×10^{20} cm^{-3}. Here, it should be mentioned that in our model the trapping site density is equivalent to the cluster density. This may help to answer the question how differently clusters are ordered in the two phases.

4.3. MgZnSc QC

Figure 8 shows the S parameters for the QC $Zn_{80}Mg_5Sc_{15}$ (open circles) and its crystalline approximant $Zn_{85}Sc_{15}$ (full circles) as a function of positron incident energy. The S parameters for the two phases rapidly increase in the energy region from 0 to 5 keV and become saturated at higher incident energies. The rapid increase indicates low positron diffusivities due to the high concentration of positron trapping sites, i.e., vacancies surrounded by Zn atoms.

The S parameter increases more gradually for the QC than for the approximant in the energy range from 0 to 5 keV. The similar surface S parameters at the lowest incident energy for the two phases indicate that surface effect is not the origin of the dissimilar variations of the S parameters with incident energy. Therefore, the slower increase of the S parameter means longer positron diffusion length for the QC, which originates from lower vacancy density in QC.

Here, we assume that the trapping site density in the approximant is given by $2/a^3$, where a is the lattice constant (a = 13.852 Å). By calculating the probability of the thermalized positron diffusing without being trapped in the dodecahedral shell, we obtain the trapping radius r_d. A solid line in Figure 8 gives the result of a fit performed by the weighted non-linear-square method, which provides a trapping radius rd of 3 Å. This is the same trapping radius as the approximant Cd_6Yb, which supports the existence of Zn atoms in the dodecahedral first shell of the approximant $Zn_{85}Sc_{15}$.

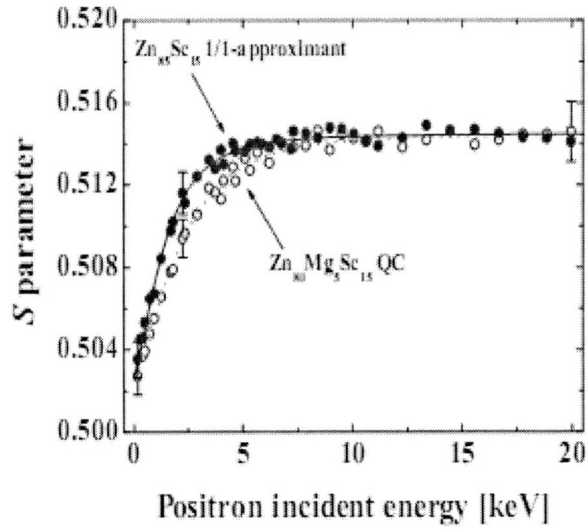

Figure 8. Measured *S* parameter data for the icosahedral QC $Zn_{81}Mg_6Sc_{13}$ (open circles) and cubic crystalline 1/1-approximant $Zn_{85}Sc_{15}$ (full circles) as a function of positron incident energy. The solid and dashed lines are results of the weighted nonlinear least square fit for the cubic crystalline 1/1-approximant $Zn_{85}Sc_{15}$ and icosahedral QC $Zn_{81}Mg_6Sc_{13}$, respectively.

Since the same local atomic structure as that in the crystalline approximant $Zn_{85}Sc_{15}$ is observed for the QC $Zn_{80}Mg_5Sc_{15}$ by the positron lifetime and coincident Doppler broadening spectroscopy, the same trapping radius of 3 Å was employed for QC. The result of the fit is shown by a dashed line in Figure 8. The lattice constant for the QC is determined to be 16.1 Å, resulting in a structural vacancy density of 4.8×10^{20} cm^{-3}. Table 1 lists the structural vacancy densities (C_d) estimated for a number of QC's and approximants.

Table 1. Structural vacancy concentration estimated for QC's and approximants

sample	C_d [cm^{-3}]
AlMnSi 1/1-approximant	9.8×10^{20} [31]
AlPdRe quasicrystal	7.7×10^{20} [15]
AlPdMn quasicrystal	5.0×10^{20} [15]
AlCuFe quasicrystal	4.7×10^{20} [18]
MgZnHo quasicrystal	4.0×10^{20} [17]
MgZnY quasicrystal	3.0×10^{20} [17]
CdYb quasicrystal	4.1×10^{20} [24]
CdYb 1/1-approximant	5.2×10^{20} [32]
ZnMgSc quasicrystal	4.8×10^{20} [25]
ZnSc 1/1-approximant	7.5×10^{20} [33]

4. DETECTION OF HIGH-TEMPERATURE THERMAL VACANCY

High-temperature thermal vacancies in metals and intermetallic compounds have been specifically studied by positron lifetime spectroscopy [32,51-52]. However, in QC's and crystalline approximants thermal vacancies may be invisible in the change of positron lifetimes due to the high concentration of structural vacancies ($C_V \sim 10^{-2}$) with a similar positron lifetime. If, however, the chemical environment of the thermally formed vacancies differs significantly from that of the structural vacancy, this can be visible in high momentum coincident Doppler broadening spectrum of the electron-positron annihilation photons.

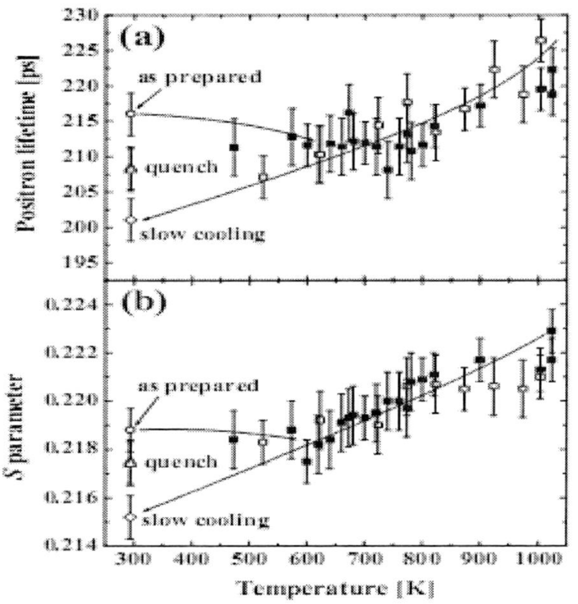

Figure 9. Temperature variations of the positron lifetime τ1 (a) and *S* parameter (b) in AlPdMn QC (full squares: heating run, open squares: cooling run). Data measured at ambient temperature in the as-prepared state (open circle), after slow cooling (-7×10^{-4} Ks^{-1}) from 1023 K (open diamond), and after quenching (-10 Ks^{-1}) from 1023 K (open triangles) are indicated additionally.

The temperature variations of the positron lifetime and of the *S* parameter, which are mainly determined by annihilation with valence electrons, are plotted in Figure 9 for slow cooling ($\sim -7\times10^{-4}$ Ks^{-1}). The enhanced positron lifetimes are found at ambient temperature due to the preparation-induced defects in the as-prepared specimen (see Figure 9 (a)) and after more rapid cooling (quenching by ~ -10 Ks^{-1}). They are annealed out at about 600 K. A similar behavior is observed in the high momentum Doppler broadening *W* parameter (Figure 10).

The *W* parameter exhibits an S-shaped curve at high temperatures, as characteristic for thermal vacancy formation, superimposed to a linear slope at lower temperatures. The existence of two types of vacancies, namely structural vacancies at ambient temperatures as well as thermally formed vacancies at high temperatures can also be seen in the *W-S* representation (see Figure 11) with the data taken from Figure 9 (b) and Figure 10. The change from a shallow slope in Figure 11 to a steep slope demonstrates the transition of positron trapping from one type of vacancy trap to another one.

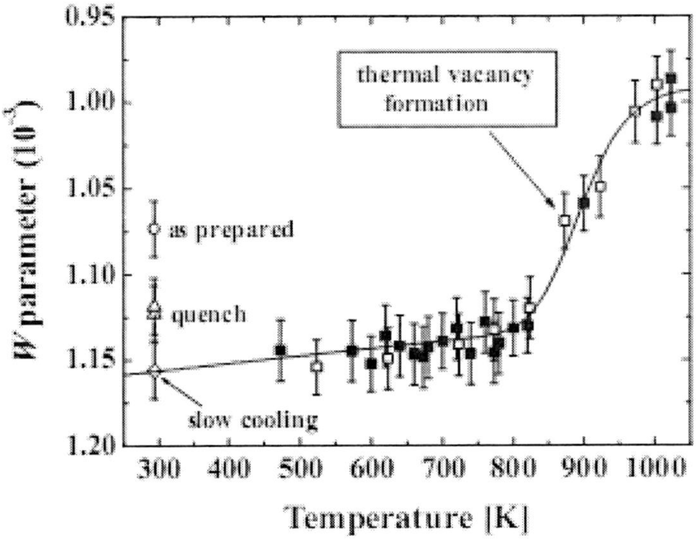

Figure 10. Temperature variation of the W parameter in the AlPdMn QC (full squares: heating run, open squares: cooling run) together with the line fit of Eq. 6 to the experimental data (solid line). The W parameters measured at ambient temperature in the as-prepared state (open circle), after slow cooling (-7×10^{-4} Ks^{-1}) from 1023 K (open diamond), and after quenching (-10 Ks^{-1}) from 1023 K (open triangles).

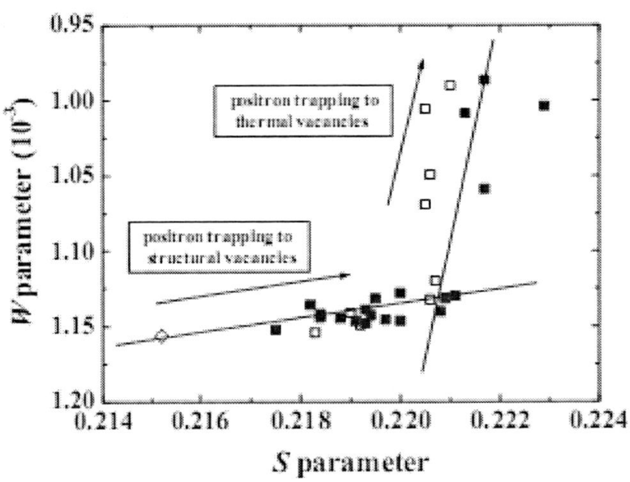

Figure 11. Correlation of the W and S parameter in the AlPdMn QC (full squares: heating run, open squares: cooling run) including data after slow cooling from 1023 K (open diamond). The solid lines are guides to the eye. The evidence for two types of vacancies (structural vacancies or thermal vacancies) is visible from the shallow or steep slope, respectively.

It should be emphasized here that the S-shaped high-temperature curve typical for thermal vacancy formation is only visible in the W parameter derived from the high momentum part of the Doppler broadening spectra but that this can be hardly seen in the temperature dependence of the positron lifetime or the S parameter (Figure 9). This means that the formation of thermal vacancies in addition to structural vacancies in complex solids as AlPdMn QC is visible by a change of the chemical surrounding of the vacancy which, however, does not substantially change the valence electron density. This is compatible with the calculations for ordered binary intermetallic compounds yielding practically the same positron lifetimes for vacancies on the two sublattices in entirely different atomic surroundings [53].

There arises the question whether we can derive from the high-momentum Doppler broadening spectra or from the ratio representation of these spectra (see Figure 12) more specific information on the chemical surroundings of the thermally formed vacancy by a comparison with the ratio spectra of the pure components. The ratio spectrum measured for AlPdMn QC at 1023 K is shifted towards the spectrum reported for vacancies in pure Al [54], however, with a maximum of 30×10^{-3} m_0c as characteristic for the transition metals Pd and Mn. This behavior demonstrates that the positron wave function in the high-temperature vacancies probes Al, Pd, and Mn atoms adjacent to the vacancy. These vacancies are highly mobile at high temperatures as concluded from the initial data on the migration of quenched-in vacancies (see Figure 10). This type of vacancies favors high-temperature atomic diffusion processes as, e.g., that of Fe [55] or Au [56] with high diffusion enthalpies which fit to the present vacancy formation and migration enthalpies (see below).

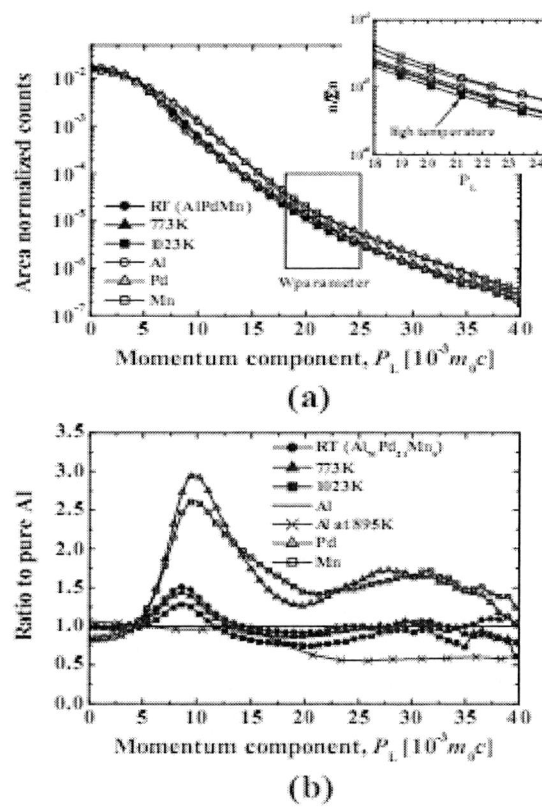

Figure 12. (a) Coincident Doppler broadening spectra at ambient temperature after slow cooling (full circles), measured at 773 K (full triangles), and measured at 1023 K (full squares) for the AlPdMn QC together with those for pure Al (open circles), pure Pd (open triangles), and pure Mn (open squares) measured at ambient temperature. Each spectrum is normalized to the total number of counts. The inset shows the blown-up section selected for the high-momentum W parameter (Figure 10). (b) Doppler broadening ratio spectra taken from Figure 11 (a). Each spectrum is normalized to the Doppler broadening spectrum of pure Al at ambient temperature (horizontal line). The ratio spectrum for thermal vacancies in pure Al according to Ref. 54 is additionally plotted (×××).

For an estimate of the vacancy formation enthalpy H_V^F in AlPdMn QC's, the S-shaped high-temperature change of the W parameter (Figure 10) can be modeled by

$$W(T) = \frac{W_1 + \frac{\sigma_2}{\sigma_1 C_1} W_2 C_2}{1 + \frac{\sigma_2}{\sigma_1 C_1} C_2} \quad (6)$$

with the temperature-dependent concentration

$$C_2 = \exp\left(\frac{S_V^F}{\kappa_B}\right)\exp\left(-\frac{H_V^F}{\kappa_B T}\right) \quad (7)$$

of thermal vacancies. Here, $W_1 = W_{1,0}(1+\beta_2 T)$ is the temperature-dependent characteristic W parameter of structural vacancies, W_2 the W parameter characteristic for thermal vacancies, $\sigma_2/\sigma_1 \sim 1$ the ratio of the specific positron trapping rates of thermal and structural vacancies, $C_1 \sim 10^{-3}$ the concentration of structural vacancies and S_V^F the vacancy formation entropy. From the fit of Eqs.(6) and (7) to the data in Figure 10 we obtain

$$H_V^F = (2.3 \pm 0.5) eV, \quad (8)$$

$$\frac{\sigma_2}{\sigma_1 C_1}\exp\left(\frac{S_V^F}{\kappa_B}\right) = (3.8 \pm 0.5) \times 10^{12} s^{-1}, \quad (9)$$

$$W_{1,0} = (1.17 \pm 0.016) \times 10^{-3}, W_2 = (0.99 \pm 0.016) \times 10^{-3}, \quad (10)$$

and the temperature coefficient $\beta_2 = (3.7 \pm 0.2) \times 10^{-5}\ K^{-1}$. Only a slight change of H_V^F within the uncertainty limits is found when instead of a temperature-independent behaviors of W_2 a temperature dependence as that of W_1 is used.

The value $H_V^F = (2.3 \pm 0.5)$ eV of the apparent vacancy formation enthalpy is considerably higher than in the pure metals and intermetallic compounds and agrees with the estimate from positron lifetime measurements for QC $Al_{70.2}Pd_{21.3}Mn_{8.5}$ [13]. This high value on the one hand may indicate that thermal vacancy formation in AlPdMn QC is difficult and complex. A complex process of vacancy formation with a higher number of atoms involved may also be anticipated from the high value of the apparent vacancy formation entropy $S_V^F = 22\kappa_B$ derived from the data of Eq.(4). On the other hand, the high H_V^F and S_V^F values may indicate deficiencies in the simple model employed here.

High-temperature atomic diffusion processes are carried by thermally formed vacancies as demonstrated by the present observation of the formation process of these vacancies by coincident Doppler broadening techniques (Figs. 10 and 11). We tentatively may employ the relationship $Q^{SD} = H_V^F + H_V^M$ which in the case of pure metals correlates the activation enthalpy Q^{SD} of self-diffusion to the enthalpies of vacancy formation (H_V^F) and migration (H_V^M). The high diffusion enthalpies reported for ^{59}Fe (2.61 eV) [55] or ^{195}Au (2.57 eV) [56] in AlPdMn QC may then be understood within the uncertainty limits in terms of the present value for $H_V^F = 2.3$ eV and a low value of $H_V^M = (0.8 \pm 0.2)$ eV anticipated from time-differential dilatometry experiments [13] and the annealing of quenched-in vacancies.

For the high-temperature plastic deformation of AlPdMn QC's with a brittle-to-ductile transition above 850 K [57-58] vacancy formation and migration appear to play a central role. The high-temperature ductility of AlPdMn QC with a mainly steady-state range of flow stress requires dislocation recovery by dislocation climb [57]. For the understanding of this process vacancy formation and migration as detected here is essential.

5. Detection of Electron-Irradiation Induced Vacancy

As discussed in the previous section, the detection of additional formation of lattice vacancies is difficult by positron lifetime spectroscopy for QC's and crystalline approximants where high concentration of structural vacancies gives rise to saturation trapping. To overcome this problem, we have employed high momentum Doppler broadening measurements and successfully detected a second type of vacancy: high temperature thermal vacancy [16]. Here, a third type of vacancy: electron irradiation-induced vacancy is addressed.

The electron irradiation was performed for the $Mg_{26}Zn_{64}Ho_{10}$ QC with the electron energy of 3 MeV in the temperature range from 243 K to 253 K and $Al_{70}Pd_{21}Mn_9$ QC with electron energies of 0.5 MeV or 3 MeV in the temperature range from 170 K to 200 K via cold nitrogen gas flow at the DYNAMITRON accelerator of the University of Stuttgart. Under this experimental condition, doses are estimated to 1.8×10^{19} m^{-2} and 1.5×10^{19} m^{-2} for 0.5 MeV and 3 MeV electron irradiation. It should be pointed out here that the radiation-induced vacancies are practically homogeneously distributed within the measuring range of the positrons. The positrons emitted from a ^{22}Na source with a mean energy of about 0.25 MeV are annihilated, e.g., in Al within a range of 0.3 μm from the specimen surface [59]. By irradiation with electrons of the energies of 0.5 MeV or 3.0 MeV, vacancies are induced in Al by atomic displacements unto a depth of about 0.6 mm or 6.7 mm, respectively [60-61] which is much deeper than the positron annihilation range.

In unirradiated $Mg_{26}Zn_{64}Ho_{10}$ QC, only a single component with a lifetime $\tau_1 = 203$ ps which is substantially longer than anticipated for the defect free alloy indicates a dense distribution of structural vacancies. This structural vacancy is surrounded by Mg, Zn, and Ho atoms as estimated from the high-momentum Doppler broadening spectra characteristic for the various core electron momentum distributions of the various atoms (Figure 13).

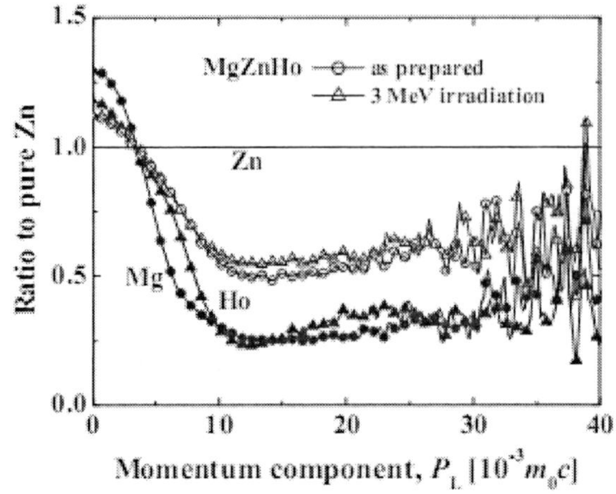

Figure 13. Coincident Doppler broadening spectra of as-prepared (open circles) and 3 MeV electron irradiated $Mg_{26}Zn_{64}Ho_{10}$ (open triangles) together with those of pure Mg (full circles), pure Zn (solid line, Y=1), and pure Ho (full triangles).

The ratio curve of unirradiated $Mg_{26}Zn_{64}Ho_{10}$ in the ratio spectra is located between those of Zn, Mg, and Ho. Upon 3 MeV electron irradiation the curve is significantly shifted toward that of pure Zn (Y = 1) in the core electron momentum range of 10-20×10^{-3} m_0c. Furthermore a long-lived component with τ_2 = 578 ps and I_2 = 8 % appears in the positron lifetime spectrum (Figure 14), whereas the main component τ_1 = 202 ps is the same as the lifetime prior to irradiation.

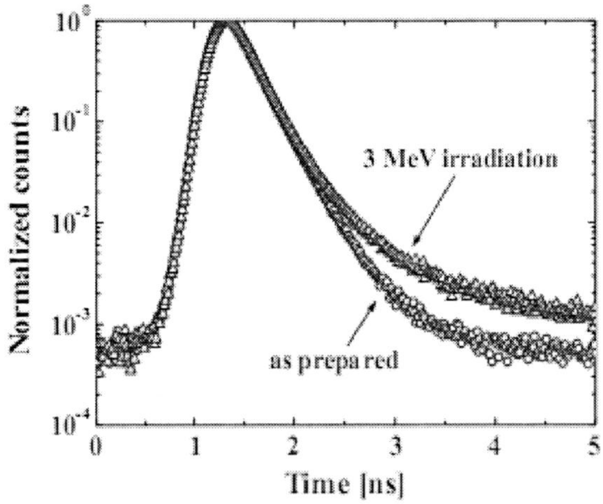

Figure 14. Positron lifetime spectra of as-prepared (open circles) and 3 MeV electron irradiated $Mg_{26}Zn_{64}Ho_{10}$ (open triangles).

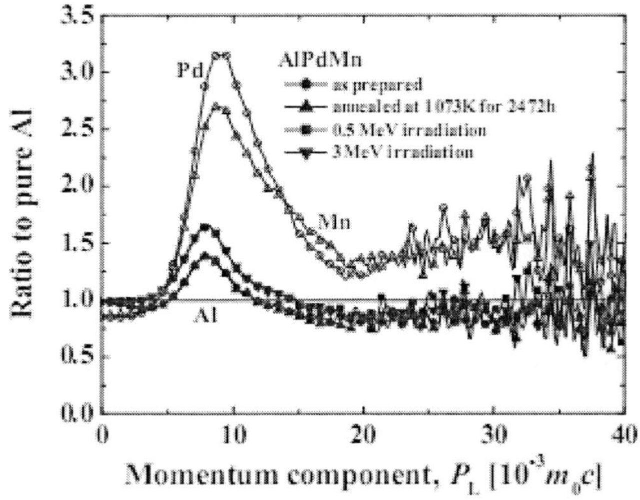

Figure 15. Coincident Doppler broadening spectra of the coincident Doppler broadening spectra of as-prepared (full circles), long-term annealed (full triangles), 0.5 MeV electron irradiated (full squares), and 3 MeV electron irradiated AlPdMn (full inverse triangles) together with those of pure Al (solid line, Y=1), pure Pd (open circles), and pure Mn (open triangles).

The long-lived component can be attributed to the agglomeration of radiation-induced vacancies which may be mobile below ambient temperature. The effect of electron irradiation can be also seen from the significant change of the *W* parameter from 4.15×10^{-3} to 4.38×10^{-3}.

In unirradiated AlPdMn again structural vacancies were available with the lifetime τ_1 = 206 ps. The chemical surroundings of the vacancies are dominated by Al atoms as demonstrated in the Doppler broadening spectra in Figure 15. There is no change of this type of vacancies visible in the Doppler broadening spectra by annealing at 1073 K for 2472 h (Figure 15). However after 0.5 MeV or 3 MeV electron irradiation the Doppler broadening spectra indicate that now the population of the nearest-neighbor atoms of the radiation-induced vacancies is changed versus the transition metal atoms Pd and Mn (Figure 15) without a significant change of the positron lifetime. In contrast to MgZnHo QC (see Figure 13) no vacancy agglomeration is visible in the case of AlPdMn QC after electron irradiation. From this one may conclude that in AlPdMn QC radiation-induced vacancies are immobile at ambient temperature. As demonstrated above, atomic diffusion processes at high temperatures are strongly correlated to Al paths within the AlPdMn QC. In contrast to that, atomic diffusion processes during the electron irradiation may be correlated to the paths of transition metal atoms Pd and Mn.

6. Application of Positron to Structural Phase Transition

Decagonal Al-Ni-Co QC's exhibit a wealth of structural modifications in dependence of composition, temperature, and time-temperature history as compiled in the phase diagram of Ritsch et al. [62] and modeled by Hiraga et al. [63] in dependence of composition. Chemical ordering between Al and transition metal (TM) atoms, which is detected by electron microscopy [63-64] as well as x-ray diffraction [65-66] and which is favored theoretically

[64,67], play a pivotal role for understanding why QC's with their aperiodic structure form. The ordered quasi-unit-cell decagons exhibit a central cluster rich in transition metals which eventually give rise to a symmetry breaking of the decagon structure [64,67-68]. This is a prerequisite for perfect quasi-periodic tiling [64,69] since it yields the same overlap rules as in the Gummelt coverage model [70]. In this sense specific investigation of chemical order-disorder processes in AlNiCo QC's, as reported earlier [71] is of particular interest. For a better understanding of this transformation, atomic-scale structural information is desirable. In this section, we apply a series of positron annihilation spectroscopy to the study of structural phase transition in decagonal AlNiCo QC.

A cylinder of the decagonal $Al_{71.5}Ni_{14}Co_{14.5}$ QC with a length of 7 mm, an outer diameter of 3.5 mm and a 1.5 mm diameter axial borehole was prepared by ultrasonic techniques. A 3.7×10^5 Bq positron source of $^{44}TiSO_4$ was deposited in the borehole, oxidized, and covered by a AlNiCo cap with subsequent evacuation and sealing in a quartz tube.

Positron lifetime spectroscopy yields one dominant component with the lifetime τ_1 which is characteristic for the $Al_{71.5}Ni_{14}Co_{14.5}$ QC and an additional weak $\tau_2 \sim 600$ ps source component in the whole temperature range. The positron lifetime $\tau_1 = 198$ ps at ambient temperature, which is substantially higher than the positron lifetime in the defect-free case as expected from the mean valence electron density (see Figure 1), demonstrates that structural vacancies with an atomic concentration > 10^{-4} are available.

The temperature variation of the positron lifetime τ_1 is shown in Figure 16. Two narrow reversible changes in dependence of temperature are observed centered at 650 K and 1140 K. The temperature dependence of the Doppler broadening *W* parameter is shown in Figure 17.

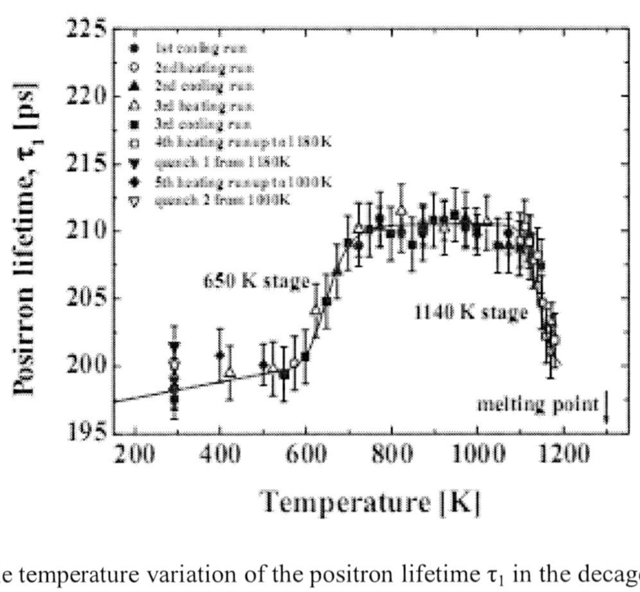

Figure 16. Reversible temperature variation of the positron lifetime τ_1 in the decagonal AlNiCo QC with various measuring runs. Positron lifetimes measured at ambient temperature after quenching from 1180 K or 1000 K are given additionally together with isochronal annealing data (full diamonds). The solid line is drawn for guiding the eye.

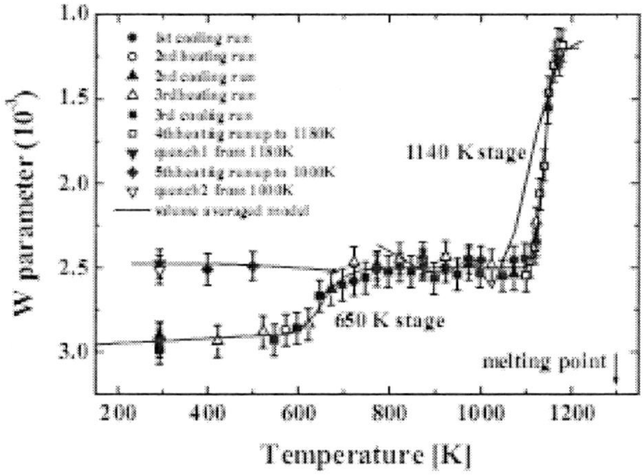

Figure 17. Reversible temperature variation of the W parameter in the decagonal AlNiCo QC with data from various measuring runs. W parameters measured at ambient temperature after quenching from 1180 K and 1000 K are given additionally together with isochronal annealing data (full diamonds). The bold line indicates a fit of the volume averaged positron annihilation in the ordered domains and in the disordered structure (see text) taking the temperature dependent domain size from neutron scattering (Ref. 74) into account. The solid line is drawn for guiding the eye.

We first consider the prominent change of the W parameters at 1140 K (see Figure 17). In this temperature range structural changes were observed for d-AlNiCo QC's with compositions similar to the present one by dilatometry [72], by x-ray diffraction [73], and by neutron scattering [74]. From the FWHM of the diffuse neutron scattering peak in decagonal $Al_{72}Ni_{12}Co_{16}$ [74] ordered domains of the width d = 2π/FWHM = 3.7 nm within the decagonal planes and a length exceeding 1 μm along the periodicity axis can be derived at ambient temperature [75]. In the 1140 K stage chemical disordering is observed by a shrinkage of the domains and the disappearance of scattering peaks. In the simple model comprising both the atomic-scale positron data as well as the temperature-dependent domain size from neutron scattering [74-75], we denote the Doppler broadening parameters characteristic for the vacancies in the ordered domain by W_0 and that for the vacancies in the disordered structure by W_d. The mean volume-averaged Doppler broadening parameter is then given by

$$W = f_0 W_0 + f_d W_d \qquad (11)$$

taking into account the volume fractions

$$f_0 = \frac{d^2}{d_{\max}^2} \qquad (12)$$

of the ordered domains and $f_d = 1 - f_0$ of the disordered structure. For the fully ordered state the relations $d = d_{max}$ and $f_0 = 1$ hold.

A fit of this model (Eqs. 11 and 12) to the 1140 K stage of the W parameter (Figure 17) with the temperature-dependent domain size $d(T)$ from neutron scattering data [74-75] demonstrates that the 1140 K stage of W in Figure 17 coincides with disordering. This fit yields $W_0 = 2.5\times10^{-3}$ below the 1140 K stage, which is intermediate between the values of pure Ni (4.9×10^{-3}) and of pure Co (1.7×10^{-3}). In the above comparison with pure metals, we neglect that the W parameter in vacancies is by about 15 % lower than in the defect free pure metal [54], because this difference is much smaller than that of the W parameters between the transition metals and of Al taken from the data in Figure 18.

Based on the present data, specific information on phase transitions in complex systems can be deduced by a comparison to the structural information derived for decagonal AlNiCo QC's from Z-contrast electron microscopy [63-64] and x-ray diffraction [66]. These studies yield in the 2.0 nm decagon a central cluster with a transition metal concentration exceeding that of the mean value of the QC. This may be the very environment selected by the present positron annihilation W parameter between the 650 K stage and the 1140 K stage. At high temperatures, upon disordering, the atomic composition of the central cluster appears to be enriched in Al [76] as expected according to the mean composition of the $Al_{71.5}Ni_{14}Co_{14.5}$ QC. This is specifically detected by the positron annihilation Doppler parameter W (Figure 18) characteristic for pure Al.

We want to point out here, that the change of the W parameter in the 1140 K stage due to chemical changes around the vacancies is accompanied by a decrease of the positron lifetime (Figure 16). This may originate from the Al rich high-temperature vacancy environment contributing a high number of valence electrons, thus increases the mean valence electron density.

At 650 K an additional reversible transition unknown from the literature is observed by the positron lifetime (Figure 16) and the W parameter (see Figure 17) signifying a change from a Ni rich vacancy surrounding (see Figure 18) to more Co or some small addition of Al. Thermal vacancy formation can be excluded for an explanation of this stage or of the 1140 K stage as discussed briefly in the following.

For an interpretation by thermal vacancy formation the two stages are by far too narrow. This can be demonstrated by a model of positron trapping of thermal vacancies formed with the atomic concentration $C_2(T)$ at high temperatures in addition to structural vacancies with the constant concentration C_1. This model yields the temperature variation $W(T)$ described by Eqs. 11 and 12.

For 650 K stage, the apparent parameters $H_V^F = 1.56$ eV and $(\sigma_2/\sigma_1 C_1)\exp(S_V^F/\kappa_B) = 10^{12}$ are obtained. With the values $\sigma_2/\sigma_1 \approx 1$ and $C_1 \approx 10^{-3}$ as derived from positron diffusion experiments in QC's [10] a value is obtained for $S_V^F/\kappa_B = 21$ which is far beyond the values $S_V^F/\kappa_B = 2$ to 5 usually obtained for intermetallic compounds [77] and therefore excludes this type of process for the 650 K stage. For interpreting the 1140 K stage thermal vacancy formation is even more inappropriate as evidenced by the apparent parameters $H_V^F = 11-15$ eV and $(\sigma_2/\sigma_1 C_1)\exp(S_V^F/\kappa_B) = 10^{50}-10^{65}$ leading to a value $S_V^F/\kappa_B = 110-150$ unreasonably high for this type of process.

About the quenching behavior of the 650 K and 1140 K effects only little information is available. After fast cooling (~ 1 K/s) from 1180 K a W parameter (see Figure 17) similar to that above the 650 K stage is observed at 295 K. By this treatment obviously the 650 K effect but not the 1140 K effect can be quenched in. As demonstrated in Figure 2 the quenching effect is still visible after heating to 500 K so that annealing is expected in the 650 K stage. In this stage disordering between Ni and Co atoms may occur since the Ni-Co ordering energy is predicted to be much smaller than the Al-TM ordering energy [64] involved in the 1140 K stage. Short-range atomic transport for disordering is well available as concluded from the Co diffusivity D_{Co} [78-89] which yields within the present measuring time of t = 10^5 s a mean diffusion length $L = \sqrt{2D_{Co}t} = 1 - 2\,\text{nm}$ at 650 K. Atomic mobility in this stage is also evidenced by internal friction studies [80]. In addition, pairs of Al atoms in the Startile section of the basic decagons are predicted to rotate between five equivalent orientations [67-68].

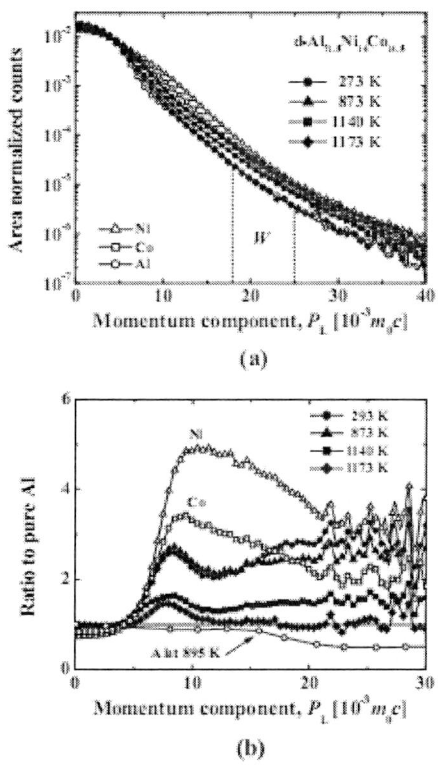

Figure 18. (a) Coincident Doppler broadening spectra of the decagonal AlNiCo QC measured at various temperatures, together with the spectra for pure Al, Ni, and Co. Each spectrum is normalized to the total number of counts. (b) Ratio curves of the coincident Doppler broadening spectra of the decagonal AlNiCo QC measured at various temperatures together with those for pure Al, pure Ni, and pure Co. All spectra are normalized to that of pure Al.

Structural phase transitions in condensed matter have been studied by scattering experiments or by macroscopic techniques as dilatometry, electrical resistivity, magnetization, etc. The characterization of phase transitions becomes more challenging in structurally complex systems as, e.g., aperiodic QC's. The present results demonstrate that the local probe with positrons yield specific information on structural phase transition on an atomic scale and can be complementary to conventionally used diffraction experiments with large coherence lengths.

Acknowledgments

The author would like to thank I. Kanazawa (Tokyo Gakugei University), H. Murakami (Tokyo Gakugei University), M. Nakata (Tokyo Gakugei University), S. Takeuchi (Science University of Tokyo), R. Tamura (Science University of Tokyo), K. Kimura (The University of Tokyo), Y. Kobayashi (AIST), H.-E. Schaefer (Universität Stuttgart), W. Sprengel (Graz University of Technology), and R. Würschum (Graz University of Technology).

References

[1] D. Shechtman, I. Blech, D. Cratias, and J.W. Cahn, *Phys. Rev. Lett.* 53, 1951 (1984).

[2] T. Kizuka, I. Kanazawa, Y. Sakurai, S. Nanao, H. Murakami, and T. Iwashita, *Phys. Rev. B* 40, 796 (1989).

[3] I. Kanazawa, T. Ohata, Y. Yamazaki, Y. Sakurai, S. Nanao, and T. Iwashita, *Phys. Rev. B* 43, 9482 (1991).

[4] Y. Nakao, T. Shibuya, S. Takeuchi, W. Liu, Xue-Song, and S. Berko, *Phys. Rev. B* 46, 3108 (1992).

[5] I. Kanazawa, C. Nakayama, J. Takahashi, T. Ohata, T. Iwashita, T. Kizuka, *Phys. Rev. B* 49, 3573 (1994).

[6] D.W. Lawther and R.A. Dunlap, *Phys. Rev. B* 49, 3183 (1994).

[7] R. Würschum, T. Troev, and B. Grushko, *Phys. Rev. B* 52, 6411 (1995).

[8] E. Hamada, N. Oshima, T. Suzuki, K. Sato, I. Kanazawa, M. Nakata, and S. Takeuchi, *Materials Science Forum* 255, 451 (1997).

[9] I. Kanazawa, E. Hamada, T. Saeki, K. Sato, M. Nakata, S. Takeuchi, and M. Wollgarten, *Phys. Rev. Lett.* 79, 2269 (1997).

[10] K. Sato, Y. Takahashi, H. Uchiyama, I. Kanazawa, R. Tamura, K. Kimura, F. Komori, R. Suzuki, T. Ohdiara, and S. Takeuchi, *Phys. Rev. B* 59, 6712 (1999).

[11] E. Hamada, K. Sato, Y. Takahashi, H. Uchiyama, I. Kanazawa, N. Oshima, T. Suzuki, M. Nakata, T. Yoshida, and S. Takeuchi, *Jpn. J. Appl. Phys* 40, 259 (2001).

[12] K. Sato, H. Uchiyama, Y. Takahashi, I. Kanazawa, R. Suzuki, T. Ohdaira, T. Takeuchi, T. Mizuno, and U. Mizutani, *Phys. Rev. B* 64, 024202 (2001).

[13] F. Baier and H.-E. Schaefer, *Phys. Rev. B* 66, 064208 (2002).

[14] K. Sato, F. Baier, W. Sprengel, and H.-E. Schaefer, *Phys. Rev. B* 66, 092201 (2002).

[15] K. Sato, H. Uchiyama, K. Arinuma, I. Kanazawa, R. Tamura, T. Shibuya, and S. Takeuchi, *Phys. Rev. B* 66, 052201 (2002).

[16] K. Sato, F. Baier, A.A. Rempel, W. Sprengel, and H.-E. Schaefer, *Phys. Rev. B* 68, 214203 (2003).

[17] H. Uchiyama, T. Takahashi, K. Arinuma, K. Sato, I. Kanazawa, T. Suzuki, K. Kirihara, and K. Kimura, *J. of Physics: Condensed Matter* 16, 1899 (2004).

[18] K. Sato, F. Baier, W. Sprengel, and H.-E. Schaefer, *Phys. Rev. Lett.* 92, 127403 (2004).

[19] K. Sato, Y. Kobayashi, K. Arinuma, I. Kanazawa, R. Tamura, T. Shibuya, and S. Takeuchi, *Phys. Rev. B* 70, 094107 (2004).

[20] K. Sato, Y. Kobayashi, I. Kanazawa, R. Tamura, and S. Takeuchi, *Phys. Rev. B* 70, 184204 (2004).

[21] F. Baier, M.A. Müller, B. Grushko, and H.-E. Schaefer, *Materials Science and Engineering* 294–296, 650 (2000).

[22] Positron Spectroscopy of Solids, *Proc. International School of Physics <Enrico Fermi>*, Course CXXV, Varenna 1993, edited by A. Dupasquier and A.P. Mills, Jr., (IOS Press, Amsterdam, 1995).

[23] P. Kirkegaard and M. Eldrup, *Comput. Phys. Commun.* 7, 401 (1974).

[24] P. Asoka-Kumar, M. Alatalo, V.J. Ghosh, A.C. Kruseman, B. Nielsen and K.G. Lynn, *Phys. Rev. Lett.* **77**, 2097 (1996).

[25] K. Sato, K. Ito, K. Hirata, R.S. Yu, and Y. Kobayashi, *Phys. Rev. B* 71, 012201 (2005).

[26] K. Sato, D. Shanai, Y. Hotani, T. Ougizawa, K. Ito, K. Hirata, and Y. Kobayashi, *Phys. Rev. Lett.* 96, 228302 (2006).

[27] Y. Kobayashi, I. Kojima, S. Hishita, T. Suzuki, E. Asari, and M. Kitajima, *Phys. Rev. B* 52, 823 (1995).

[28] V. J. Ghosh, D. O. Welch, and K. G. Lynn, in *Proc. 5th Inter. Workshop on Slow Positron Beams for Solids and Surface*, edited by E. H. Ottewitte (AIP, New York, 1992), p. 937.

[29] V. J. Ghosh, *Appl. Surf. Sci.* 85, 187 (1995).

[30] G. C. Aers, P. A. Marshall, T. C. Leung, and R. D. Goldberg, *Appl. Surf. Sci.* 85, 196 (1995).

[31] G. C. Aers, *J. Appl. Phys.* 76, 1622 (1994).

[32] H.-E. Schaefer, *Phys. Status Solidi (a)* 102, 47 (1987).

[33] R. Wurschum, K. Badura-Gergen, E.A. Kummerle, C. Grupp, and H.-E. Schaefer, *Phys. Rev. B* 54, 849 (1996).

[34] R. Wurschum, W. Bauer, K. Maier, A. Seeger, and H.-E. Schaefer, *J. Phys.: Condensed. Matter* 1, 33 (1989).

[35] M.A. Muller, A.A. Rempel, K. Reichle, W. Sprengel, J. Major, and H.-E. Schaefer, *Mater. Sci. Forum* 363-365, 70 (2001).

[36] H.-E. Schaefer, M. Forster, R. Wurschum, W. Kratschmer, and D. R. Huffman, *Phys. Rev. B* 45, 12164 (1992).

[37] V. Elser and C. L. Henley, *Phys. Rev. Lett.* 55, 2883 (1985).

[38] A.P. Tsai, J.Q. Guo, E. Abe, H. Takakura, and T.J. Sato, *Nature* 408, 537 (2000).

[39] J.Q. Guo, E. Abe, and A.P. Tsai, *Phys. Rev. B* 62, 14605 (2000).

[40] A. Palenzona, *J. Less-Common Metals* 25, 367 (1971).

[41] D. Herlach, H. Stoll, W. Trost, H. Metz, T.E. Jackman, K. Maier, H.-E. Schaefer, and A. Seeger, *Appl. Phys.* 12, 59 (1977).

[42] J.M. Campillo and F. Plazaola, *Mater. Sci. Forum* 363-365, 594 (2001).

[43] Y. Kaneko, Y. Arichika, and T. Ishimasa, *Philos. Mag. Lett.* 81, 777 (2001).

[44] M.J. Puska and R.M. Nieminen, *Rev. Mod. Phys.* 66, 841 (1994).

[45] I.K. MacKenzie, T.L. Khoo, A.B. McDonald, and B.T.A. MacKee, *Phys. Rev. Lett.* 19, 946 (1967).

[46] R.I. Andrusyak, B.Y.A. Kotur, and V.E. Zavodnik, *Kristallografiya* 34, 996 (1989).

[47] K. Sato, E. Hamada, M. Tashiro, T. Koizumi, I. Kanazawa, F. Komori, Y. Ito, and S. Takeuchi, in *Proc. 6th Int. Conf. on Quasicrystals*, edited by S. Takeuchi and T. Fujiwara (World Scientific, Singapore, New Jersey, London, Hong Kong, 1997), pp. 425.

[48] K. Sato, H. Uchiyama, Y. Takahashi, I. Kanazawa, R. Tamura, K. Kimura, F. Komori, R. Suzuki, T. Ohdaira, and S. Takeuchi, *Materials Science Forum* 363-365, 481 (2001).
[49] A.C. Larson and D.T. Cromer, *Acta Crystallogr.* B 27, 1875 (1970).
[50] R. Tamura, Y. Maruo, S. Takeuchi, M. Ichihara, M. Isobe, and Y. Ueda, *Jpn. J. Appl. Phys.* 41, L524 (2002).
[51] I.K. Mackenzie, T.L. Khoo, A.B. McDonald, and B.A.T. McKee, *Phys. Rev. Lett.* 19, 146 (1967).
[52] R. Würschum, C. Grupp, and H.-E. Schaefer, *Phys. Rev. Lett.* 75, 97 (1995).
[53] Y. Jirásková, O. Schneeweiss, M. Šob, I. Novotný, I. Procházka, F. Bečvář, B. Sedlák, F. Šebesta, and M.J. Puska, *J. Phys. IV* 5, 157 (1995).
[54] K.G. Lynn, J.E. Dickman, W.L. Brown, M.F. Robbins, and E. Bonderup, *Phys. Rev. B* 20, 3566 (1979).
[55] Th. Zumkley, H. Mehrer, K. Freitag, M. Wollgarten, N. Tamura, and K. Urban, *Phys. Rev. Lett.* 54, R6815 (1996).
[56] P. Blüher, P. Scharwaechter, W. Frank, and H. Kronmüller, *Phys. Rev. Lett.* 80, 1014 (1998).
[57] U. Messerschmidt, M. Bartsch, B. Geyer, M. Feuerbacher, and K. Urban, *Philos. Mag. Lett.* 80, 1165 (2000).
[58] D. Brunner, D. Plachke, and H.D. Carstanjen, *Mater. Sci. Eng.* A234-A236, 310 (1997).
[59] W. Braudt, *Appl. Phys.* 5, 1 (1974).
[60] M.J. Berger and S.M. Seltzer, *Tables of Energy Losses and Ranges of Electrons and Positrons* (NASA, Washington, 1964).
[61] P. Vajda, *Rev. Mod. Phys.* 49, 481 (1977).
[62] S. Ritsch, C. Beeli, H.-U. Nissen, T. Gödecke, M. Scheffer, and R. Lück, *Philos. Mag. Lett.* 78, 67 (1998).
[63] K. Hiraga, T. Ohsuna, W. Sun, and K. Sugiyama, *Mater. Trans.* 42, 2354 (2001).
[64] Y. Yan and S. Pennycook, *Phys. Rev. Lett.* 86, 1542 (2001).
[65] H. Takakura, A. Yamamoto, and A.P. Tsai, *Acta Crystallogr.* A 57, 576 (2001).
[66] A. Cervellino, T. Haibach, and W. Steurer, *Acta Crystallogr.* B 58, 8 (2002).
[67] C.L. Henley, M. Mihalkovič, and M. Widom, *J. Alloys Comp.* 342, 221 (2002).
[68] M. Mihalkovič, I. Al-Lehyani, E. Cockayne, C.H. Henley, N. Moghadam, J.A. Moriarty, Y. Wang, and M. Widom, *Phys. Rev. B* 65, 104205 (2002).
[69] Y. Yan and S. Pennycook, *Nature* 403, 266 (1999).
[70] P. Gummelt, *Geometriae Dedicata* 62, 1 (1996).
[71] A. Baumgarte, J. Schreuer, M.A. Estermann, and W. Steurer, *Philos. Mag. A* 75, 1665 (1997).
[72] R. Lück, M. Scheffer, T. Gödecke, S. Ritsch, and C. Beeli, *Mat. Res. Soc. Symp. Proc.* Vol 553, 25 (1999).
[73] W. Steurer, A. Cervellino, K. Lemster, S. Ortelli, M.A. Estermann, *Chimia* 55, 528 (2001).
[74] F. Frey, E. Weidner, K. Hradil, M. de Boissieu, A. Letoublon, G. McIntyre, R. Currat, and A.-P. Tsai, *J. Alloys Comp.* 342, 57 (2002).
[75] F. Frey and E. Weidner, *Z. Kristallogr.* 218, 160 (2003).
[76] E. Abe et al., *8th Int. Conf. on Quasicrystals*, Bangalore, India, (2002).
[77] H.-E. Schaefer, K. Frenner, and R. Würschum, *Intermetallics* 7, 277 (1999).

[78] C. Khoukaz, R. Galler, H. Mehrer, P.C. Canfield, I.R. Fisher, and M. Feuerbacher, *Mat. Sci. Eng.* 294-296, 697 (2000).

[79] T. Zumkley, H. Nakajima, J.Q. Quo, and A.P. Tsai, *Philos. Mag. A* 82, 205 (2002).

[80] M. Weller and B. Damson, *in Quasicrystals: Structure and Physical Properties*, Wiley-VCH, Weinheim, p.523 (2003).

In: Quasicrystals: Types, Systems, and Techniques
Editor: B. E. Puckermann, pp. 107-126
ISBN 978-1-61761-123-0

Chapter 4

STRUCTURE MODELS OF QUASICRYSTAL APPROXIMANTS DEDUCED FROM THE STRONG-REFLECTIONS APPROACH

Junliang Sun, Xiaodong Zou[1] and Sven Hovmöller
Inorganic and Structural Chemistry Unit, Department of Materials and Environmental Chemistry,
Stockholm University, SE-106 91 Stockholm, Sweden

ABSTRACT

The structures of many quasicrystals have still remained unknown since the publication of the first icosahedral quasicrystal in rapidly solidified Al-Mn alloys in 1982. The main obstacle is that the quasicrystals always contain defects and it is difficult to synthesize high quality single crystals which are needed for a good structure determination by single crystal X-ray diffraction. In most cases, quasicrystals coexist with several complex quasicrystal approximants. These approximants have similar local atomic structures as the quasicrystals and many of them also contain defects that make diffraction spots from quasicrystals and different approximants overlapped and the whole diffraction pattern blurred. Meanwhile, some less complicated approximants in the same system can be synthesized as large single crystals with fewer defects, and their atomic structures can be determined. Due to the similar local atomic structures, a quasicrystal and its approximants always show similar intensity distribution and phase relationships for the strong reflections in reciprocal space. Thus, the structure factors with both amplitudes and phases can be calculated from a known approximant for strong reflections and after re-indexing them, they can be used to calculate a 3D electron density map for more complex approximants by inverse Fourier transformation. The structure model can be deduced from this 3D electron density map since the strongest reflections mainly determine the atomic positions in a structure. In principle, the perfect quasicrystal structure model can be obtained by this approach. The strong reflections approach avoids a direct structure determination from quasicrystals containing defects but takes the

1 Email: xzou@mmk.su.se.

advantage of using the common features of quasicrystals and approximants. The model deduced from this approach will be an ideal model for the quasicrystal, free of defects.

1. INTRODUCTION

Where all the atoms are in quasicrystals and their approximants has been an open question for more than two decades since the discovery of the icosahedral quasicrystal in rapidly solidified Al-Mn alloys (Shechtman *et al.*, 1984). This is a fundamental interesting problem for understanding the formation of this kind of materials and their special properties, such as mechanical properties, thermo-electric power and so on (Xing et al., 1999; Dubois, 2000; Pope, 1999). One way to deal with this problem is to use the same method as used for normal crystal structures, but instead of using normal 3D space groups, high-dimensional superspace groups are applied to solve the atomic structures through single crystal X-ray diffraction techniques. This is a very effective way to solve the structure if large high quality single crystals are available, but unfortunately, it is very difficult to obtain large single crystals for most quasicrystals and their approximants. It often happens that several quasicrystal approximants coexist, causing lots of defects. In very many cases, it is impossible to obtain a pure phase in a single small particle. This is why only a few of these structures have yet been solved by X-ray diffraction.

A possible solution for this situation is to use electron crystallography. Then a sufficiently strong signal can be obtained from a few nm area. In this technique, electrostatic potential maps can be obtained by combining the structure factor phases from high resolution transmission electron microscopy (HRTEM) images and amplitudes from HRTEM images or electron diffraction patterns. A successful application was on the complicated quasicrystal approximant ν-AlCrFe with space group $P6_3/m$, a = 40.687 Å and c = 12.546 Å (Zou *et al.*, 2003). A 3D electrostatic potential map was calculated by combining the structure factor phases from HRTEM images and amplitudes from selected-area electron diffraction (SAED) patterns of 13 zone axes. 124 of the 129 unique atoms in the unit cell were found from the electrostatic potential map. However, this method requires extensive experimental work for determination of complicated structures such as quasicrystal approximants. A sufficient number of high quality HRTEM images and electron diffraction patterns need to be collected along different orientations. Obtaining high quality data requires correct sample preparations and an experienced operator. The post treatment of the data includes many correction terms, such as defocus, astigmatism, crystal thickness and so on. All of these limit the application of electron crystallography for structure determination of quasicrystals and their approximants.

Quasicrystals and their approximants exist as a series of compounds in most cases, all of them with similar local atomic structures with different packing. Some less complicated approximants in the same system can be synthesized as large single crystals or pure phases with fewer defects, and their atomic structures can be well-determined. From knowing related approximants in the same series, structure modeling in real space is an efficient approach. (Takakura et al., 2006; Lord et al., 2001; Fu et al., 2003) From the tiling and the decoration of each tile, the whole structure model with atomic positions can be obtained.

The *strong-reflections approach* is an approach in reciprocal space. The structure similarity of approximants in real space results in similar structure factor intensity distribution and phase relationships for the strong reflections in reciprocal space. Structural models of six

Table 1. List of quasicrystal approximants solved by the strong reflections approach

Known structure						Deduced structure						
Phase	Space group	Lattice parameters (Å)				Phase	Space group	Lattice parameters (Å)				Ref.
		a	*b*	*c*	*β* °			*a*	*b*	*c*	*β* °	
m-$Al_{13}Co_4$	*C*2/*m*	15.2	8.1	12.4	108	τ^2-$Al_{13}Co_4$	*P*2/*m*	39.9	8.1	32.2	108	Christensen et al., 2004
μ_3-ZnMgRE	$P6_3/mmc$	14.6		8.6		μ_5-ZnMgRE	$P6_3/mmc$	23.5		8.6		Zhang, Zou et al., 2006
						μ_7-ZnMgRE	$P6_3/mmc$	33.6		8.9		
λ-Al_4Mn	$P6_3/m$	28.4		12.4		τ(μ)-AlCrSi	$P6_3/mmc$	32.3		12.4		Zhang, He et al, 2006
						μ′-AlCrSi	$P6_3/mmc$	20.1		12.4		He et al. 2006
ε_6	*Pnma*	23.5	16.8	12.3		ε_{16}	*B*2*mm*	23.5	16.8	32.4		Li et al. 2010

compounds in four groups of approximants (Table 1) have been successfully deduced from their known related structures by the strong-reflections approach. This approach is based on two facts: one is that the strongest reflections mainly determine the atomic positions in a structure; the other is that the strong reflections that are close to each other in reciprocal space have similar structure factor amplitudes and phase relations for all the approximants in a series. Thus, the structure factor amplitudes and phases of strong reflections for an unknown approximant can be estimated from those of a known related approximant. Atomic positions in the unknown approximant are then obtained directly from the 3D electron density map calculated by inverse Fourier transformation of the structure-factor amplitudes and phases of the strong reflections.

2. Why the Strong-Reflections Approach Works

The strong-reflections approach is based on two basic facts: that the strong reflections determine the atomic positions and the similarity between different approximants in the same series. In this section, we will discuss these two facts.

2.1. Strong Reflections Determine the Atomic Positions

If both structure factor phases and amplitudes are known, the electron density map can be obtained through the inverse Fourier transform (Equation 1). From this electron density map, sphere-like peaks can be assigned as atomic positions.

$$\rho(x,y,z) = \sum_{h,k,l} F_{hkl}\, e^{2\pi(hx+ky+lz)+\varphi_{hkl}} \qquad (1)$$

To assign non-hydrogen atomic positions, not all reflections are needed during the Fourier transformation. A resolution of 1.2Å for these structure factors is typically enough since the atomic distances for non-hydrogen atoms are normally larger than 1.2Å. Moreover, within this resolution, it is not necessary to include all reflections for the Fourier transformation, because the weak reflections only slightly modify the electron density map. Taking chemical knowledge (such as characteristic interatomic distances, angles and coordination numbers) into account, the requirement for completeness of structure factors can be further relaxed. A good example is the structure determination of IM-5 (one of the most complicated zeolites) by electron crystallography (Sun et al. 2010). In this work, only the structure factors from three main zone axes were included in the Fourier transformation. Although the completeness was only about 22% at the resolution of 2.5 Å, all Si positions were located from the potential map with the help of chemical knowledge (Figure 1). All oxygen atoms could be inserted between each Si-Si pair afterwards.

For the quasicrystal approximants, the situation becomes even easier because the intensity distribution is very uneven in this kind of materials. As shown in Figure 2, the red, green and blue curves present the amplitude distributions for three approximants, ν-$Al_{80.61}Cr_{10.71}Fe_{8.68}$, λ-$Al_4Mn$ and κ-$Al_{76}Cr_{18}Ni_6$ phases, respectively (Sato et al., 1997; Li et al., 1997; Marsh, 1998 ; Kreiner and Franzen, 1997 ; Mo et al., 2000; Zou et al., 2003). They are closely related quasicrystal approximants with almost the same *c*-axis (≈12.5 Å) but different *a*-axes. The ν-phase is the most complicated one with the longest *a*-axis (40.687 Å), and its amplitude distribution is the most uneven as shown by the red curve in Figure 2. κ-$Al_{76}Cr_{18}Ni_6$ is the simplest structure in this series, with the shortest *a*-axis (17.674 Å).

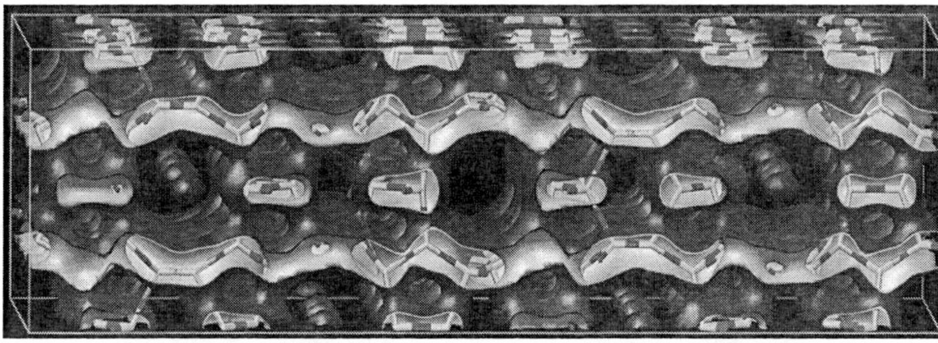

Figure 1. Potential map of IM-5 reconstructed from 144 unique reflections along three main zone axes. From this potential map, all 24 unique Si atoms could be located and one oxygen atom inserted between each Si-Si pair. (Sun et al. 2010).

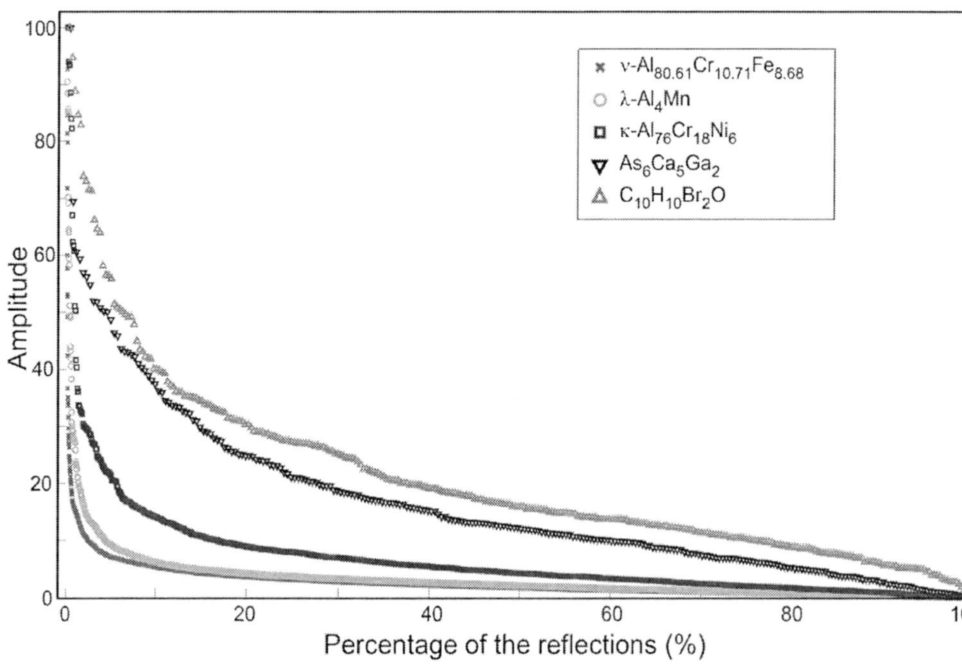

Figure 2. The structure factor amplitude distributions of five structures. The quasicrystal approximants (ν-$Al_{80.61}Cr_{10.71}Fe_{8.68}$, λ-$Al_4Mn$, κ-$Al_{76}Cr_{18}Ni_6$) show very uneven amplitude distributions (Red, green and blue curves respectively). Black and purple curves represent amplitude distributions of $As_6Ca_5Ga_2$ and $C_9H_{10}Br_2O$, a typical normal intermetallic compound and an organic crystal structure.

It is striking how uneven amplitude distribution these three approximants have, when compared to a normal intermetallic compound and an organic crystal structure respectively (the black and purple curves). This kind of uneven amplitude distribution makes it possible to determine atomic positions from just a few strong reflections.

As a rule-of-thumb, more than 90% of all atomic positions can be located in a crystal structure, if the summed amplitudes of the included reflections reaches 50% of the total amplitude. For quasicrystal approximants, always having a few exceptionally strong reflections, this can be obtained by including a very small fraction of all reflections. For example, in the ε_6-Rh-Al system, the 256 strongest reflections among all 2640 independent reflections sum up to 57% of the total amplitude. From the Fourier transform of these strong reflections, all atomic positions can be obtained (Boundard et al., 1996; Li 2010).

Another example is μ_5-ZnMgRE which is one of a series of approximants with the same *c*-axis but different *a*-axes (Abe et al., 1999). Figure 3 shows one slice of the electron density map at $z = 0.25$ calculated from different numbers of strong reflections.

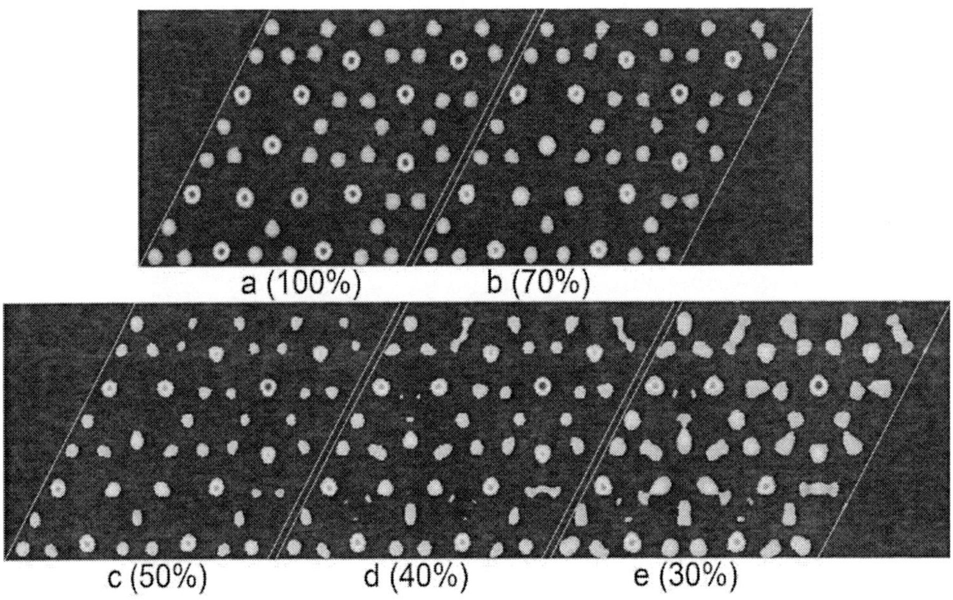

Figure 3. Electron density map at z = ¼ for the μ_5-ZnMgRE structure calculated from different amount of the strongest reflections within 1.2 Å resolution. a) all 514 reflections; b) 70% of the total amplitude (150 reflections); c) 50% of the total amplitude (75 reflections); d) 40% of the total amplitude (49 reflections); d) 30% of the total amplitude (30 reflections). Note that not only the atomic positions can be seen, but also the heavier elements (red) are easily distinguished from the lighter ones (light blue).

With 50% of the total amplitude, all atomic position can be indentified. With 40% and 30% of the total amplitude, the electron density map is not clean any more, two atoms start merging into one position. But, even with only 30% of the total amplitude (30 out of total 514 reflections), 21 out of 23 atomic positions can be obtained easily and the remaining two can be located from the elongated peaks.

The main tendency is that with less and less strong reflections, atomic positions become less accurate and even merge into one or split into two. Finally, the element types become less obvious and ghost peaks appear in the electron density map. When very few strong reflections are available, chemical knowledge can help to assign atomic positions and element types. An important fact is that the structure factor phases of the strong reflections are more important than the amplitudes for obtaining a correct electron density map. An average amplitude deviation of 30% will normally not change the feature of the whole density map very much, while a few strong reflections with wrong phases may totally mess up the density map. It is thus of paramount importance to obtain the correct phases of the strongest reflections.

2.2. Similarity

For quasicrystals and their approximants, the intensity distributions are very uneven and a few strong reflections with both phases and amplitudes can already represent the whole structure very well as discussed in the previous subsection. The problem now is how to obtain the structure factor phases and amplitudes for these strong reflections.

It is not easy to obtain the amplitudes and phases of those strong reflections directly, if we don't have high quality single crystal X-ray diffraction data or HRTEM images. Fortunately, for the quasicrystal approximants, they always appear as a series of compounds and all of them have similar local atomic structures with slightly different packing in a large scale. As shown in Figure 4, two quasicrystal approximants (μ_5 and μ_7) have the same local cluster but with different arrangements (Abe et al., 1999; Sugiyama et al. 1999).

This similarity in real space leads to similarity also in reciprocal space. To show the similarity for both amplitudes and phases, the simulated structure factors of μ_3 (Sugiyama et al., 1998), μ_5 (Abe et al., 1999) and μ_7 (Sugiyama et al. 1999) are shown in Figure 5a-c, as well as κ-$Al_{76}Cr_{18}Ni_6$ (Sato et al., 1997; Li et al., 1997; Marsh, 1998), λ-Al_4Mn (Kreiner and Franzen, 1997) and ν-$Al_{80.61}Cr_{10.71}Fe_{8.68}$ phases (Mo et al., 2000; Zou et al., 2003) in Figure 5d-f.

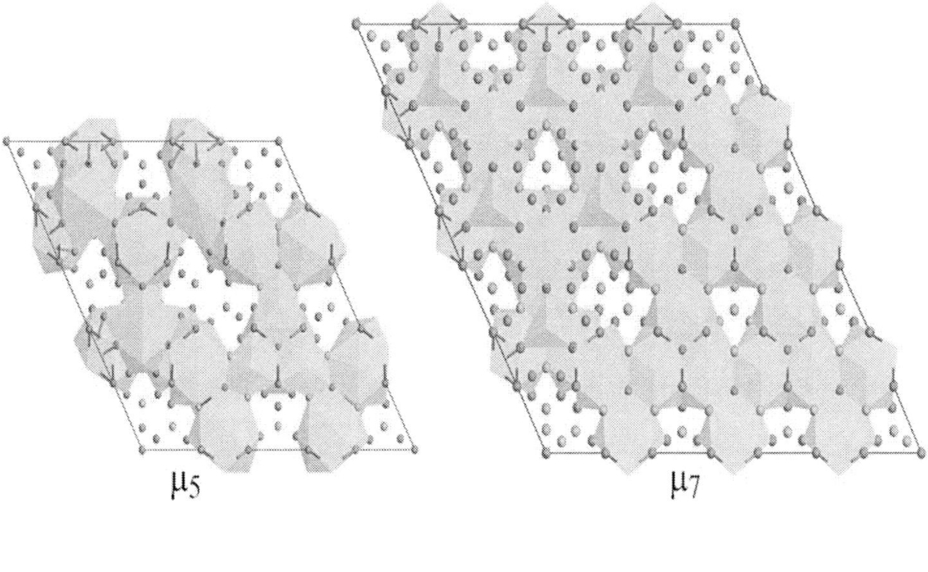

Figure 4. Structure projections along the *c*-axis of μ_5 and μ_7. (Abe et al., 1999; Sugiyama et al. 1999).

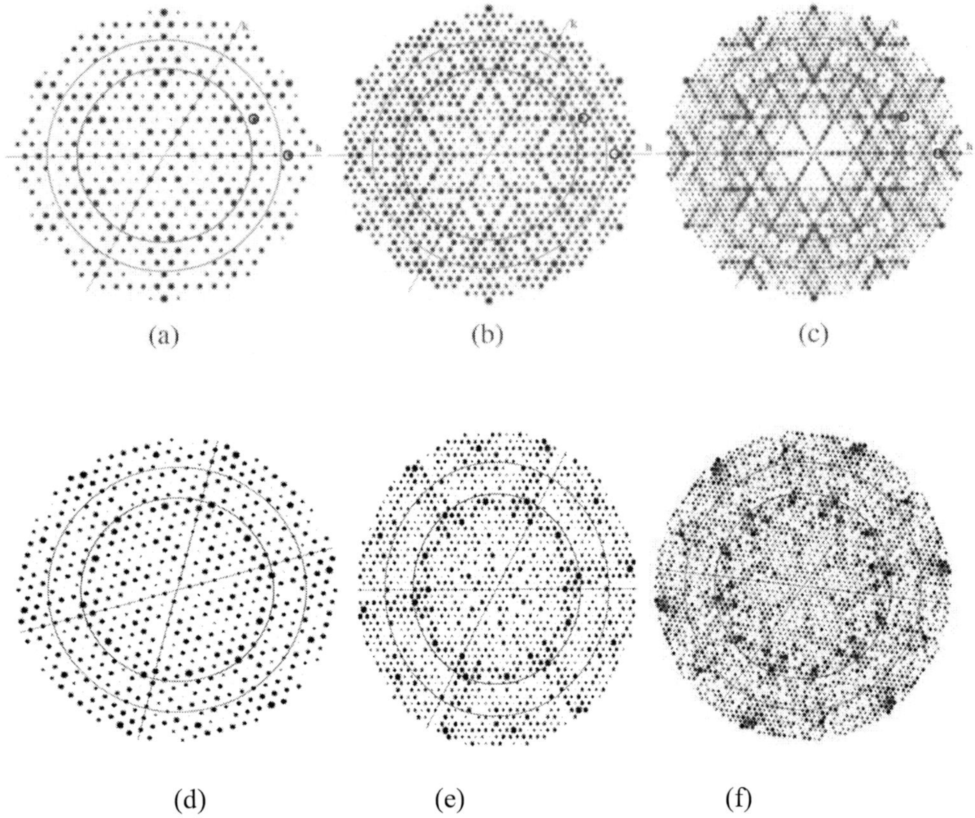

Figure 5. Simulated structure factors of six quasicrystal approximants with both amplitudes and phases. The size of each spot represents the amplitude and the colour represents the phase (red for 0° and blue for 180°). (a), (b) and (c) are μ_3, μ_5 and μ_7 in the same Zn–Mg–rare-earth quasicrystal approximant series (see Table 1). (d), (e) and (f) are κ-$Al_{76}Cr_{18}Ni_6$, λ-Al_4Mn and ν-$Al_{80.61}Cr_{10.71}Fe_{8.68}$ phases. They all have space group $P6_3/m$ and essentially the same c-axis dimension (≈12.5 Å). However, the a-axes differ, ranging from 17.674 Å for κ, over 28.389 Å for λ to 40.687 Å for ν. (Zhang, Zou et al. 2006; Zhang, Zou et al. unpublished work).

The size and color of each spot represents the amplitude and phase respectively. All strong reflections locate at more or less the same positions in reciprocal space, and with the same colour (i.e. phase). See for example those reflections marked by black circles.

The deviations in reciprocal space for the strongest reflections of μ_5 and μ_7 (>3% of the largest amplitude) within 1.2 Å resolution are plotted in Figure 6. The strongest reflection with a deviation larger than 0.025 $Å^{-1}$ is about 10% of the largest amplitude. Moreover, the strongest peak with different phases is not more than 10% of the largest amplitude. These great similarities in both amplitudes and phases make it possible to deduce the atomic structure from one approximant to another by the strong-reflections approach.

For unknown phases, the similarity of structure factors can be checked also by experimental electron diffraction patterns and HRTEM images. As the authors did for ε_{16} and ε_6 (Li et al., 2010), the structure factor amplitudes extracted from electron diffraction patterns were compared, using different numbers of strong reflections (Figure 7). The intensities of the corresponding reflections are very similar in these two structures, with an R-value of 0.19 for the 30 strongest reflections.

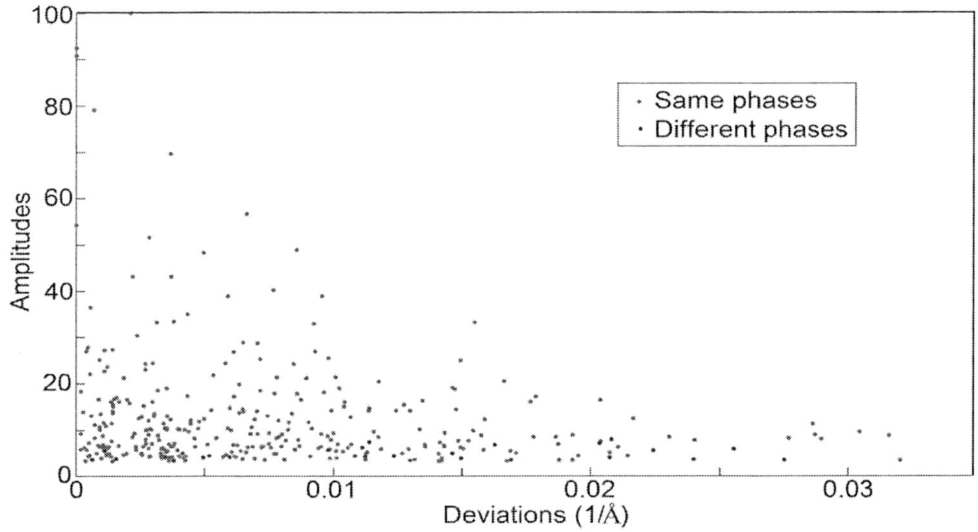

Figure 6. Deviations for the strong reflections of μ_5 and μ_7 in reciprocal space. The Y-axis is the amplitudes of the corresponding peaks in μ_5. The red spots represent corresponding peaks in μ_5 and μ_7 having the same phases, while the blue spots represent different phases. All reflections whose amplitude in μ_5 is stronger than 3% of the strongest one are included (341 out of 514). The few reflections with different phases (30 out of 514) all have amplitudes below 10% of the highest amplitude.

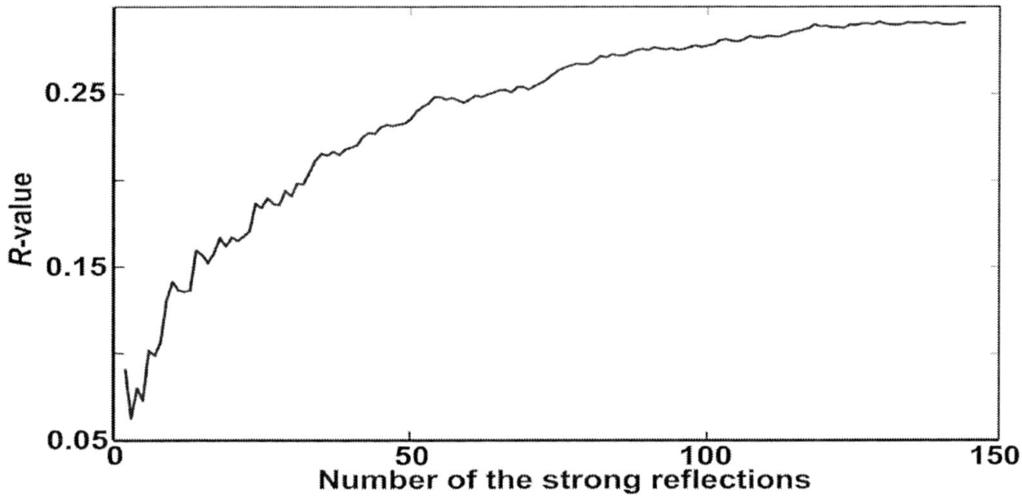

Figure 7. R-values plotted against the number of the corresponding strongest reflections in ε_6 and ε_{16}. For the 30 strongest reflections, an R-value of 0.19 shows a good correspondence between ε_6 and ε_{16}. As more and more moderately strong reflections are included, the R-value increases, reaching 0.29 for the 146 strongest reflections. (Li et al., 2010).

As more and more moderately strong reflections are included, the R-value increases, reaching 0.29 for the 146 strongest reflections. This kind of R-value is close to a typical internal R-value for electron diffraction data obtained from different particles of the same structure. Thus, there is obvious similarity in the ε_{16} and ε_6 structures. The similarity of the structure factor phases were checked in these two structures by a more complicated way as we will discuss later.

In summary, the structure factors for strong reflections in the same series of quasicrystal approximants are shown to be similar. Normally, the similarity between two high order quasicrystal approximants with large unit cells are bigger than that between two low order approximants with small unit cells as we can see in Figure 5. Thus, if possible, it is better to use a known approximant with a large unit cell as the starting structure for the strong-reflections approach.

3. Strong-Reflections Approach

In the previous section, we discussed the similarity of different approximants from the same series in reciprocal space. To execute the strong-reflections approach, the first thing which must be done is to find a known approximant related to the unknown structure.

Normally the composition, space groups and unit cell parameters give hints about this similarity. Sometimes one or more of the unit-cell dimensions of different approximants within the same series are related by the golden mean τ as 1: τ: τ^2: τ^3 etc, such as a series of approximants in the Al–Co system i.e. *m*-$Al_{13}Co_4$ (*C2/m*, $a = 15.173$, $b = 8.109$, $c = 12.349$Å and $\beta = 107.90°$), τ^2-$Al_{13}Co_4$ (*P2/m*, $a = 39.863$, $b = 8.139$, $c = 32.208$ Å and $\beta = 107.94°$), τ^3 ($a = 64$, $b = 8.1$, $c = 52$Å and $\beta = 108°$) and τ^4 ($a = 104$, b is unknown, $c = 8.4$Å and $\beta = 108°$). (Christensen et al., 2004; Hudd et al., 1962; Ma et al., 1995).

Note that even if two approximants have similar composition, the same space group and related unit cell parameters, they are not necessarily similar in reciprocal space. In the structure deduction of μ'-$(Al,Si)_4Cr$ (*P*6_3*/mmc*, $a = 20.1$ and $c = 12.4$ Å), the authors did not select μ-$(Al,Si)_4Cr$ (*P*6_3*/mmc*, $a = 20.0$ and $c = 24.7$ Å) as the starting structure, because the phase relations of the symmetry-related reflections are different due to the space group (He et al., 2007). For example, according to the space group, the phases of (5 0 5) and (5 0 -5) in μ' must differ by 180°, while the corresponding reflections (5 0 10) and (5 0 -10) in μ should have the same phases. Instead, λ-$(Al,Si)_4Cr$ was selected as the starting structure since their strong reflections are more closely related. It is recommended to check this similarity experimentally through some simple techniques such as electron diffraction and HRTEM.

After confirming the similarity, the phase and amplitude information from the known structure can be used for the structure determination of the others. This can be done mainly by three steps:

a) Re-indexing and phasing
b) Identifying atomic positions
c) Structure verification

3.1. Re-Indexing and Phasing

This is a key step for the strong-reflections approach. The structure factors with both phases and amplitudes for a known structure can be directly calculated from the structure model. When the *hkl* list of the strong reflections with indices, amplitudes and phases is transformed into the structure factors of an unknown structure, the structure factor amplitudes are normally kept in the new *hkl* list but the indices and phases might be changed. We will address these two problems separately here.

In many cases, the re-indexing is rather straight-forward. An example is shown in Figure 5a-c with simulated structure factor phases and amplitudes of μ_3, μ_5 andμ_7 in the Zn-Mg-RE system. Starting from any of them, simply rescaling the index with a new set of reciprocal axes without any rotation, the new index is obtained. For example, using μ_7 as the starting known structure to deduce the structure model of μ_5 (Zhang, Zou et al., 2006), all *h* and *k*-indexes will be multiplied by 0.718 and *l*-indexes kept unchanged, since the a^* and b^*-axes in μ_5 are longer while c^* is almost the same as that of μ_7. After this operation, most indices may become non-integers, but most indices can easily be rounded to the nearest integer index. Normally they are already close to certain integers due to their similarity in reciprocal space. In Table 2 the 20 strongest reflections in μ_7 and their corresponding reflections in μ_5 are listed, with simulated intensities and phases. All of them show a one-to-one correspondence.

Table 2. The 20 strongest reflections in μ_7 and their corresponding ones in μ_5

	μ_7					μ_5			
h	*k*	*l*	Amp.	Phase	*h*	*k*	*l*	Amp.	Phase
7	7	0	1000	0	5	5	0	1000	0
0	0	4	925	0	0	0	4	907	0
0	0	6	795	180	0	0	6	923	180
7	0	3	761	0	5	0	3	790	0
21	0	0	674	0	15	0	0	696	0
11	0	3	568	0	8	0	3	566	0
14	7	3	503	0	10	5	3	516	0
11	0	1	424	180	8	0	1	488	180
11	0	5	389	180	8	0	5	482	180
0	0	2	381	180	0	0	2	542	180
11	0	2	380	0	8	0	2	403	0
11	7	0	351	0	8	5	0	329	0
18	0	2	335	0	13	0	2	389	0
14	0	2	334	0	10	0	2	432	0
7	4	2	321	0	5	3	2	389	0
10	0	3	321	180	7	0	3	431	180
7	4	6	303	0	5	3	6	332	0
11	0	6	298	0	8	0	6	350	0
7	7	6	288	180	5	5	6	271	180
14	4	3	274	180	10	3	3	288	180

In more complicated cases, not only the length of the axes, but also the orientations of the axes are different from one approximant to another, in one quasicrystal approximant series. In this case, a 3×3 orientation matrix is needed to relate the strong reflection indexes. The orientation matrix ***A*** is defined as

$$(h\ k\ l)' = (h\ k\ l)\ \mathrm{A}.$$

By finding several related strong reflections experimentally, the nine terms in the orientation matrix can be obtained and subsequently refined. This approach was applied in the structure deduction of τ(μ)-$Al_{3.82-x}CrSi_x$ and μ'-$(Al,Si)_4Cr$ from the known phase λ-$(Al,Si)_4Cr$ (Zhang, He et al., 2006; He et al., 2007). In these three structures, the c* axes are orientated in the same direction and with the same length. Thus four terms in the orientation matrix are zero (a_{13}, a_{23}, a_{31} and a_{32}) and a_{33} is always equal to one. To determine the other four terms, experimental SAED patterns were collected to find the corresponding strong reflections as shown in Figure 8. In Figure 8, the patterns are orientated based on the strong reflections, rather than making the a* or b*-axes parallel. Obviously, the strong reflections are one-to-one corresponding, such as $(13\ {-5}\ 0)_\lambda \Leftrightarrow (13\ 0\ 0)_{\tau(\mu)}$ and $(11\ 2\ 0)_\lambda \Leftrightarrow (8\ 8\ 0)_{\tau(\mu)}$. From this correspondence, the orientation matrix was determined as

$$A = \begin{pmatrix} \tau/2 & 1/2 & 0 \\ -1/2 & \tau^2/2 & 0 \\ 0 & 0 & 1 \end{pmatrix}.$$

Considering the simpler cases, approximants related without rotation can also be understood as having an orientation matrix with only a_{11}, a_{22} and a_{33} non-zero. Even more complicated orientation matrices with all nine terms non-zero may appear when 3D quasicrystal approximant systems are considered, such as icosahedral quasicrystals.

After obtaining the new indices, we need to consider the phases. In most cases, the unit cell origins of the known and unknown structures are at corresponding positions. Then the phases of the unknown structure can be inferred directly from the known structure and used for the Fourier transform in the next step. This is normally true when all related structures are centrosymmetric, such as τ(μ)-$Al_{3.82-x}CrSi_x$, μ'-$(Al,Si)_4Cr$ and λ-$(Al,Si)_4Cr$, m-$Al_{13}Co_4$ and τ^2-$Al_{13}Co_4$, and μ_3, μ_5 and μ_7 in the Zn-Mg-RE system.

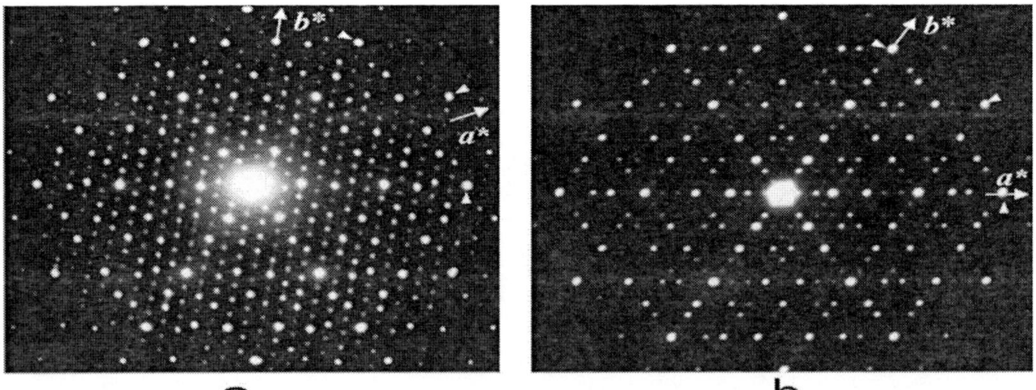

Figure 8. Electron diffraction patterns from (a) λ-$(Al,Si)_4Cr$ and (b) τ(μ)-$Al_{3.82-x}CrSi_x$ approximants taken at 100 kV along the c-axis. Both are on the same scale and with similar orientation. The corresponding strongest diffraction spots in the two approximants are marked by arrows. The diffraction pattern of λ in (a) shows only $p6$ symmetry, while that of τ(μ) in (b) shows $p6m$ symmetry. Note, however, that the strongest diffraction spots in λ (a) have approximate $p6m$ symmetry. (Zhang, He et al., 2006).

The situation is different when deducing ε_{16} from ε_6 in Al-Rh alloys. The space group of ε_6 is *Pnma* with $a = 23.6$ Å, $b = 16.8$ Å and $c = 12.3$ Å while the space group of ε_{16} is *B2mm* with $a = 23.5$ Å, $b = 16.8$ Å and $c = 32.4$ Å (deduced from ED patterns and HRTEM images). The phase relations between the symmetry-related reflections in the two space groups are different. For the space group *Pnma*, the relations are:

(i) if $h + l = 2n$ and $k = 2n$:
$\varphi_{hkl} = \varphi_{-hk-l} = \varphi_{h-kl} = \varphi_{-h-k-l} = \varphi_{hk-l} = \varphi_{-h-kl} = \varphi_{-hkl} = \varphi_{h-k-l}$ (φ: structure factor phase)

(ii) if $h + l = 2n$ and $k = 2n + 1$:
$\varphi_{hkl} = \varphi_{hk-l} = \varphi_{-h-k-l} = \varphi_{-h-kl} = \pi + \varphi_{h-k-l} = \pi + \varphi_{h-kl} = \pi + \varphi_{-hkl} = \pi + \varphi_{-hk-l}$

(iii) if $h + l = 2n + 1$ and $k = 2n$:
$\varphi_{hkl} = \varphi_{h-kl} = \varphi_{-h-k-l} = \varphi_{-hk-l} = \pi + \varphi_{hk-l} = \pi + \varphi_{h-k-l} = \pi + \varphi_{-h-kl} = \pi + \varphi_{-hkl}$

(iv) if $h + l = 2n + 1$ and $k = 2n + 1$:
$\varphi_{hkl} = \varphi_{h-k-l} = \varphi_{-h-k-l} = \varphi_{-hkl} = \pi + \varphi_{hk-l} = \pi + \varphi_{h-kl} = \pi + \varphi_{-h-kl} = \pi + \varphi_{-hk-l}$

For the space group *B2mm*, the relations are simpler:
$\varphi_{hkl} = \varphi_{h-kl} = \varphi_{hk-l} = \varphi_{h-k-l} = -\varphi_{-hkl} = -\varphi_{-h-kl} = -\varphi_{-hk-l} = -\varphi_{-h-k-l}$

Apparently there are lots of equivalent strong reflections that have different phase relations in the two space groups. For example, the third strongest reflections $(8\ 4\ 3)_{\varepsilon 6}$ and $(8\ 4\ {-3})_{\varepsilon 6}$ must have phases differing by 180° while their corresponding reflections in ε_{16} $(8\ 4\ 8)_{\varepsilon 16}$ and $(8\ 4\ {-8})_{\varepsilon 16}$ must have the same phases. Fortunately, the situation here is different from the one we discussed above. Here the space group of ε_{16} is non-centrosymmetric, so it is possible to get the same phases for $(8\ 4\ 8)_{\varepsilon 16}$ and $(8\ 4\ 8)_{\varepsilon 16}$ by moving the origin of the unit cell.

A simple way to find the origin shift is using *P*1 symmetry for both ε_6 and ε_{16}. The 256 strongest reflections in ε_6, were expanded into 1590 reflections using *Pnma* symmetry. Based on the strong reflection approach, each of these strong reflections in ε_6 has one corresponding reflection in the ε_{16} structure in *P*1 symmetry with the same phases and amplitudes but different indices. The 3D electron density map of ε_{16} in *P*1 symmetry then can be calculated by Fourier transformation as shown in Figure 9. Approximately, the symmetry operations can be identified from this 3D electron density map as follows: *B*-centering, $2 \parallel \mathbf{a}$, $m \perp \mathbf{b}$, $m \perp \mathbf{c}$, which agrees with the space group *B2mm* with the origin at (0, 0.25, 0.15625). Then, the origin was shifted to this position and the new structure factor phases were calculated using the following equation for all these 1590 reflections:

$$\phi'(h\,k\,l) = \phi(h\,k\,l) + 360^\circ \times (h \times 0 + k \times 0.25 + l \times 0.15625).$$

Now the symmetry related reflections almost have the correct phase relations according to the space group *B2mm,* with an average deviation for all reflections about 7.8°. Note that although the phase relations between the symmetry-related reflections in the two space groups *Pnma* and *B2mm* are different, the phase relations for the strong reflections in the two structures are almost the same. It is similar to the electron diffraction patterns in Figure 8, where Figure 8a does not have *p6m* symmetry, but the strong reflections follow *p6m* symmetry, just as all the reflections in Figure 8b.

In this section, we have discussed the method of transforming the *hkl* list from a known structure to that of an unknown structure. The most important thing here is that only strong reflections can be transformed. Weak reflections may not have corresponding reflections in the unknown structure at all or they may have quite unrelated amplitudes and phases. To select those strong reflections, first the strong reflections are included to sum up to 50-75% of the total amplitude. Then, the reflections with large deviations in reciprocal space are removed, since the more a reflection pair of two structures deviates from each other in reciprocal space, the less their phases will be related. Those that do not follow the space group symmetry of the unknown phases should also be deleted, such as those in Figure 8a which don't follow *p6m* symmetry and those in ε_{16} with large phase errors in *B2mm* symmetry after the origin shift. If the amplitudes of the transformed reflections sum up to 40-50% of the total amplitude, the similarity of the known and unknown structures are very high and the structure model of the unknown phases can be easily established in the next step.

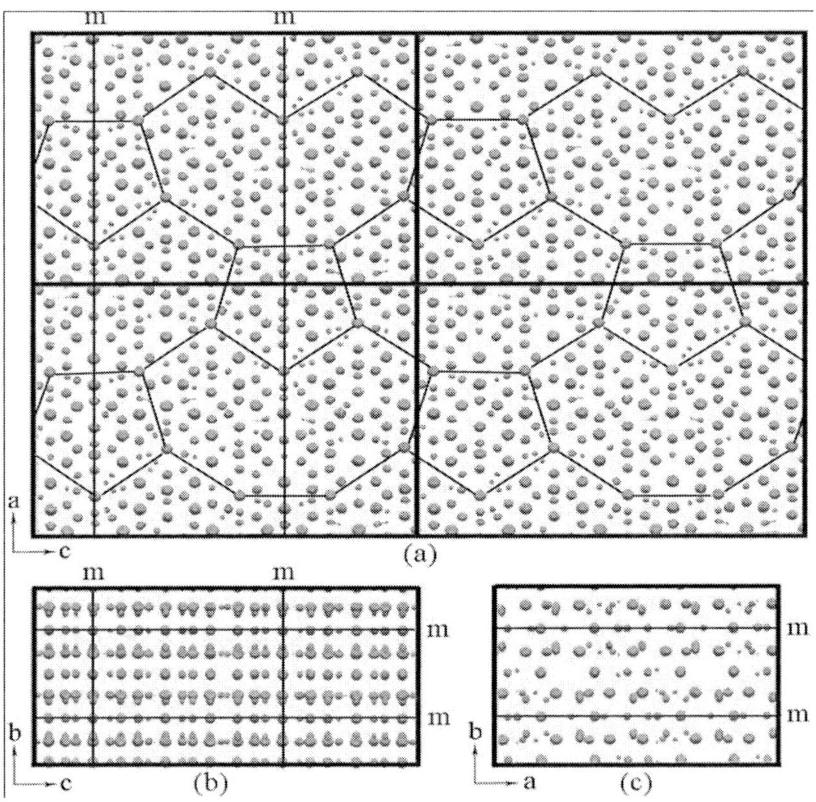

Figure 9. A three-dimensional electron-density map of ε_{16} calculated from 1590 (256 independent) strong reflections using the space group *P*1. (a), (b) and (c) are three-dimensional density maps viewed along the *b*, *a* and *c* axes, respectively. Four unit cells are outlined in (a). The symmetry elements can be identified from the density map, with mirrors perpendicular to the *b* and *c* axes (marked). The new origin is set on 2*mm*. The origin shifts obtained from the electron-density maps are $\Delta x = 0$, $\Delta y = 0.25$ and $\Delta z = 0.15625$. Banana-shaped clusters and pentagonal clusters are outlined in (a) to show the symmetry. (Li et al., 2010).

3.2. Identifying Atomic Positions

As discussed in the previous subsection, through the orientation matrix and the origin shift, both indices and phases can be transformed into the corresponding one in the unknown structure. In the next step we will use these derived structure factors, with both amplitudes and phases, to locate the atomic positions in the unknown structure.

The 3D electron density map can be calculated from the structure factors by Fourier transformation. In most cases, many spherical peaks can be found in the map (Figure 9). The independent fractional positions can be obtained by the programs eMap (Oleynikov, 2006) or EDMA (van Smaalen, 1995) automatically or manually.

Due to the missing reflections and errors in amplitudes, the peak heights will not accurately represent the electron density in the real structure. This makes it difficult to assign element type for each peak, especially for elements with similar atomic numbers. For example, in Figure 2 the peak heights of the two peaks at the center of the unit cell become higher and higher with fewer reflections included in the Fourier transform. In Figure 2e, they almost have similar heights as the three around the 3-fold axes with 30% of the total amplitude included. By checking with other related approximants (μ_3 and μ_7) shown in Figure 10, rare earth atoms are never found so close to each other, and thus they should be assigned as Zn atomic positions.

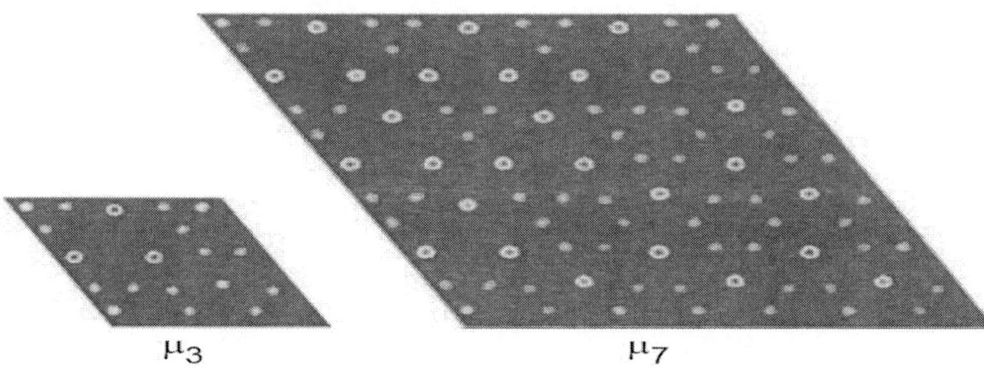

Figure 10. Density maps of layers at z=1/4 for μ_3 calculated with 108 unique reflections and μ_7 calculated with 279 unique reflections. (Zhang, Zou et al., 2006).

In other cases, higher peaks can even correspond to light elements in so heavily distorted electron density maps. For example, in the structure deduction of $\tau(\mu)$ from λ in the Al-Cr-Si system (Zhang, He et al., 2006), the 13 strongest peaks and the 18th strongest peak were assigned to be Cr atoms instead of the first 14 peaks, while all the other peaks except peak 46 were assigned as Al or Si (Al and Si are too similar to be distinguished). Peak 46 is a ghost peak which is very close to its neighboring atoms.

In more complicated cases, peaks for light elements might be missing. When ε_{16} was deduced from ε_6 in the Rh-Al system, 150 unique peaks were found from the 3D density map (Li et al, 2010). 33 of them were assigned as Rh, and the other 117 as Al. These 150 didn't form similar local environments as in ε_6 and some empty voids could also be found. Thus, three Al atoms were added to complete the structure based on the geometry and similarity to ε_6.

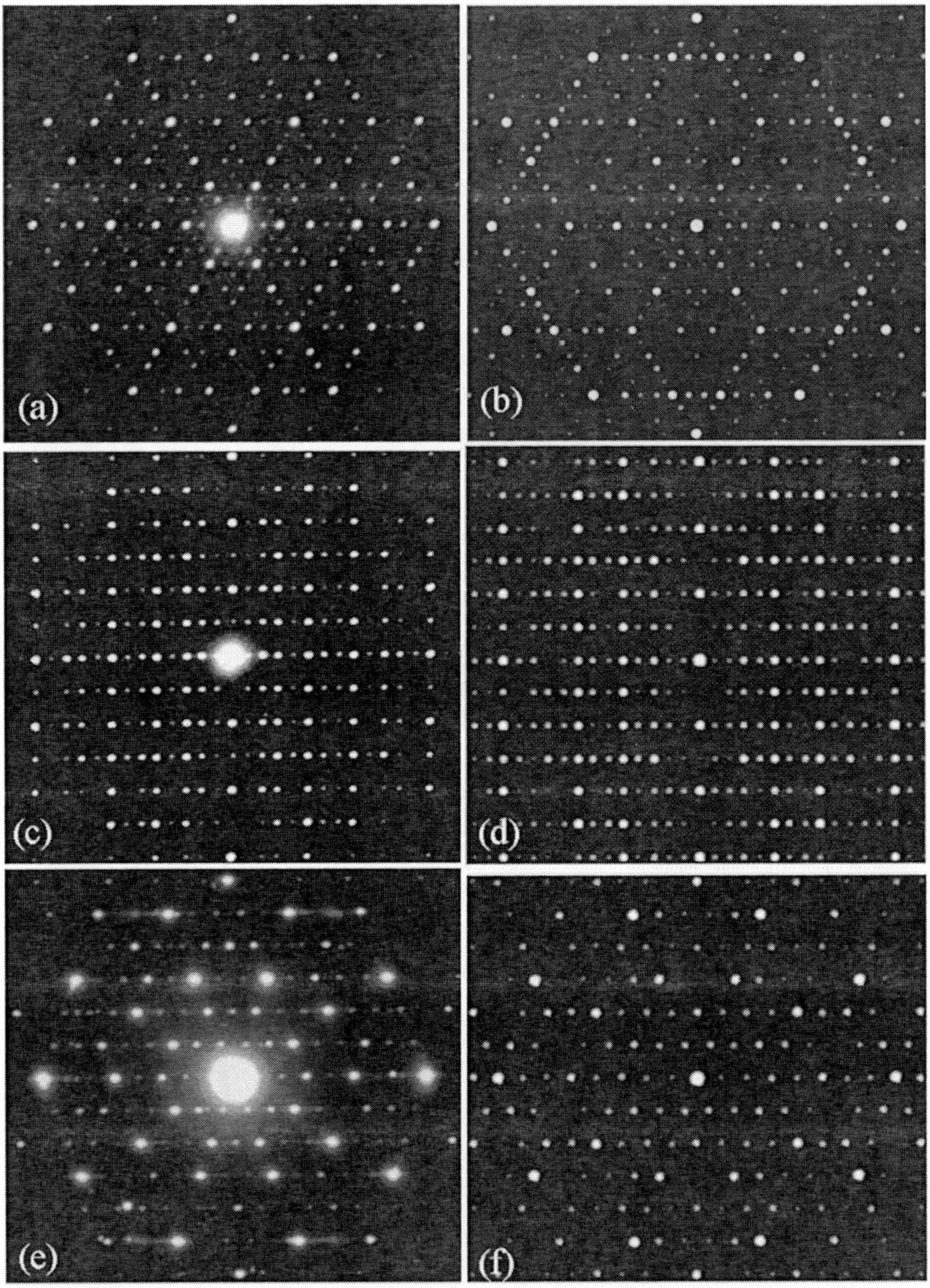

Figure 11. Experimental electron diffraction patterns of τ(μ) taken along the (a) [001], (c) [010] and (e) [211] zone axes on a Philips CM12 transmission electron microscope at 100 kV are compared with those simulated from the structure model using the program MacTempas, shown in (b), (d) and (f), respectively. The specimen thicknesses used in the simulations were 100, 50 and 50 Å for (b), (d) and (f). (Zhang, He et al., 2006).

In summary, due to the limited accuracy of amplitudes and missing peaks, after getting fractional positions from the 3D density map, it is necessary to combine the real space approach to assign element types and missing positions. The accuracy of atomic positions strongly relies on the similarity of the known and unknown structures, and normally 90% of all atoms can be determined within 0.2Å deviation from their correct positions (Zhang, Zou et al. 2006). This is actually better than one can expect from the first density map obtained in X-ray crystallography, based for example on phasing by direct methods. The structure model can be further improved by structural optimization through energy minimization or refinement based on the complete set of experimental electron diffraction data.

3.3. Structure Verification

The final structure model obtained in the previous step can be verified using any kind of experimental data. A pictorial way is to compare their simulated ED patterns with the experimental ones. As shown in Figure 11, the simulated electron diffraction patterns of $\tau(\mu)$ show a great agreement with the experimental ones. To quantify this agreement, the electron diffraction intensities can be extracted from the patterns by the program ELD (Zou *et al.*, 1993) and then diffraction patterns from different orientations are merged into a 3D *hkl* list. A least-squares refinement based on this *hkl* file can be done by the SHELXL program (Sheldrick, 2008). Normally an $R1$ value of ~0.3 indicates good agreement, since the dynamic effects and other distortions are not considered yet. Experimental HRTEM images can also be compared with simulated images as was done for the ε_{16} structure (Li et al., 2010).

4. Extending to Ideal Quasicrystals

The strong-reflections approach can be extended straightforwardly to the structure prediction of quasicrystals. The strong reflections in the electron diffraction patterns have great similarity between the quasicrystals and their approximants. Three patterns from the Al-Co-Ni decagonal quasicrystal system are shown in Figure 11. All the strong reflections show great similarity for both positions and intensities in the approximants and the quasicrystal. It is clear that the strongest 10 reflections have slightly different intensities in low order approximants (Figure 11a-b) while they are equally strong in the quasicrystal due to the 10-fold symmetry (Figure 11c). Thus, the amplitudes of these reflections calculated from a known approximant must be averaged instead of using their amplitudes directly, for the deduction of quasicrystal structures.

To deduce the quasicrystal atomic structure, it is essential to deduce the atomic structures for several approximants. As shown in Figure 12, three approximant structures are deduced with the same list of reflections using the strong-reflections procedure. The same amplitudes and phases are used for each structure while the indices are changed, based on the unit cell parameters. In the structure of PD9, all fragments can be found in PD2 or PD3, which means as long as we know the rough arrangement of clusters, all atomic positions can be obtained from PD2 or PD3. With larger and larger unit cell parameters ($a = \tau^n a_{PD2}$, $b = \tau^n b_{PD2}$, $n \rightarrow \infty$), more complicated approximants and finally the quasicrystal structure can be obtained.

Due to the similar local atomic structures, the atomic positions in the quasicrystal can be located by finding the same fragments in PD2 and PD3.

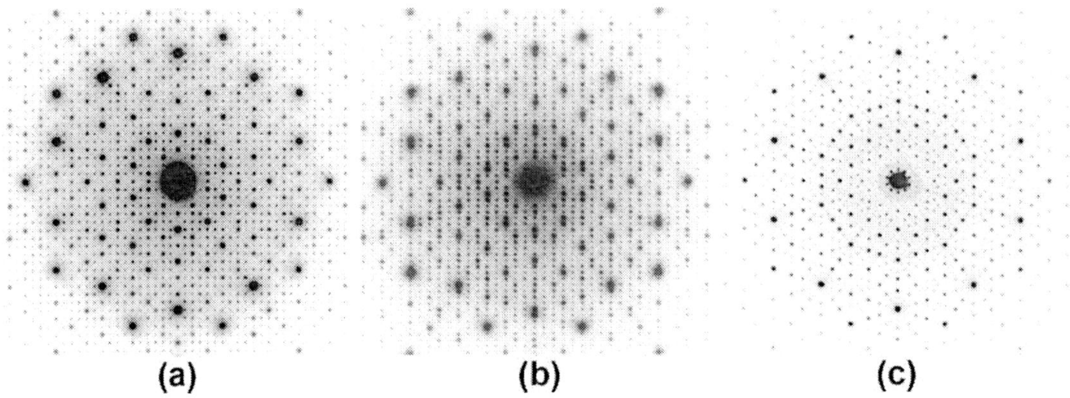

Figure 11. Electron diffraction patterns of a quasicrystal and its approximants in the Al-Co-Ni system along the ten-fold [001] direction. (a) PseudoDecagonal-1 (PD1, $a = 37.7$, $b = 39.7$, $c = 8.2$ Å) (Grushko et al., 2002); (b) PD4 (a =50.8, $b = 32$, $c = 8.2$ Å) (Oleynikov et al., 2006); (c) quasicrystal of $Al_{72}Ni_{20}Co_8$ (Abe et al., 2004).

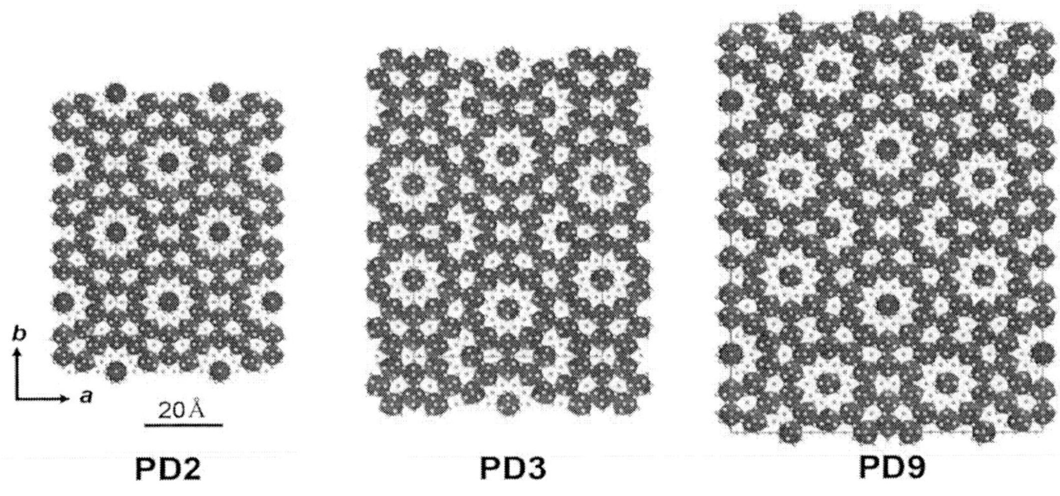

Figure 12. Projection of the approximant structures in the Al-Co-Ni system along the [001] direction. All structures are deduced by the strong-reflections approach with the same amplitudes and phases but different indices, based on the unit cell parameters. The unit cell parameters are $a = 23.2$, $b = 32.0$ Å for PD2, $a = 37.7$, $b = 51.8$ Å for PD3 and $a = 60.9$, $b = 83.3$ Å for PD9. All of them have the c-axis about 8.2 Å with all angles equal to 90°. (Zou and Hovmöller, unpublished).

The last step of the structure deduction of quasicrystals by the strong-reflections approach is to combine it with the structure modeling in real space where the atomic structure is obtained by decorating each tile. The key point here is that the tile arrangement (i.e. the fragment arrangement) and the decoration of each tile are deduced from the strong-reflections approach. Thus, from any known approximant, it should be possible to deduce its corresponding quasicrystal structure.

REFERENCES

Abe, E.; Takakura, H.; Singh, A.; Tsai, A.P. "Hexagonal superstructure in the Zn-Mg-rare-earth alloys" (1999) *J. Alloys Compd.* 283, 169-172.

Abe, E.; Yan, Y.F.; Pennycook, S.J. "Quasicrystals as cluster aggregates" (2004) *Nature Mater.* 3, 759-767.

Christensen, J.; Oleynikov, P.; Hovmöller, S. and Zou, X. D. "Solving Approximant Structure Using a 'Strong Reflection' approach " (2004). *Ferroelectrics*, 305, 273-277.

Dubois, J.M. "New prospects from potential applications of quasicrytalline materials" (2000) *Mater. Sci. Eng.* 4, 294-296.

Fu, X.J.; Yang, Y.T.; Hou, Z.L.; Liu, Y.Y. "Configuration correlations in decagonal covering structures" (2003) *Phys. Lett. A*, 315, 156-161.

Grushko, B.; Doblinger, M.; Wittmann, R.; Holland-Moritz, D. "A study of high-Co Al-Ni-Co decagonal phase" (2002) *J. Alloys Compd.*, 342, 30-34.

He, Z. B.; Zou, X. D.; Hovmöller, S.; Oleynikov, P. and Kuo, K. H. Structure determination of the hexagonal quasicrystal approximant μ'-$(Al,Si)_4Cr$ by the strong reflections approach *Ultramicroscopy*, (2007). 107, 495-500.

Hudd, R.C.; Taylor, W.H. "The structure of Co_4Al_{13}" (1962) *Acta Cryst.* 15, 441-442.

Kreiner. G.; Franzen, H.F. "The crystal structure of λ-Al_4Mn" (1997). *J. Alloys Compd.* 261, 83-104.

Li, M.R.; Sun, J.L.; Oleynikov, P.; Hovmöller, S.; Zou, X.D.; Grushko, B. "A complicated quasicrystal approximant ε_{16} predicted by the strong-reflections approach" (2010) *Acta Cryst.* B66, 17-26.

Li, X.Z.; Hiraga, K.; Yamamoto, A. "Icosahedral cluster in the structure of an Al-Cr-Ni κ phase" (1997) *Philos. Mag.* A76, 657-666.

Lord, E.A.; Ranganathan, S.; Kulkarni, U.D. "Quasicrystals: tiling versus clustering" (2001) *Philosophical Magazine A*, 81, 2645-2651.

Ma, X.L.; Li, X.Z.; Kuo, K.H. "A family of τ-inflated monoclinic $Al_{13}Co_4$ phases" (1995) *Acta Cryst.* B51, 36-43.

Marsh, R.E. "Concerning the κ Phases of Al-Cr-Ni" (1998). *Acta Cryst.* B54, 925-926.

Mo, Z.M.; Zhou, H.Y.; Kuo, K.H. "Structure of ν-$Al_{80.61}Cr_{10.71}Fe_{8.68}$, a giant hexagonal approximant of a quasicrystal determined by a combination of electron microscopy and X-ray diffraction" (2000) *Acta Cryst.* B56, 392-401.

Oleynikov, P. (2006). eMap, http://www.analitex.com/eMap.html.

Oleynikov, P.; Demchenko, L.; Christensen, J.; Hovmöller, S.; Yokosawa, T.; Döblinger, M.; Gruschko, B.; Zou, X.D. "Structures of the pseudodecagonal Al-Co-Ni approximant PD4" (2006) *Philosophical Magazine* 86 457-462.

Pope, A.L.; Tritt, T.M.; Chernikov, M.A.; Feuerbacher, M. "Thermal and electrical transport properties of the single-phase quasicrystalline materials: $Al_{70.8}Pd_{20.9}Mn_{8.3}$" (1999) *App. Phys. Lett.*, 75, 1854-1856.

Sato, A.; Yamamoto, A.; Li, X.Z.; Hiraga, K.; Haibach, T.; Steurer, W. "A new hexagonal κ phase of Al-Cr-Ni" (1997) *Acta Cryst.* C53, 1531-1533.

Shechtman, D.; Blech, I.; Gratias, D. and Cahn, J.W. "Metallic Phase with Long-Range Orientational Order and No Translational Symmetry" (1984). *Phys. Rev. Lett.* 53, 1951-1953.

Sheldrick, G.M. "A short history of SHELX" (2008). *Acta Cryst.* A64, 112-122.

Sugiyama, K.; Yasuda, K.; Horikawa, Y. Ohsuna. T.; Hiraga, K. "Crystal structure of μ_7-MgZnSm" (1999), *J. Alloys Compd.* 285, 172-178.

Sugiyama, K.; Yasuda, K.; Ohsuna, T.; Hiraga, K. "The structures of hexagonal phases in Mg-Zn-RE (RE = Sm and Gd) alloys" (1998) *Z. Krist.* 213, 537-543.

Sun, J.L.; He, Z.B.; Hovmöller, S.; Zou, X.D.; Gramm, F.; Baerlocher, Ch.; McCusker, L.B. "Structure determination of the zeolite IM-5 using electron crystallography" (2010) *Z. Kristallogr.* 225, 77-85.

Takakura, H.; Gómez, C.P.; Yamamoto, A.; De Boissieu, M.; Tsai, A.P. "Atomic structure of the binary icosahedral Yb-Cd quasicrystal" (2006) *Nature Mater.* 6, 58-63.

van Smaalen, S. "Incommensrate crystal strctures" (1995) *Crystallogr. Rev.* 5, 79-202.

Xing, L.Q.; Eckert, J.; Löser, W.; Schultz, L. "High-strength materials produced by precipitation of icosahedral quasicrystals in bulk Zr-Ti-Cu-Ni-Al amorphous alloys" (1999) *Appl. Phys. Lett.* 74, 664-666.

Zhang, H.; He, Z. B.; Oleynikov, P.; Zou, X. D.; Hovmöller, S. and Kuo, K. H. "Structure model for the $\tau(\mu)$ phase in Al–Cr–Si alloys deduced from the λ phase by the strong reflections approach" (2006). *Acta Cryst.* B62, 16-25.

Zhang, H.; Zou, X. D.; Oleynikov, P. and Hovmöller, S. "Structure relations in real and reciprocal space of hexagonal phases related to i-ZnMgRE quasicrystals" (2006). *Philos. Mag.* B86, 543-548.

Zou, X. D., Sukharev, Y. and Hovmöller, S. "ELD- A computer program system for extracting intensities from electrong-diffraction patterns" (1993). *Ultramicroscopy*, 52, 436-444.

Zou, X.D.; Mo, Z.M.; Hovmöller, S.; Li, X.Z.; Kuo, K.H. "Three-dimensional reconstruction of the ν-AlCrFe phase by electron crystallography" (2003) *Acta Cryst.* A59, 526-539.

In: Quasicrystals: Types, Systems, and Techniques
Editor: B. E. Puckermann, pp. 127-146 ISBN 978-1-61761-123-0

Chapter 5

HYDROGEN STORAGE IN Ti-Zr/Hf-Ni QUASICRYSTAL AND RELATED CRYSTAL POWDERS SYNTHESIZED BY MECHANICAL ALLOYING

***Akito Takasaki*[1] *and K. F. Kelton**[*2]
Department of Engineering Science and Mechanics, Shibaura Institute of Technology, Toyosu, Koto-ku, Tokyo 135-8548, Japan,
* Department of Physics, Washington University, St. Louis, Missouri 63130, USA

ABSTRACT

The dominant cluster in the Ti/Zr-based quasicrystals is a Bergman-type cluster possessing a large number of tetrahedral interstitial sites; this makes these quasicrystals attractive as potential hydrogen storage materials. This paper summarizes our recent research results on the hydrogen absorption and desorption properties of the Ti-Zr-Ni and Ti-Hf-Ni quasicrystals and related amorphous or crystal phases produced by a combination of mechanical alloying and subsequent annealing. The effects on the microstructures and hydrogenation properties of the substitution of Zr for either Ti or Hf in alloys based on the $Ti_{45}Zr_{38}Ni_{17}$ compositions are investigated. Comparisons between results reported for samples prepared by rapid quenching or annealing are also made.

1. INTRODUCTION

Numerous quasicrystals have been discovered since the report of the first quasicrystal in an aluminum alloy by Shechtman et al [1]. All of these have a new type of translational long-range order and display non-crystallographic rotational symmetry. There are several types of quasicrystals; most prominent are the icosahedral phase (i-phase), the octagonal phase, the

1 takasaki@sic.shibaura-it.ac.jp.
2 kfk@wustl.edu.

decagonal phase, and several one dimensional quasicrystals. The earliest quasicrystals were metastable, normally produced by rapid solidification of the melt. However, some stable quasicrystals have now been discovered; those can be produced by isothermal heat treatments. I-phase formation by mechanical alloying (MA) has also been reported in alloys such as Fe-Cu-Al [2] and Al-Mg-Zn [3-5]. MA could yield powders that have high chemical and structural internal energy, leading to the formation of extended solid solutions or metastable phases with crystalline or amorphous structures.

Research on quasicrystals has mostly focused on fundamental scientific questions, such as the search for new quasicrystal alloy systems, attempts to understand their rules of formation, and an identification of their local atomic structures. However, several industrial applications were suggested for i-phase quasicrystals shortly after their discovery [6,7], including uses as (a) surface coating materials or thermal barriers due to their low friction and low thermal conductivity, (b) reinforcing particles or precipitates in metal matrix composites and (c) thermoelectric materials. A few examples of products incorporating quasicrystals are now on the market [8].

The chemical elements in the i-phases first discovered and studied had a low chemical affinity for hydrogen. However, the production of Ti-Zr-Ni quasicrystals by rapid quenching [9-11] led to extensive studies of their hydrogenation properties, due to the high affinity of Ti and Zr for hydrogen. These studies have demonstrated favorable properties that make these quasicrystals viable materials for hydrogen-storage materials. We discuss here our recent research results on the formation of the i-phase and related phases in powders produced by a combination of MA and subsequent annealing in vacuum for the Ti-Zr-Ni and Ti-Hf-Ni ternary systems. Their hydrogen absorption, either from the gas phase or by electrochemical loading, and hydrogen desorption properties are also discussed. These results are also compared with those obtained in Ti/Zr-based quasicrystals produced by rapid quenching or annealing. The formation, stability and structures of the i-phase and related crystal and amorphous phases in Ti/Zr/Hf-based alloys, produced either by rapid quenching or annealing, have been reviewed elsewhere [12].

2. Hydrogen in Quasicrystals

The ability to absorb hydrogen in metals or alloys is greatly dependent upon the chemical affinity of the host atoms for hydrogen. Since hydrogen occupies the interstitial sites in the host, it also depends on the types of interstitial sites and the total number and sizes of these sites. There are two common types of interstitial sites in crystal structures, octahedral and tetrahedral ones. Hydrogen has a tendency to occupy tetrahedral sites first.

Hydrogen molecules in the gas phase that approach a metal surface are first dissociated at the gas/metal interface and then adsorbed at appropriate surface or near-surface sites (physical and chemical adsorption). The hydrogen atoms diffuse further into the solid metal and finally occupy the interstitial sites of the host metal. If the local concentration of hydrogen exceeds a certain limit, a hydride phase precipitates. In the desorption reaction, the same steps occur in the reverse sequence, *i.e.* diffusion, transfer through the surface, recombination of adsorbed hydrogen atoms into hydrogen molecules and desorption of molecular hydrogen.

Although the local structure of the Ti-Zr-Ni i-phase is very complicated and is not yet fully determined, NMR and neutron diffraction measurements [10,17,18], indicate that it is built up from Bergman two-shell atomic clusters. These clusters contain 20 tetrahedral interstitial sites within the inner shell and 120 between the inner and outer shells [7]. Simple crystal structures, such as face centered cubic (fcc) or body centered cubic (bcc), contain tetrahedral sites but also contain a large number of octahedral sites that are less favorable for hydrogen storage; there are no octahedral sites in the Bergman cluster. Comparing with these structures, the large number of tetrahedral interstitial sites in the i-phase suggests that it could absorb a larger number of hydrogen atoms per metal atom. Furthermore, the introduction of hydrogen atoms into the i-phase may allow a clearer view of the local structure to be obtained because the hydrogen atom can act as a micro/nano structural probe. Recently, a realistic model of the Ti-Zr-Ni i-phase was proposed, based on a canonical cell tiling of a high order rational approximant that was determined by invoking similarities to the lower order 1/1 crystal approximant and from fitting X-ray and neutron diffraction data [19]. The structure was based on decorated rhombic dodecahedral, prolate rhombohedral and oblate rhombohedral tiles. These decorated tilings were relaxed under realistic electronic potentials to further refine the structure.

3. Formation of the Quasicrystal Phase by Mechanical Alloying

3.1. Ti-Zr-Ni System

Rapidly quenched $Ti_{45}Zr_{38}Ni_{17}$ alloys have been studied extensively. Since it is well known that the i-phase can be produced by rapid quenching, we have started attempting to produce the $Ti_{45}Zr_{38}Ni_{17}$ i-phase by MA. The elemental powder mixture of chemical composition $Ti_{45}Zr_{38}Ni_{17}$ (the purity of all elemental powders was 99.9%) was mechanically alloyed in a Fritsch Pulverisette 7 planetary ball mill. The initial mass of the powder mixture before MA was about 8.5g, and the ball-to-powder weight ratio was approximately 8:1. The ball acceleration was 15g (where g is the gravitational acceleration). X-ray diffraction patterns from $Ti_{45}Zr_{38}Ni_{17}$ powder mixtures that were mechanically alloyed for different times are shown in Figure 1.

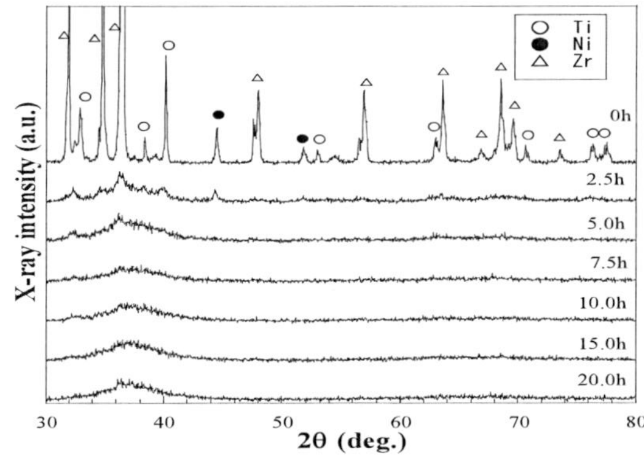

Figure 1. X-ray diffraction patterns for the $Ti_{45}Zr_{38}Ni_{17}$ powder mixtures that were mechanically alloyed for different lengths of time.

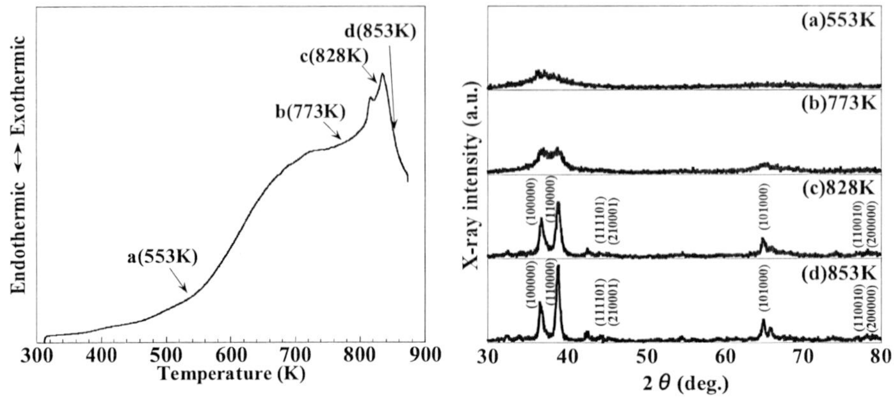

Figure 2. DSC curve for the $Ti_{45}Zr_{38}Ni_{17}$ amorphous powder obtained after MA for 20h, as well as the XRD patterns corresponding to the heating temperatures labeled with(a) to (d) on the DSC curve.

The elemental powders are alloyed well at an early stage of MA and an amorphous state appears after MA for more than 7.5h. The i-phase could not be obtained directly by this MA process. Figure 2 shows a DSC curve for the amorphous powder obtained after MA for 20h, as well as the XRD patterns corresponding to the temperatures labeled on the DSC curve. Exothermic features are observed between 300 and 900 K.The broad exothermic peak probably corresponds to a release of stored energy introduced during the MA process and to the formation of the i-phase. Three weak XRD peaks corresponding to the i-phase can be seen after heating to 773K (Figure 2 (b)). These peaks become sharper and more intense upon heating to 828K (Figure 2 (c)). The unlabelled (weak) X-ray diffraction peaks observed in Figure2 (c) and (d) (beside the (111101) or (101000) peaks of the i-phase) are due to the formation of a Ti_2Ni type crystal phase (fcc structure, lattice parameter, a= 1.21 nm). The X-ray diffraction peaks for the i-phase shown in Figure 2(c) and (d) were indexed using the scheme suggested by Bancel et al [20]. Different from the solidification route, MA processes the elemental powders by mechanical forces (the temperature is far below the liquidus temperature of the alloy or elements), which leads to a chemical inhomogeneity in the final products. This may make it difficult to produce powders that contain only the quasicrystal phase. Furthermore, because the Ti_2Ni-type phase is stabilized by the presence of small amounts of oxygen [21] the oxygen concentration during MA makes it difficult to eliminate this phase in the final product.

To investigate the microstructural dependence on the substitution of Ti/Zr, $Ti_{45+x}Zr_{38-x}Ni_{17}$ powders were also mechanically alloyed. Even if Ti were substituted for Zr or Zr substituted for Ti, while keeping the Ni concentration constant at 17 at.%, the final products were amorphous after longer MA. However, the broad X-ray diffraction peak of the amorphous phase shifts linearly with the Ti concentration to higher angle (*c.f.* Figure 9), indicating a decreasing average nearest neighbor separation that reflects the smaller size of the Ti relative to Zr. The nearest neighbor distance in the $Ti_{29}Zr_{54}Ni_{17}$ amorphous phase (before hydrogenation) is about 0.244 nm, whereas it is about 0.235 nm in amorphous $Ti_{61}Zr_{22}Ni_{17}$ (*c.f.* Figure 9).

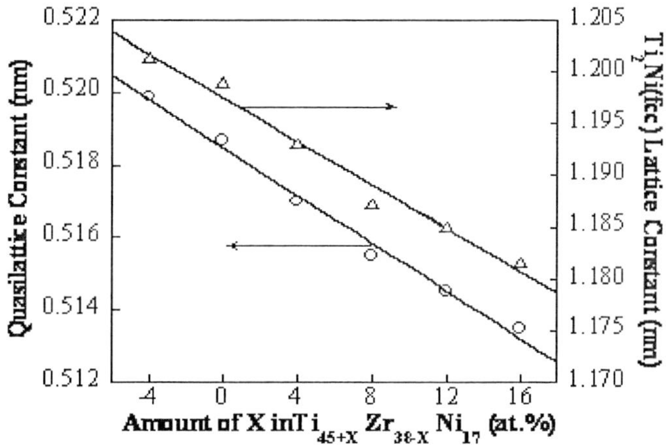

Figure 3. The quasilattice constant of the i-phase and the lattice parameter of the Ti_2Ni-type crystal phase as a function of the amount of Ti substituted for Zr in the powders.

Subsequent annealing of the amorphous $Ti_{45+x}Zr_{38-x}Ni_{17}$ powders produced the i-phase and a Ti_2Ni-type crystal phase, as for the $Ti_{45}Zr_{38}Ni_{17}$ powders mentioned earlier. The α-Ti phase (hexagonal closed pack structure) also appeared in the Ti-rich powders. The ratio of the volume fractions of these phases appears to depend on the chemical compositions of the powders. The dominant equilibrium phases in the as-cast and annealed Ti-Zr-Ni alloys near the i-phase forming composition are the C-14 like Laves phase, a Ti_2Ni-like phase, α-Ti/Zr, the approximant 1/1 W-phase, and the i-phase [22]. The W-phase [23] has a bcc-type cubic lattice structure composed of large tetrahedrally-coordinated Bergman clusters at each lattice site. The local structure of the approximant phase is believed to be similar to that of the i-phase [19]. Interestingly, the C-14 Laves and W phases were not detected by X-ray diffraction in the mechanically alloyed powders. The quasilattice constant [24] of the i-phase and the lattice parameter of the Ti_2Ni-type crystal phase as a function of the amount of Ti substituted for Zr in the powders are shown in Figure 3. The quasilattice constant for the i-phase and the lattice

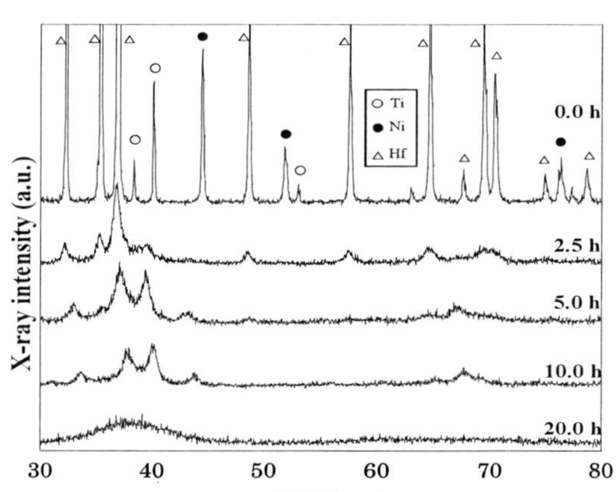

Figure 4. X-ray diffraction patterns for the $Ti_{40}Hf_{40}Ni_{20}$ powder mixtures mechanically alloyed for different lengths of times.

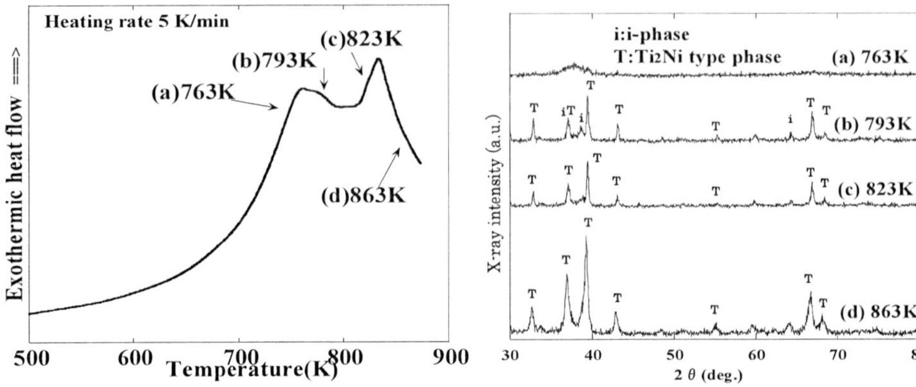

Figure 5. DSC curve for the $Ti_{40}Hf_{40}Ni_{20}$ amorphous powder obtained after MA for 15h, and the X-ray diffraction patterns corresponding to the heating temperatures labeled by (a) to (d) on the DSC curve.

parameter of the Ti_2Ni-type crystal phase in the Zr-rich powder, $Ti_{41}Zr_{42}Ni_{17}$, are about 0.520 nm and 1.20 nm respectively. Both decrease monotonically with increasing Ti concentration indicating that the Ti substituted for Zr in both phases. As already mentioned, the structure of the i-phase is presumed to be dominated by local tetrahedral order, which provide suitable sites for the location of interstitial hydrogen. Since the quasilattice of the i-phase decreased as Ti was substituted for Zr, the volume of the tetrahedral sites presumably decreases, but the total number of the tetrahedral sites in the i-phase should not vary.

3.2. Ti-Hf-Ni System

A high-order rational approximant phase that has more a complicated structure than the W-phase (1/1) observed in Ti-Zr-Ni alloys has been reported in rapidly quenched Ti-Hf-Ni alloys [25]. Although the X-ray diffraction patterns from this phase appear to come from an i-phase, systematic shifts of the electron diffraction spots from their expected positions for the i-phase demonstrate that the dominant phase is actually a 3/2 approximant phase [25]. This 3/2 approximant phase transforms to a strongly phason-disordered i-phase with annealing at temperatures between 623 and 773 K and then crystallizes to a Ti_2Ni-type phase at 893 K [25].

X-ray diffraction patterns from MA $Ti_{40}Hf_{40}Ni_{20}$ powders are shown in Figure 4. Unlike the Ti-Zr-Ni powders, some unalloyed elements remained even after MA for about 10 hours, but as for the Ti-Zr-Ni powders longer MA leads to the formation of an amorphous powder. Figure 5 shows a DSC curve for the $Ti_{40}Hf_{40}Ni_{20}$ amorphous powder obtained after MA for 15h, as well as the X-ray diffraction patterns corresponding to the temperatures labeled on the DSC curve. As for the Ti-Zr-Ni powders, several exothermic features are observed. Several X-ray diffraction peaks are observed after heating at 793 K (Figure 5(b)). The X-ray diffraction peaks labeled as "T" in Figure 5 correspond to the Ti_2Ni-type phase, but some peaks labeled as "i" match those expected for the i-phase. As mentioned, X-ray diffraction alone cannot distinguish the i-phase from the 3/2 approximant phase. However, the annealing temperature for the amorphous powders in this study was 793 K, which is a sufficient temperature for the i-phase formation during the annealing of a rapidly quenched alloy, suggesting that the peaks labeled as "i" in Figure 5 arise from the i-phase and not the 3/2 approximant. Additional heating at temperatures greater than 823 K decomposes the i-phase, and only the Ti_2Ni type phase is observed after annealing at 863 K (Figure 5 (d)). These observations agree fairly well with those reported for the rapidly quenched alloys [25]. The quasilattice constant of the i-phase and the lattice parameter of the Ti_2Ni-type phase in the $Ti_{40}Hf_{40}Ni_{20}$ powder were 0.518nm and 1.19 nm respectively; these are almost the same as those in the $Ti_{45}Zr_{38}Ni_{17}$ powder.

4. Hydrogen Absorption

4.1. Ti-Zr-Ni System

The $Ti_{45}Zr_{38}Ni_{17}$ amorphous and i-phase powders obtained by MA and subsequent annealing were hydrogenated in a high-pressure stainless steel vessel. The vessel containing the powder was evacuated by a rotary pump and then back-filled with pure (99.99999%) hydrogen gas. It was heated by an electric heater to a constant temperature of 573 K; the initial hydrogen pressure was 3.8 MPa. The hydrogen concentration in the powder was determined from the hydrogen pressure change in the vessel, which was measured by a pressure transducer. The maximum hydrogen concentration achieved was almost the same for both powders, approximately 60 at.% and corresponding to a hydrogen-to-metal ratio (H/M) of 1.50. A long induction time, about 200 h, was required to reach this concentration, presumably due to a very thin oxide barrier that was present on the powder surface. The hydrogen concentration levels for the powders produced by MA agree well with those obtained for the rapidly-quenched i-phase [11, 26]. Figure 6 (a) shows an X-ray diffraction pattern from the $Ti_{45}Zr_{38}Ni_{17}$ i-phase powders after the initial hydrogen gas loading and Figure 6 (b) shows the X-ray diffraction pattern after the second loading, which followed the desorption of the initially loaded hydrogen No new prominent phase appears in the i-phase powder after the initial loading, however the quasilattice expanded by about 6.6%. This indicates that the hydrogen was absorbed into the quasilattice and suggests that the i-phase is stable against the formation of hydrides. The lattice parameter of the Ti_2Ni-type crystal phase expanded by a maximum of 2.5%, but this tended to vary from experiment to experiment. It has been reported previously that oxygen can dissolve in a Ti_2Ni phase by up to 14 at.% and that the presence of oxygen reduces the maximum hydrogen uptake capacity of the Ti_2NiO_x compound [27]. This may account for some of the variability from sample to sample. The difference in the expansion between the quasilattice and the crystal lattice after the first loading of hydrogen gas (Figure 6(a)) causes the appearance of the (511) diffraction peak for the Ti_2Ni-type phase. This peak overlaps with the (110000) peak of the i-phase in the nonhydrogenated samples. After the second gas phase loading of hydrogen (Figure 6 (b)), the i-phase is still present, but it now coexists with an fcc (Ti, Zr)H_2 hydride phase. It is not certain whether this hydride is due to a partial decomposition of the i-phase or the transformation of the Ti_2Ni-type phase.

On the other hand, the (Ti, Zr)H_2 hydride phase formed directly from the amorphous powder produced by MA after gas phase loading of hydrogen at a temperature of 573 K and at an initial hydrogen pressure of 3.8 MPa. This hydride has also been reported in the i-phase produced by rapid quenching after loading hydrogen from the gas phase [11] and by electrochemical loading [28].

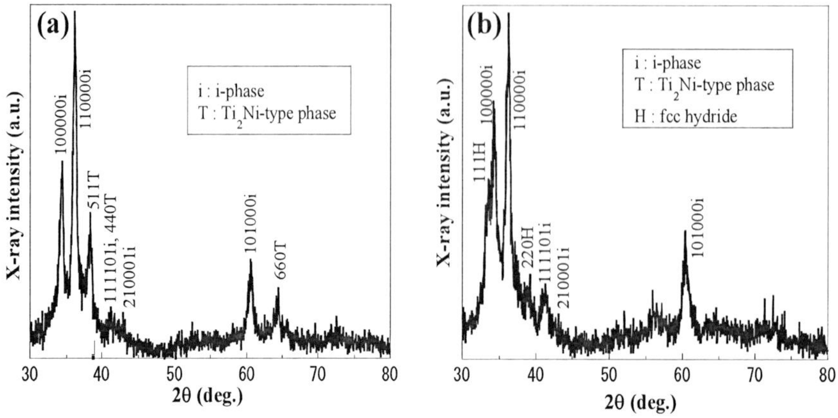

Figure 6. X-ray diffraction patterns of the $Ti_{45}Zr_{38}Ni_{17}$ i-phase powders (a) after the initial loading and (b) the second loading of hydrogen gas.

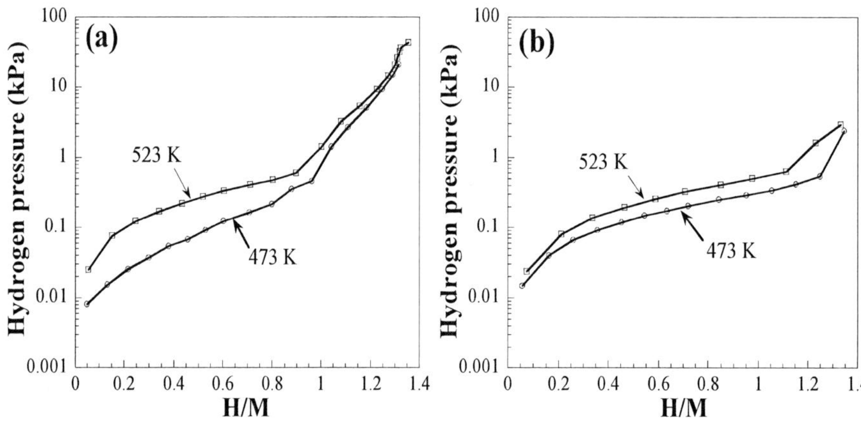

Figure 7. Low-pressure absorption pressure-composition isotherms for (a) the i-phase and (b) the amorphous $Ti_{45}Zr_{38}Ni_{17}$ powders at temperatures of 473 K and 523 K.

Low-pressure absorption pressure-composition isotherms for the i-phase and the amorphous $Ti_{45}Zr_{38}Ni_{17}$ powders at temperatures of 473 K and 523 K are shown in Figure 7(a) and (b) respectively. In contrast with the high pressure results mentioned earlier, no phase change was observed even for the amorphous powder, probably because of the temperatures for hydrogen loading and the comparatively low hydrogen absorption concentration. Plateau-like regions are observed in both powders and both temperatures, but they have a small slope, at hydrogen pressure lower than 1 kPa. The equilibrium hydrogen pressure measured at 623 K is low (less than 0.5 kPa) below H/M $\approx$ 1 in rapidly-quenched $Ti_{45}Zr_{38}Ni_{17}$ i-phase ribbons [29], which is almost the same as the present results. A slight difference between the isotherms of the i-phase and the amorphous powders implies that the local atomic structures are slightly different. The distribution of hydrogen site energies for the i-phase powder, estimated from fitting to the low-pressure isotherms, was slightly wider than that for the amorphous powder, indicating that some hydrogen atoms in the i-phase are more tightly bound and some other hydrogen are more weakly bound than in the amorphous phase [16].

High-pressure desorption isotherms at 573 K for the i-phase and the amorphous $Ti_{45}Zr_{38}Ni_{17}$ powders are shown in Figure 8 (a) and (b) respectively. Two plateaus are observed for the i-phase powder during absorption while only one plateau is observed for the amorphous powder. The plateau observed at higher hydrogen pressure for the i-phase powder is almost the same as the one for the amorphous phase, occurring at about 0.9 MPa. During desorption, on the other hand, only one plateau is observed for both the amorphous and i-phase powders.

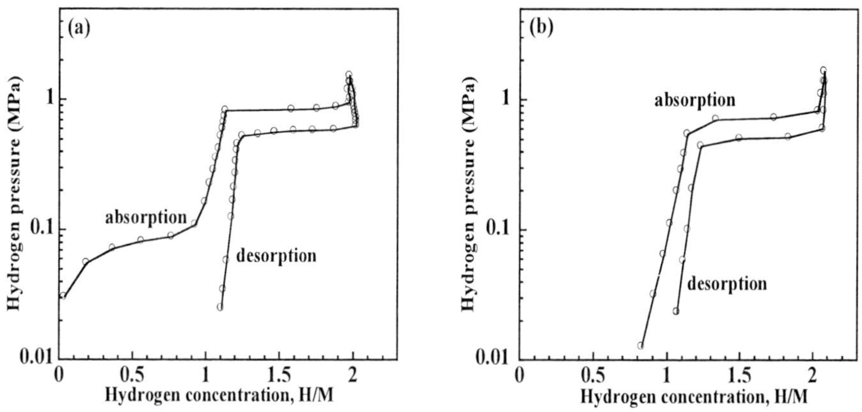

Figure 8. High-pressure isotherms for (a) the i-phase and (b) the amorphous $Ti_{45}Zr_{38}Ni_{17}$ powders at a temperature of 573 K.

The desorption plateau pressure is almost the same, 0.5 MPa, and the maximum hydrogen concentration is about 67 at.% (H/M $\approx$ 2.02) for both powders. For the isothermal measurements, hydrogen concentration level in sample is generally assumed to be zero at the beginning of the measurement, but the initial vacuum level depends on the experimental equipment. Because the equipment used for the high-pressure measurement was different from one used for low-pressure one, it is difficult to compare these data with the low-pressure absorption isotherms shown in Figure 7. Surprisingly, however, each desorption curve connects fairly well with the absorption isotherm at a reasonable hydrogen pressure and hydrogen concentration. After absorption for the high-pressure isotherm measurements, the (Ti, Zr)H_2 hydride formed in the amorphous powder, although the i-phase remained in the i-phase powder. These results also agree well with the hydrogen loading experiments using a high-pressure vessel (573K, 3.8 MPa) as mentioned earlier.

Electrochemical hydrogenation studies of the Ti-Zr-Ni powders were also performed. After the substitution of Ti for Zr, the i-phase powders were equiaxially pressed with a die and a punch at a pressure of 90 MPa, and compacted into a disk shape with diameter and thickness of about 5 mm and 1 mm respectively. The disk compacts were electrochemically hydrogenated in a 6M KOH solution at room temperature using a platinum counter-electrode. The current density was maintained at a constant value of 400 mA/g (1 kA/m^2). Although as mentioned earlier, a long induction time was required to reach the maximum hydrogen concentration during gas phase loading (200 h), electrochemical hydrogenation required only 6 h. The maximum H/M in the i-phase was estimated from an equation [11] derived from a regression analysis of the change in the quasilattice constant before and after gas phase loading of hydrogen in the Ti-Zr-Ni i-phases. Maximum hydrogen concentrations in the i-phases of different composition (the effect of substitution of Hf for Zr will be discussed later) are summarized in Table 1 and compared with results obtained for several metal hydrides. The maximum hydrogen concentrations in the powders where Ti was substituted for Zr were approximately 63 at.%, independent of the chemical composition of the i-phase. This concentration level is almost the same as that after the gas phase loading of hydrogen, exceeding that for $LaNi_5$ and TiFe hydrides. It is suggested that the total number of interstitial sites preferred by the hydrogen atoms does not change if the original Zr sites in the i-phase are occupied by Ti atoms. Hydrogen concentrations shown in the unit of wt. %, however, depend on the chemical composition of the i-phase because of the atomic weight difference between Ti and Zr (Ti is lighter). The i-phase and the Ti_2Ni-type phase remain even after electrochemical hydrogenation, but the quasilattice and the crystal lattice are expanded due to the hydrogen absorption. An extended X-ray absorption fine structure (EXAF) study of the Ti-Zr-Ni i-phase has indicated that hydrogen atoms sit preferentially near Ti and Zr neighbors [30, 31]. The TiH_2-type hydride (fcc) is also observed for Ti-rich powder after electrochemical hydrogenation, suggesting that the α-Ti phase (solid solution phase) formed in the Ti-rich powders might transform to the hydride.

The $Ti_{45+x}Zr_{38-x}Ni_{17}$ amorphous powders were also hydrogenated electrochemically. It was found that all of the powders remained amorphous after electrochemical hydrogenation, which differs from the gas phase loading results. It was difficult to estimate the hydrogen concentration by weighing because the pellet samples were immersed in a KOH solution. Hydrogen concentrations were, therefore, estimated from changes in the positions of the X-ray diffraction peaks. The nearest-neighbor distances in the amorphous $Ti_{45+x}Zr_{38-x}Ni_{17}$ powders as a function of the Ti concentration before and after electrochemical hydrogenation are shown in Figure 9. After hydrogenation, the nearest-neighbor distance increases slightly with increasing Ti concentration, while the distance before hydrogenation decreases by a comparatively larger amount, indicating that the amorphous phase with a higher Ti concentration absorbs more hydrogen.

Table 1. Maximum hydrogen concentrations in the i-phases for Ti-Zr/Hf-Ni i-phase powders and several metal hydrides.

		Maximum hydrogen concentration		
	Hydrogenation	H/M	at.%	wt.%
$Ti_{41}Zr_{42}Ni_{17}$	Electrochemical	1.74	63.5	2.52
$Ti_{45}Zr_{38}Ni_{17}$	Gaseous*	2.02	66.9	2.98
$Ti_{49}Zr_{34}Ni_{17}$	Electrochemical	1.73	63.3	2.63
$Ti_{53}Zr_{30}Ni_{17}$	Electrochemical	1.70	63.0	2.66
$Ti_{61}Zr_{22}Ni_{17}$	Electrochemical	1.69	62.9	2.80
$Ti_{40}Hf_{40}Ni_{20}$	Gaseous**	1.20	54.5	1.17
$LaNi_5$		1.0	50.0	1.38
TiFe		0.98	49.5	1.87
Mg_2Ni		1.33	57.1	3.61

* measured during high-pressure isotherm measurement at 573 K.

** measured in high-pressure vessel at temperature of 573 K.

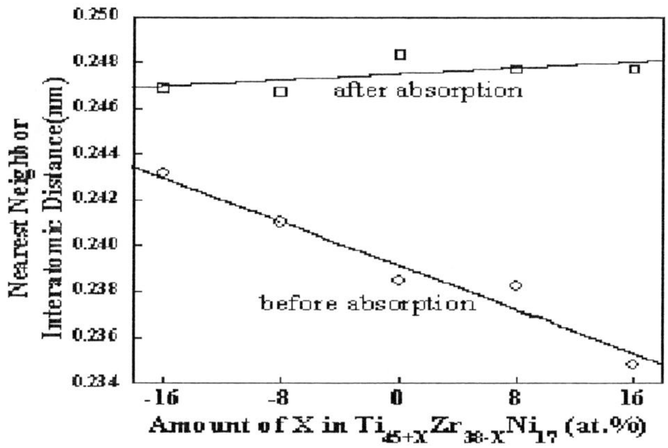

Figure 9. Nearest-neighbor distance in the $Ti_{45+x}Zr_{38-x}Ni_{17}$ amorphous powders as a function of the amount of Ti substituted for Zr in the powders.

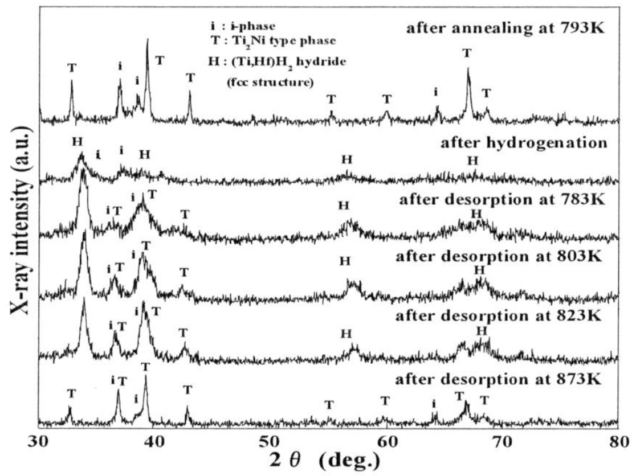

Figure 10. X-ray diffraction patterns for the annealed $Ti_{40}Hf_{40}Ni_{20}$ powder before and after gas phase loading of hydrogen, and after hydrogen desorption at several temperatures.

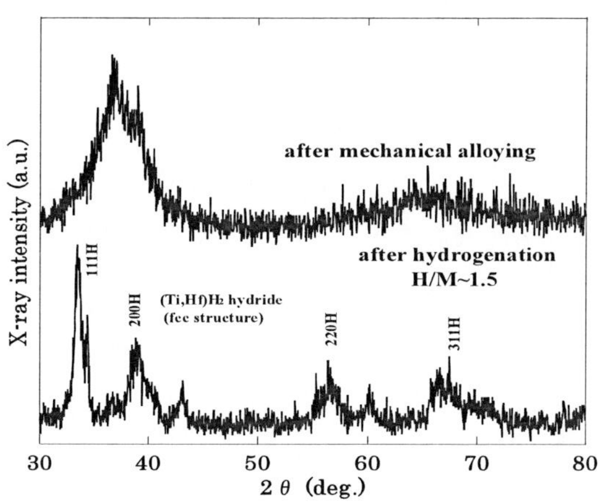

Figure 11. X-ray diffraction patterns for the amorphous $Ti_{40}Hf_{40}Ni_{20}$ powder before and after gas phase loading of hydrogen.

4.2. Ti-Hf-Ni System

The i-phase powders obtained after annealing MA powders where Hf had been substituted for Zr ($Ti_{40}Hf_{40}Ni_{20}$) also contained a small amount of the Ti_2Ni-type crystal phase. The maximum hydrogen concentrations in these powders were about 55 at.%, for gas phase loading at 573 K and an initial hydrogen pressure of 3.8 MPa. This is slightly lower than for the Ti-Zr-Ni i-phase powders discussed earlier. A high-order rational approximant (3/2) phase was recently discovered in a rapidly quenched $Ti_{40}Hf_{40}Ni_{20}$ alloy with a structure that is believed to be similar to that of the i-phase. This alloy was reported to absorb hydrogen up to H/M=1.2 [25], similar to that observed for these $Ti_{40}Hf_{40}Ni_{20}$ i-phase powders and consistent with the similarity in the local structures of the i-phase and the 3/2 approximant phases. Figure10 shows X-ray diffraction patterns for the $Ti_{40}Hf_{40}Ni_{20}$ powder before and after gas phase loading of hydrogen, as well as after hydrogen desorption at several heating temperatures. Although the weak XRD peaks corresponding to the i-phase, which shift to higher angles with hydrogen absorption, are still observed after hydrogenation, the major phase is a fcc TiH_2-type ((Ti, Hf)H_2) hydride phase. This is similar to the observations in the Ti-Zr-Ni powders after gas phase hydrogenation. Interestingly, the i-phase and the Ti_2Ni-type crystal phase are again observed after hydrogen desorption at 873 K.

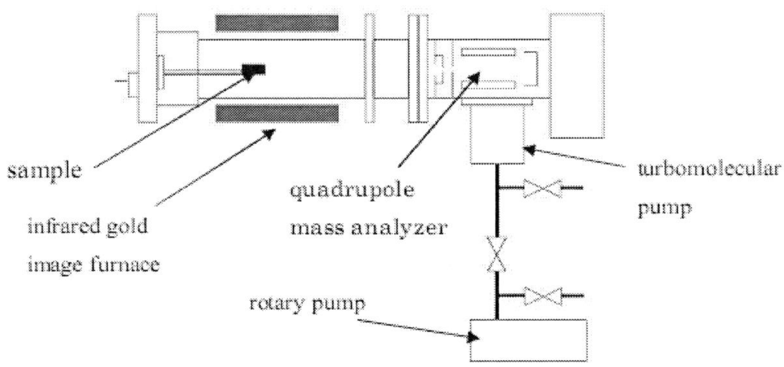

Figure 12. Schematic diagram of the thermal desorption spectroscopy system.

The amorphous $Ti_{40}Hf_{40}Ni_{20}$ powder obtained after MA was also hydrogenated with hydrogen gas at a temperature of 573 K at an initial hydrogen pressure of 3.8 MPa. The X-ray diffraction patterns for the amorphous $Ti_{40}Hf_{40}Ni_{20}$ powder before and after gas phase loading of hydrogen are shown in Figure 11. As for the Ti-Zr-Ni amorphous powder hydrogenated from the gas phase, the $Ti_{40}Hf_{40}Ni_{20}$ amorphous powder transformed almost completely to the fcc TiH_2-type ((Ti,Hf)H_2) hydride.

5. Hydrogen Desorption

5.1. Ti-Zr-Ni System

The hydrogen desorption kinetics were measured by thermal desorption spectroscopy (TDS). A schematic diagram of the TDS system used in this study is shown in Figure 12. The sample powder was mounted in an infrared gold-image furnace, and the furnace was evacuated to 10^{-6} Pa using a turbomolecular pump; the sample was then heated up to 1000 K at a constant heating rate. During heating, the hydrogen partial pressure (ion current of H_2^+), which corresponds directly to the rate of hydrogen desorption, was monitored by a quadrupole mass analyzer.

The thermal desorption spectra for the $Ti_{45}Zr_{38}Ni_{17}$ i-phase and amorphous powders that had been hydrogenated from the gas phase at 573 K (at a heating rate of 5 K/min) are shown in Figure 13(a) and (b) respectively. A single peak at about 700K was found for both powders. The spectrum for the i-phase powder also contained a broad shoulder on the lower temperature side of the peak; as will be discussed later, this is probably due to the Ti_2Ni-type phase contained in the i-phase powder. The onset temperature for hydrogen evolution from the i-phase powder was about 400 K, lower than for the amorphous powder (actually the amorphous powder had transformed to the (Ti, Zr)H_2 hydride), for which it occurred at approximately 500 K.

The activation energy for hydrogen desorption is a measure of the thermal energy needed for a hydrogen atom to overcome the barrier for hydrogen desorption (enthalpy for hydrogen desorption). It can be estimated from the shift in the peak temperature obtained from the thermal desorption spectroscopy measurements as a function of the heating rate using the following equation, which was initially proposed by Kissinger [32].

$$\ln\left(\frac{Q}{T^2}\right) = -\left(\frac{E}{RT}\right) + A \qquad (1)$$

Here, Q is heating rate (K/min), T is the peak temperature (K), R is the gas constant (J/(mol·K)) and E is the activation energy for hydrogen desorption (J/mol). The activation energies estimated from eq. (1) were 127 kJ/mol for the $Ti_{45}Zr_{38}Ni_{17}$ i-phase powder and 168 kJ/mol for the $Ti_{45}Zr_{38}Ni_{17}$ amorphous powder, showing that hydrogen in the i-phase can be desorbed more easily than from the amorphous phase.

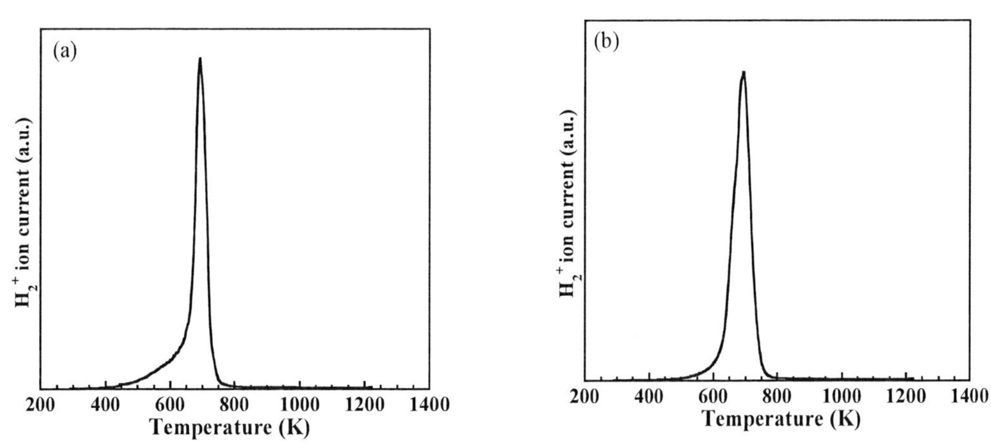

Figure 13. Thermal desorption spectra of hydrogen for $Ti_{45}Zr_{38}Ni_{17}$ (a) i-phase and (b) amorphous powders hydrogenated from gas phase hydrogen at 573 K.

After hydrogen desorption of the i-phase powder at a temperature of 800 K, the quasilattice constant and the lattice parameter of the Ti_2Ni-type phase returned to their values before hydrogenation, showing good reversibility for the absorption and desorption of hydrogen. On the other hand, the fcc hydride $(Ti,Zr)H_2$, which was formed in the amorphous powder after the gas phase loading of hydrogen, was stable to temperatures up to 803 K. It began to decompose at 853 K and completely transformed to the Ti_2Ni-type crystal phase at 903 K.

The thermal desorption spectra for hydrogen in the $Ti_{61}Zr_{22}Ni_{17}$ and $Ti_{45}Zr_{38}Ni_{17}$ i-phase powders that had been electrochemically loaded are shown in Figure 14 (the heating rate was 5 K/min). The data shown with open circles in Figure 14 are the measured signals and the solid curves are fits of a Gaussian to the peaks; the resultant curve from the combined Gaussian fits is also shown. Although as mentioned earlier, a single peak was observed in the samples loaded from the gas phase, two or three peaks were observed in each spectrum of the samples that were loaded electrochemically, arising from the different ordered phases present in the hydrogenated powders. The onset temperature for hydrogen desorption appears to be lower in the Ti-rich powder ($Ti_{61}Zr_{22}Ni_{17}$) since a broad weak peak appears at a low temperature, but the primary hydrogen desorption starts at about 570 K for all powders. It is also found that the main features of the spectrum shift to higher temperatures with increasing Ti concentration, suggesting that some hydrogen atoms are more strongly bound in the quasilattice. To investigate the origin of the thermal desorption peaks, X-ray diffraction studies were made on the $Ti_{61}Zr_{22}Ni_{17}$ i-phase powder after annealing at several temperatures. The X-ray diffraction patterns before and after hydrogenation are shown in Figure 15. Upon heating the powder sample at 683K, a terminal temperature for the first broad desorption peak shown in Figure 14, the strong X-ray diffraction peak corresponding to the Ti_2Ni-type phase (observed at $2\theta \approx 38^\circ$ after hydrogenation) shifts to higher angles because of the hydrogen desorption, eventually returning to its pre-hydrogenation value ($2\theta \approx 40^\circ$). The X-ray diffraction peaks corresponding to the i-phase, on the other hand, do not shift to higher angles until 683 K, mostly returning to their pre-hydrogenation positions at 733 K or 833 K. Those temperatures are very close to the terminal temperature for the second desorption peak shown in Figure 14. From these results, it is suggested that the first thermal desorption peak shown in Figure 14, particularly observed in the Ti-rich powder, is due to desorption from the Ti_2Ni-type phase, and the second peak is due to desorption from the i-phase. The third peak shown in Figure14 is probably due to desorption from the hydride. However, the amount of hydride formed after hydrogenation was so small that X-ray diffraction measurement could not detect it. To investigate the hydrogen desorption properties of the Ti_2Ni-type crystal in more detail, we measured the thermal desorption spectra for a binary Ti_2Ni crystal phase that was produced by MA and subsequent annealing. We confirmed that the desorption spectra extended over a wide range from low to high temperature, which agrees well with the thermal desorption results for the Ti_2Ni-type phase studied here.

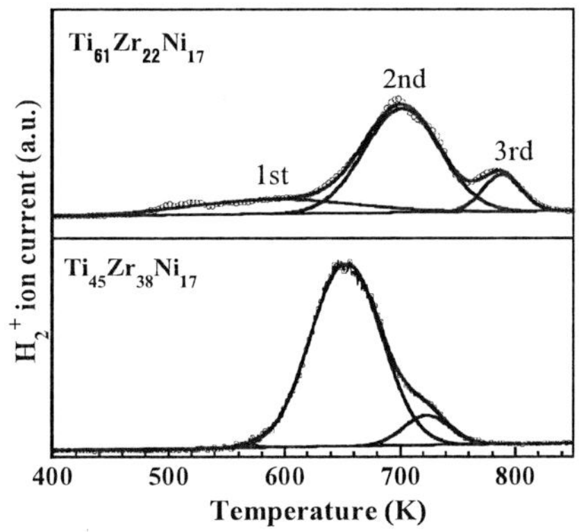

Figure 14. Thermal desorption spectra of hydrogen for $Ti_{61}Zr_{22}Ni_{20}$ and $Ti_{45}Zr_{38}Ni_{17}$ i-phase powders after electrochemical hydrogenation for 6h.

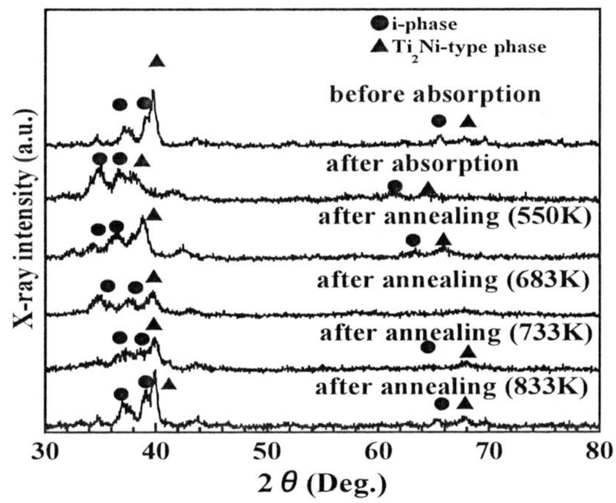

Figure 15. X-ray diffraction patterns for the $Ti_{61}Zr_{22}Ni_{17}$ i-phase powder annealed at several temperatures.

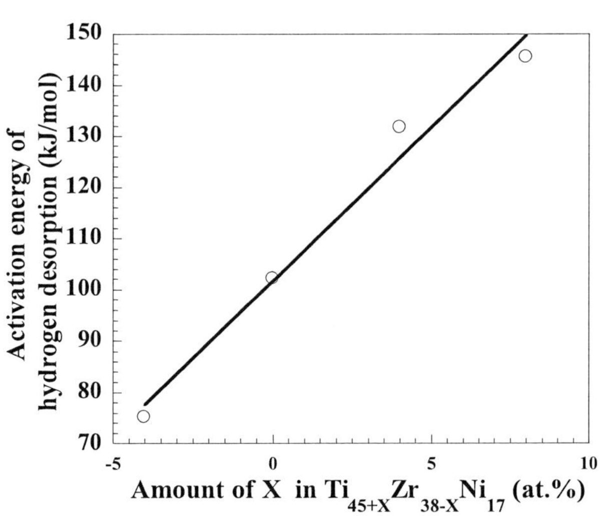

Figure 16. Activation energies for the Ti-Zr-Ni i-phase powders as a function of the amount of Ti substituted for Zr.

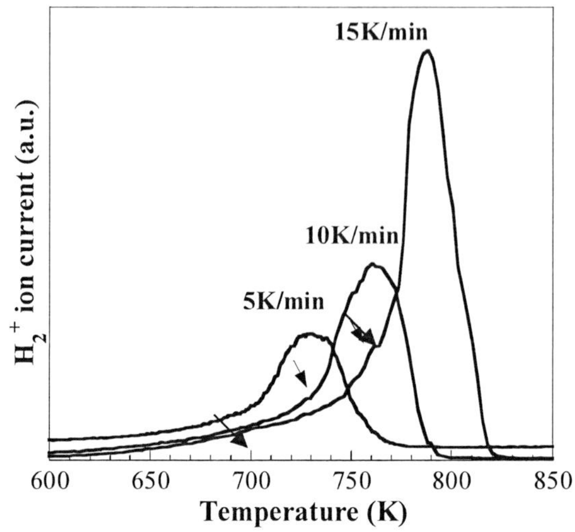

Figure 17. Thermal desorption spectra of hydrogen for $Ti_{40}Hf_{40}Ni_{20}$ i-phase powder measures at several heating rates.

Figure 16 shows the activation energy estimated from eq. (1) for the Ti-Zr-Ni i-phase powders as a function of the amount of Ti substituted for Zr. The activation energy for hydrogen desorption for the $Ti_{41}Zr_{32}Ni_{17}$ i-phase is about 75 kJ/mol, whereas for the $Ti_{53}Zr_{30}Ni_{17}$ i-phase it is about 145 kJ/mol. The activation energy increases monotonically with increasing Ti concentration in the powders. As mentioned earlier, the volume of the tetrahedral sites in the i-phase should decrease with increasing Ti concentration, evidenced by the shrinking quasilattice constant, making it harder to desorb the hydrogen. The increasing activation energy, then, suggests that the Ti strengthens the physical interaction between the hydrogen and the neighboring metal atoms.

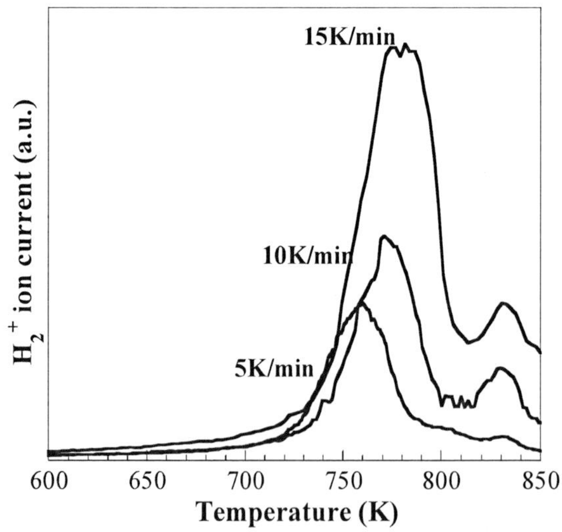

Figure 18. Thermal desorption spectra of hydrogen for $Ti_{40}Hf_{40}Ni_{20}$ amorphous powder measured at several heating rates.

5.2. Ti-Hf-Ni System

The $Ti_{40}Hf_{40}Ni_{20}$ i-phase and amorphous powders produced by MA, in which a Ti_2Ni-type crystal phase coexisted in the i-phase powder, were hydrogenated from the gas phase at 573 K at an initial hydrogen pressure of 3.8 MPa. The thermal hydrogen desorption spectra of the i-phase powder at several heating rates are shown in Figure 17. The primary peak temperature decreases with decreasing heating rate. In addition, there seems to be a shoulder (shown by the arrows in Figure 17) at a temperature below the peak temperature, particularly for the spectra obtained at heating rates of 10 and 15 K/min. The presence of this shoulder in the thermal desorption spectra for the $Ti_{61}Zr_{22}Ni_{17}$ i-phase after the electrochemical hydrogen loading shown in Figure 14, indicates that it is probably due to hydrogen desorption from the Ti_2Ni-type phase. The primary peak is at about 730 K for a heating rate of 5 K/min, which is almost same as for the $Ti_{45}Zr_{38}Ni_{17}$ i-phase powder (700 K). However, the activation energy for hydrogen desorption from the $Ti_{40}Hf_{40}Ni_{20}$ i-phase powder calculated from eq. (1) is about 74 kJ/mol, which is less than that for the $Ti_{45}Zr_{38}Ni_{17}$ i-phase powder (about 120 kJ/mol), suggesting that the substitution of Hf for Zr weakens the interaction between hydrogen atoms in the tetrahedral sites of the i-phase and the surrounding metal atoms.

As mentioned earlier, the $Ti_{40}Hf_{40}Ni_{20}$ amorphous powder obtained directly after MA transformed to the fcc (Ti, Hf)H_2-type hydride. The thermal desorption spectra measured at several heating rates for the amorphous powder after gas phase loading at 573 K at an initial hydrogen pressure of 3.8 MPa are shown in Figure 18. The main peak is broader than for the i-phase powders with a peak temperature near 770 K for a heating rate of 5 K/min. This peak temperature is higher than for the $Ti_{40}Hf_{40}Ni_{20}$ and $Ti_{45}Zr_{38}Ni_{17}$ i-phase powder. An additional weak peak at about 830 K in the desorption spectra is particularly prominent for the higher heating rates, 10 and 15 K/min, implying desorption from a minor phase. The activation energy for hydrogen desorption for the main peak is about 230 kJ/mol, which is much higher than that for the $Ti_{40}Hf_{40}Ni_{20}$ i-phase powder.

The activation energies for hydrogen desorption and phase formation before and after hydrogenation for the $Ti_{45}Zr_{38}Ni_{17}$, $Ti_{40}Hf_{40}Ni_{20}$ and binary Ti_2Ni powders produced by MA and subsequent annealing are summarized in Table 2. All of these powders were hydrogenated from the gas phase at a temperature of 573 K.

Table 2. Activation energies for hydrogen desorption from $Ti_{45}Zr_{38}Ni_{17}$, $Ti_{40}Zr_{40}Ni_{20}$ and binary Ti_2Ni powders produced by mechanical alloying. The i-phase and/or Ti_2Ni phases were obtained after subsequent annealing in vacuum before hydrogenation. All powders were hydrogenated from the gas phase at a temperature of 573 K.

Powder Type	Phase before hydrogenation	Phase after hydrogenation	Activation energy for hydrogen desorption (main peak)
$Ti_{45}Zr_{38}Ni_{17}$	i + Ti_2Ni i amorphous	i + Ti_2Ni hydride	127 kJ/mol 168 kJ/mol
$Ti_{40}Hf_{40}Ni_{20}$	i + Ti_2Ni amorphous Ti_2Ni	i + Ti_2Ni hydride hydride	74 kJ/mol 230 kJ/mol 314 kJ/mol
Ti_2Ni	Ti_2Ni amorphous	hydride hydride	96 kJ/mol -

The series of hydrogen desorption experiments for the Ti-Zr-Ni and the $Ti_{40}Hf_{40}Ni_{20}$ powders show that hydrogen desorption from the i-phase is easier than from the amorphous structure. This implies that there are differences in the local structures of these two phases in Ti-Zr-Ni or Ti-Hf-Ni alloys, even though both are believed to be dominated by a local tetrahedral order.

6. CONCLUSION

Besides the conventional solidification processes like rapid quenching and isothermal heat treatments, a combination of mechanical alloying (MA) and subsequent annealing in vacuum was shown to also produce the Ti-Zr-Ni icosahedral phase (i-phase) for a range of chemical composition. Depending upon the chemical compositions, minor phases such as the solid solution phase (α-Ti/Zr) and the Ti_2Ni-type phase (fcc structure) also formed after subsequent annealing, but no Laves or crystal approximant phases were found. The local tetrahedral structure of the i-phase provides many suitable sites for interstitial hydrogen. The Ti-Zr-Ni i-phase powders could load hydrogen to a concentration greater than 60 at.%. The maximum concentration reached was 67 at% ($H/M \approx 2.02$) which was obtained for the $Ti_{45}Zr_{38}Ni_{17}$ i-phase powder during a high-pressure isotherm measurement. The substitution of Zr atoms for Ti ones in the Ti-Zr-Ni i-phase powders decreased the activation energy for hydrogen desorption and the hydrogen desorption temperatures, indicating a decrease in the energy barrier for hydrogen desorption. Besides the plateau-like pressure observed during the low-pressure isotherm measurement for the i-phase powder an additional plateau was also observed at about 0.5 MPa (desorption process) during a high-pressure isotherm measurement. The i-phase in the Ti-Zr-Ni powder remained stable even after hydrogenation, but the amorphous powders transformed into a fcc $(Ti, Zr)H_2$ hydride phase, whose activation energy for hydrogen desorption was generally higher than that measured for the i-phase.

The i-phase could also be produced from $Ti_{40}Hf_{40}Ni_{20}$ elemental powders by MA and subsequent annealing, coexisting with a Ti_2Ni-type crystal phase. The amount of the Ti_2Ni-type phase was much larger than obtained in the Ti-Zr-Ni powders. The hydrogen concentration in the $Ti_{40}Hf_{40}Ni_{20}$ i-phase powder was about 55 at.% after loading from the gas phase at 573 K and an initial hydrogen pressure of 3.8 MPa. This was slightly lower than observed for the Ti-Zr-Ni i-phase powders. The activation energy for hydrogen desorption was also lower than for the Ti-Zr-Ni i-phases, suggesting that the substitution of Hf for Zr decreased the chemical interaction between hydrogen in the tetrahedral sites and the surrounding metal atoms. Thus, an optimization of the Hf concentration in Ti-Zr-Hf-Ni powders is one way to develop an i-phase with good hydrogen desorption properties as well as a favorable hydrogen storage capacity. The $Ti_{40}Hf_{40}Ni_{20}$ amorphous powder obtained directly after MA transformed to a fcc $(Ti, Hf)H_2$ hydride phase after gas phase loading, the same as observed for the Ti-Zr-Ni amorphous powder.

Comparing the hydrogenation properties of the i-phase and the amorphous phase in the Ti-Zr-Ni and Ti-Hf-Ni powders, it can be concluded that structural changes after gas phase hydrogenation do not depend on the chemical composition of the host alloy but on the structures of the initial phases (the i-phase or amorphous). The i-phase remained stable with hydrogenation but the amorphous powder turned into a hydride phase.

The storage capacity for hydrogen in these quasicrystals exceeds that in conventional hydrides such as $LaNi_5$ and TiFe. Further, the hydrogen can be desorbed from the i-phase in both Ti-Zr-Ni or Ti-Hf-Ni alloys and the desorption is easier than from the amorphous structure. These features show that the i-phase is a promising new hydrogen-storage material for fuel-storage and battery applications.

ACKNOWLEDGMENTS

We are grateful to Messrs. Naoki Imai, Takanobu Sato and Takahiro Tomizawa, graduate students of Shibaura Institute of Technology, for their help with the experiments carried out at Shibaura Institute of Technology. We also thank Dr. Van T. Huett, Washington University, St. Louis, USA, for the low-pressure composition isotherm measurements. This research was partially supported by a Grant-in Aid for Scientific Research (Grant No. 15560577) from the Ministry of Education, Culture, Sports, Science, and Technology of the Japanese Government, the European Commission (project Dev-BIOSOFC, FP6-042436, MTKD-CT-2006-042436), and by the National Science Foundation, USA, under grant DMR 03-07410.

REFERENCES

[1] D.Shechtman, I.Blech, D.Gratias, and J.W.Cahn, Phys.Rev.Lett. 53 (1984) 195
[2] J.Eckert, L.Schultz, and K.Urban, Appl.Phys.Lett. 52 (1989) 117
[3] E.Ivanov, B.Bokhonov, and I.Konstanchuk, J.Mater.Sci. 26 (1991) 1409
[4] U.Mizutani, T.Takeuchi, T.Fukunaga, S.Murasaki, and K.Kaneko, J.Mater.Sci.Lett. 12 (1991) 629
[5] U.Mizutani, T.Takeuchi, and T.Fukunaga, Mater.Trans.,JIM 34 (1993) 102
[6] D.J.Sordelet, and J.M.Dubois, MRS Bulletin 22 (1997) 34
[7] P.C. Gibbons, and K.F. Kelton, in *Physical Properties of Quasicrystals*, ed. Z.M. Stadnik, (Springer, 1999) pp.403
[8] J.M. Dubois, Mater. Sci. and Eng., 294-296 (2000) 4
[9] K.F. Kelton, and P.C. Gibbons, MRS Bulletin 22 (1997) 69
[10] A.M. Viano, R.M. Stroud, P.C. Gibbons, A.F. McDowell, M.S. Conradi, and K.F. Kelton, Phys. Rev. B51 (1995) 12026
[11] A.M.Viano, E.H.Majzoub, R.M.Stroud, M.J.Kramer, S.T.Misture, P.C.Gibbons and K.F.Kelton, Phil. Mag. A 78 (1998) 131
[12] K.F. Kelton, Mater.Sci. and Eng. A, 375-377 (2004) 31
[13] A.Takasaki, C.H.Han, Y.Furuya, and K.F.Kelton, Phil.Mag.Lett., 82 (2002), 353
[14] A.Takasaki, and K.F.Kelton, J.Alloys Compd., 347 (2002) 295
[15] A.Takasaki, V.T.Huett, and K.F.Kelton, Mater. Trans., 43 (2002) 1
[16] A.Takasaki, V.T.Huett., and K.F.Kelton, J.Non-Crystall.Sol., 334-335C (2004) 457
[17] A.Shastri, E.H.Majzoub, F.Borsa, P.C.Gibbons, K.F.Kelton, Phys.Rev. B57 (1998) 5148
[18] A.Shastri, E.H.Majzoub, F.Borsa, P.C.Gibbons, K.F.Kelton, Phys.Rev. B59 (1999) 14108

[19] R.G. Hennig, K.F. Kelton, A.E. Carlsson, C.L. Henley, Phys. Rev.B., 67 (2003) 134202/1

[20] P.A. Bancel, P.A. Heiney, P.W. Stephens, A.I. Goldman, and P.M. Horn, Phys. Rev. Lett. 54 (1985) 2422

[21] K.F.Kelton, W.J.Kim, and R.M.Stroud, Appl.Phys.Lett. 70 (1997) 3230

[22] J.P. Davis, E.H. Majzoub, J.K. Simmons, K.F. Kelton, Mater.Sci.Eng. 294-296 (2000) 104

[23] W.J.Kim, P.C.Gibbons, and K.F.Kelton, Phil.Mag.Lett. 76 (1997) 199

[24] V. Elser, Phys. Rev. B 32 (1985) 4892

[25] V.T.Huett, and K.F.Kelton, Phil.Mag.Lett., 82 (2002) 191

[26] J.Y.Kim, P.C.Gibbons, K.F.Kelton, J.Alloys Compd. 266 (1998) 311

[27] M.H.Mintz, Z. Hadari, and M.P. Dariel, J. Less-Common Metals 63 (1979) 181

[28] E.H.Majzoub, J.Y.Kim, R.G.Hennig, K.F.Kelton, P.C.Gibons, and W.B.Yelon, Mater. Sci. Eng. 294-296 (2000) 108

[29] J.Y.Kim, P.C.Gibbons, and K.F.Kelton, Metals Mater. 5 (1999) 589

[30] A.Sadoc, J.Y. Kim, K.F. Kelton, Phil.Mag. A, 79 (1999) 2763

[31] A.Sadoc,E.H. Majzoub, V.T. Huett, K.F. Kelton, J.Phys.: Condens. Matter, 14 (2002) 6413

[32] H.E. Kissinger, Anal.Chem. 27 (1957) 1702

In: Quasicrystals: Types, Systems, and Techniques
Editor: B. E. Puckermann, pp. 147-161 ISBN 978-1-61761-123-0

Chapter 6

FORMATION OF QUASICRYSTALS IN BULK METALLIC GLASSES AND THEIR EFFECT ON MECHANICAL BEHAVIOR

Jenő Gubicza and János Lendvai
Department of Materials Physics, Eötvös Lorand University
H-1117, Pázmány Péter sétány 1/A, Budapest, Hungary

ABSTRACT

The annealing of bulk metallic glasses (BMGs) at elevated temperatures usually leads to partial or full crystallization. The crystallization in several systems starts with the formation of metastable quasicrystalline (QC) particles and then the material can be regarded as a composite of QC and amorphous phases. The appearance of QC particles significantly affects the mechanical properties of BMGs. In this chapter, the morphology, structure and chemical composition of QC particles formed during heat-treatment of BMGs are reviewed according to the relevant literature. Special attention is paid to the influence of the formation of QC particles on the mechanical behavior at room and high temperatures. It was found that during heat-treatment of a commercial ZrTiCuNiBe BMG above the glass transition temperature nanosized spherical QC particles containing smaller grains were formed. Depending on the annealing temperature the volume fraction of the QC phase varied between 25 and 37%. The QC particles contain Ti, Zr and Ni in high concentration, while the amorphous matrix is enriched in Be. The high temperature viscosity increases mainly due to the hard QC particles but there is also a slight contribution from the compositional changes of the supercooled liquid matrix. The bending strength measured at room temperature decreases in consequence of QC formation, most probably mainly due to the loss of free volume in the amorphous matrix.

1. INTRODUCTION

Due to their unique mechanical properties the deformation behavior of bulk metallic glasses (BMGs) has been intensively studied in recent years [Waniuk, 1998; Inoue, 2001;

Wei, 2004; Schuh, 2007]. BMGs usually have lower elastic modulus, higher strength, and reduced plasticity compared to their crystalline counterparts of the same chemical composition [Schuh, 2007]. The application of BMGs at high temperatures often leads to partial or full crystallization. It was suggested [Frank, 1952; Kelton, 2006] that the ability of metals to form amorphous structure by fast cooling is enhanced by the dominance of icosahedral short-range order (ISRO) in melts that is incompatible with translational periodicity of crystallographic structures. An icosahedral packing of twenty slightly distorted tetrahedra is more dense than fcc or hcp packings, therefore although it is incompatible with translational periodicity it might be a natural choice for liquid and amorphous structures [Kelton, 2004; Kelton, 2006b]. The existence of ISRO in the supercooled liquid state brings about an extremely small interfacial free energy between an icosahedral quasicrystal phase (i-phase) and a metallic glass of the same composition [Holzer, 1991]. Consequently, the nucleation of the i-phase during annealing of BMGs is easier than the formation of the more stable crystalline phases. In support of this, the i-phase is frequently reported as the primary devitrification phase, particularly for the Zr- and Hf-based bulk metallic glasses [Xing, 2000; Saida, 2001, Chang, 2006].

The precipitation of an icosahedral quasicrystalline phase upon devitrification of Zr-based BMGs has been reported for ZrCuAl [Fan, 2006], ZrCuNiAl [Köster, 1996], ZrCuAlNiTi [Xing, 1998] and ZrTiCuNiBe alloys [Wanderka, 2000; Mechler, 2004]. It was found that the chemical composition of BMGs has a deterministic influence on the local atomic order in the glassy state and therefore on the crystallization sequence during annealing [Saida, 2007]. Comparing $Zr_{70}Ni_{30}$ and $Zr_{70}Cu_{30}$ metallic glasses, the former shows a tetragonal atomic order while the latter exhibits an icosahedral local atomic configuration. As a consequence, in $Zr_{70}Cu_{30}$ the supercooled liquid state has a higher stability (larger temperature difference upon heating between the glass transition and the primary crystallization) and as a first step quasicrystalline i-phase forms while $Zr_{70}Ni_{30}$ crystallizes into tetragonal Zr_2Ni phase during heat-treatment. It is suggested that structural differences in the glassy phase is caused by a strong chemical affinity of a Zr–Ni pair compared with that of a Zr–Cu pair. The sensitivity of local atomic order to the chemical composition was demonstrated by adding 1 at.% Pd into a $Zr_{70}Al_{10}Ni_{20}$ metallic glass [Saida, 2007]. Without the Pd addition, tetragonal Zr_2Ni was observed in the initial stage of transformation, however, the primary crystallization process changes markedly into single i-phase formation by the addition of 1 at.% Pd. Addition elements, such as Ag, Pd, Au or Pt, to ZrAlNiCu glasses are believed to generate inhomogeneous atomic configuration regions including ISRO configurations in the supercooled liquid [Inoue, 1999; Chen, 1999; Murty, 2000; Saida, 2000; Saida, 2003b], which then promote the precipitation of an icosahedral phase. It was also found that the addition of Ti to bulk amorphous $Zr_{62-x}Ti_xCu_{20}Ni_8Al_{10}$ alloys resulted in the formation of a metastable quasicrystalline phase in the first transformation step but only in the composition range of $2 \leq x < 4$ [Kühn, 2006]. Additions of oxygen or palladium to Zr-based bulk amorphous alloys may also induce quasicrystallizaion [Köster, 1996; Eckert, 1998; Murty, 2000b]. When oxygen is added to metallic glasses with high glass forming ability, it increases the stability of the icosahedral phase during crystallization. On the other hand, Pd stabilizes the amorphous phase in Zr-based metallic glasses in which icosahedral clusters are presumed to be stable [Murty, 2001]. Thus, nanoquasicrystallization occurs even from binary Zr–Pd binary metallic glass [Murty, 2001]. The size of the QC domains usually falls in the nanometer range, however formation of micrometer-sized icosahedral quasicrystalline particles has also been reported for Zr–Ti–Nb–Cu–Ni–Al alloys [Kühn, 2000; Kühn, 2005].

The local atomic structure in metallic glasses can be changed by mechanical disordering (MD). It was demonstrated that ball milling a $Zr_{65}Al_{7.5}Ni_{10}Cu_{12.5}Pd_5$ glassy alloy resulted in the formation of the fcc Zr_2Ni instead of the icosahedral QC phase in the primary crystallization step, due to the MD-induced distortion of the local structure [Saida, 2003]. On the other hand, the appearance of QC particles in the amorphous matrix results in significant changes in the mechanical properties which depend strongly on the chemical composition of BMGs as well as on the temperature of annealing [Xing, 2001]. In this chapter, the effect of QC phase formation in a commercial ZrTiCuNiBe metallic glass on the deformation behavior at room and elevated temperatures is summarized. First the evolution of the microstructure during annealing is characterized in detail. Then the viscosity as a function of QC phase content is investigated by indentation creep testing in the supercooled liquid region (the temperature range between the glass transition temperature and the onset temperature of crystallization). The deformation behavior at room temperature is studied by three-point bending tests.

2. Formation of Quasicrystalline Phase in the BMGs During Heat-Treatments

The formation of QC phase during annealing was studied in a commercial Zr-based bulk metallic glass with the composition of $Zr_{44}Ti_{11}Cu_{10}Ni_{10}Be_{25}$ (LM-1B, manufacturer: Liquidmetal Technologies, Inc). The diameter and the length of the cylindrical specimens were 9 and 85 mm, respectively. Heat effects were detected by differential scanning calorimetry (DSC) during isothermal heat-treatments at 677, 682 and 687 K which are slightly above the middle temperature of the supercooled liquid regime, $(T_g + T_{xo})/2$, where T_g = 625 K is the glass transition temperature and T_{xo} = 725 K is the onset temperature of crystallization. The heat flow versus time curves, recorded during isothermal annealings at 677, 682 and 687 K are shown in Figure 1. The curves display two exothermic peaks. The phase composition was determined by X-ray diffraction on samples cooled rapidly to room temperature after different annealing times. Figure 2 shows the X-ray diffractograms corresponding to the as-received state as well as to four annealing times at 682 K, marked by dots on the DSC thermogram in Figure 1. In the as-received state only a halo was observed indicating that the initial sample is fully amorphous. The X-ray diffraction pattern obtained after annealing for 1300 s contains an amorphous halo and the peaks of a QC phase [Gubicza, 2008]. The diffractograms corresponding to shorter annealing times (not shown here) have the same characteristics with weaker reflections of the QC phase. Consequently, the first exothermic peaks on the thermograms of Figure 1 are related to the formation of a QC phase. The X-ray diffractograms in Figure 2 shows that the second exothermic peak on the DSC thermogram corresponds to the crystallization of stable phases, namely Be_2Zr, Zr_2Ni, Zr_2Cu and NiTi. At the end of the second crystallization peak (the annealing time is 4900 s at 682 K) a significant amount of QC phase remained in the microstructure which disappeared completely only after prolonged (55000 s) heat-treatment while the fractions of Zr_2Cu and Be_2Zr increased. This transformation of the remaining QC phase is so slow and the released heat so small, that it could not been detected in the DSC measurements. It is noted that similar two-step devitrification sequence was detected for other alloys, such as $Zr_{65}Ni_{10}Cu_{7.5}Al_{7.5}Ag_{10}$ where the first reaction corresponds to the formation of a QC phase from the amorphous matrix, while the second one results from the formation of a non-stoichiometric Zr_2Cu-like phase from the previously formed quasicrystals [Liu, 2004; Liu, 2004b]. It was also observed that the DSC peak related to the formation of QC phase was shifted to a lower temparature if the $Zr_{44}Ti_{11}Cu_{10}Ni_{10}Be_{25}$ amorphous alloy had been deformed before heat-treatment [Révész, 2010].

The heat released in the first exothermic peak during isothermal annealing was used to estimate the relative fraction of the QC phase. Integrating the heat flow versus time function, the heat released during crystallization (H) can be calculated at any time of annealing. It is found that the total heat released during crystallization, i.e. the area under the two DSC peaks is the same within the experimental error for all the three temperatures, H_{total}= 60±1 J/g. The total heat was distributed between the two peaks by separating them at the time value where the heat flow is minimal after the first peak (see Figure 1). The heat fractions related to the two peaks are different for the three temperatures as it is shown in Figure 3.

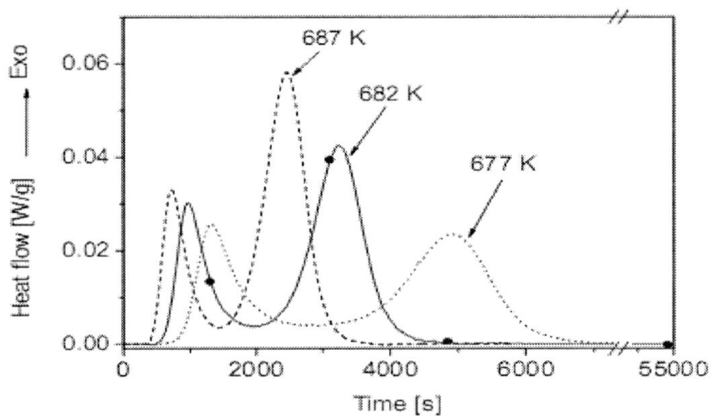

Figure 1. The heat flow as a function of time during isothermal annealing at 677, 682 and 687 K. The dots on the curve for 682 K mark the states where the crystalline phase composition following rapid cooling to room temperature was investigated by X-ray diffraction (see Figure 2).

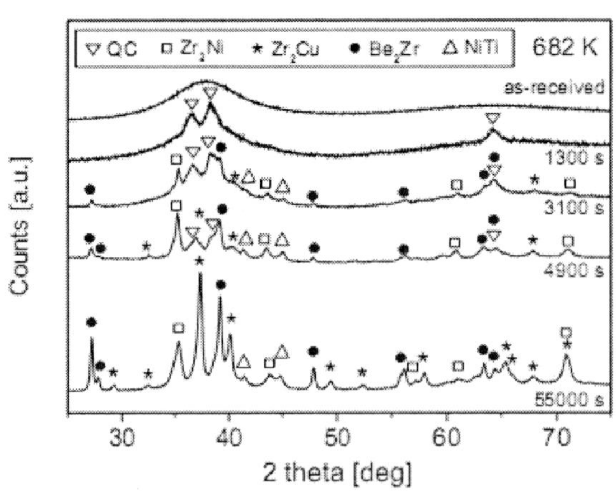

Figure 2. X-ray diffraction patterns for different times of isothermal annealing at 682 K. QC: quasicrystalline phase.

During the first peak corresponding to the formation of QC phase, 37, 30 and 25% of the total heat are released at 677, 682 and 687 K, respectively. Complementary TEM investigations have shown that the ratio of the heat released in the first exothermic peak to the total heat is very close to the volume fraction of QC phase formed up to the end of the first DSC peak. The higher the temperature of annealing, the shorter the incubation time before the formation of stable crystalline phases, therefore the maximum volume fraction of QC phase is smaller and correspondingly the fraction of the first peak in the total heat is also lower.

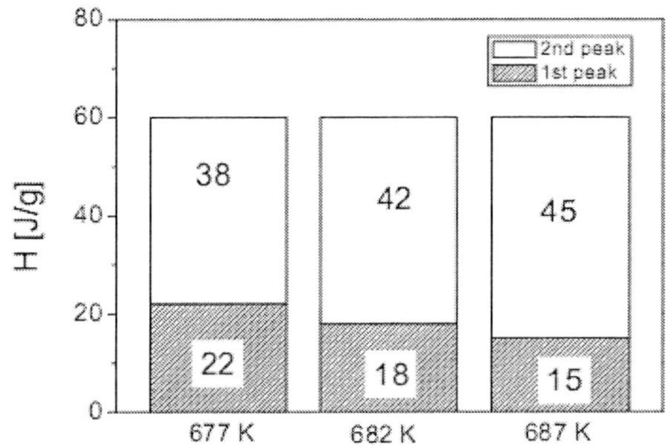

Figure 3. The total released heat and the heat fractions related to the first and second exothermic peaks in the DSC thermograms at 677, 682 and 687 K.

3. Morphology, Structure and Chemical Composition of Quasicrystals Formed During Annealing BMGs

The morphology of QC phase formed during isothermal annealing of $Zr_{44}Ti_{11}Cu_{10}Ni_{10}Be_{25}$ BMG was studied by transmission electron microscopy (TEM) [Gubicza, 2008]. Figure 4 shows TEM images of the microstructure after annealing at 682 K for 600 and 1800 s which correspond to the beginning and the end of the first exothermic DSC peak, respectively (see Figure 1). The images illustrate that the microstructure can be described as a composite consisting of amorphous matrix and spherical QC nanoparticles which were also identified in electron diffraction pattern by the software ProcessDiffraction [Lábár, 2005]. The distributions of particle diameters for 600 and 1800 s are shown in Figs. 4c and d, respectively. The size distribution of QC particles is relatively broad containing particles of 10 as well as 130 nm size. The average sizes of the QC particles are 40 and 51 nm for 600 and 1800 s, respectively, indicating that the increase of the volume fraction of QC particles can be attributed mainly to the increasing number, and not so much to the growth of particles. The average size of the quasicrystalline diffracting domains at the end of the first exothermic peak is 12 nm as estimated from the breadth of the X-ray peak profiles on the basis of the Scherrer-equation. The difference between the particle size and the diffracting domain size suggests that the particles are built up from smaller quasycrystalline grains or subgrains. This is supported by the TEM images in Figs. 4a and b where the areas having different contrasts inside nano-QC particles correspond to grains of 15-20 nm size. This value is in good agreement with the coherently scattering domain size (12 nm) estimated from the breadth of the X-ray peaks. The average size of the QC particles at the end of the first exothermic peak is found to be between 51 and 58 nm for all the three temperatures, indicating that the difference between the annealing temperatures results only in slight variation in the particle size of the QC phase.

The positions of the diffraction peaks of the QC phase in Figure 2 agree well with those of the icosahedral i-phase having the composition of $Ti_xZr_yNi_z$ (x=33-37, y=42-46 and z=21) [Kelton, 2004]. The indices of QC peaks are *100000*, *110000* and *101000* in the order of increasing diffraction angle in Figure 2. The stochiometry of the QC phase suggests that the QC particles are enriched in Ti and Ni and depleted from Be and Cu which was also supported by the analysis of their chemical composition by Electron Energy Loss Spectroscopy (EELS) studies [Gubicza, 2008]. Similar compositional differences between the amorphous and the QC phases were observed for $Zr_{46.8}Ti_{8.2}Cu_{7.5}Ni_{10}Be_{27.5}$ [Wollgarten, 2004; Van de Moortéle, 2004].

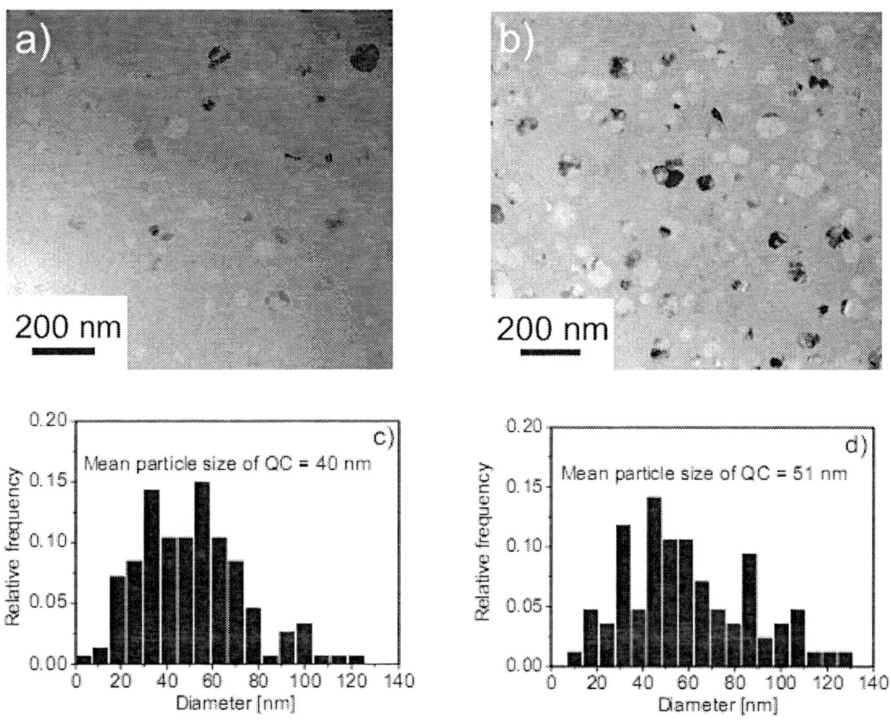

Figure 4. TEM images showing the microstructures consist of QC particles and amorphous matrices obtained after annealing at 682 K for 600 (a) and 1800 s (b). The size distributions of QC particles for 600 (c) and 1800 s (d) are also shown.

The chemical heterogeneities suggest that after the nucleation of QC particles their growth is most probably controlled by atomic diffusion supplying the increase of Ti and Ni concentration in the QC particles. At the same time, Figure 4 shows only a limited growth in the particle size when the duration of annealing is increased by a factor of three from 600 to 1800 s at 682 K. This indicates that as the phase transformation proceeds, the growth of larger QC particles is slowed down. The larger the size of a QC particle, the larger the surrounding depleted zone, and correspondingly the diffusion path of the Ti and Ni atoms necessary for its further growing. Therefore, new nucleation becomes preferred to the growth of preexisting particles which is reflected in the slowing down or stop of QC particle-growth.

4. Effect of Formation of Quasicrystalline Particles on the High-Temperature Creep of Metallic Glasses

In recent years the deformation behavior of bulk metallic glasses (BMGs) in the supercooled liquid region has been studied over a wide range of strain rates [Waniuk, 1998; de Hey, 1998; Heilmaier, 2001; Lee, 2003; Heggen, 2004; Bletry, 2004]. High temperature creep behavior is traditionally investigated in tension or compression [Heggen, 2004; Bletry, 2004]. At the same time, indentation testing has been also successfully applied in studying the creep of different crystalline materials and glasses [Yu, 1977; Han, 1990; Prakash, 1996; Cseh, 1997]. The most important advantages of this method are the ease of sample preparation and that a small piece of specimen is enough for the measurement. The latter is particularly important feature in the case of bulk metallic glasses where the dimensions of samples are often limited. Recently, it has been shown that the viscosity and the activation energy of deformation determined by compression and indentation tests are in good agreement [Fátay, 2004].

Although, the influence of crystallization on the deformation behavior at high temperature has been investigated previously [Galano, 2003; Galano, 2003], the detailed study of the effect of QC phase on creep process was only recently performed [Gubicza, 2008; Lendvai, 2008; Lendvai, 2009]. The effect of the formation of QC phase on the high-temperature creep behavior of $Zr_{44}Ti_{11}Cu_{10}Ni_{10}Be_{25}$ BMG was studied by indentation at 677, 682 and 687 K [Lendvai, 2008]. The experimental setup of the indentation test is shown schematically in Figure 5. Isothermal indentation tests were carried out in a Setaram TMA-92 thermomechanical analyzer. The indentation measurements were carried out on specimens of 3 mm height by using a flat end cylindrical punch of 1.2 mm diameter under constant load of 50 g which corresponds to 0.4 MPa pressure. It has been convincingly confirmed by experimental and theoretical investigations for different metals, alloys and ionic crystals that the equivalent stress (σ) and strain rate ($d\varepsilon/dt$) in indentation creep tests can be expressed by the applied pressure (p) and the indentation rate (dh/dt), respectively, as [Yu, 1977; Cseh, 1997]:

$$\sigma = \frac{p}{3} \tag{1}$$

and

$$\frac{d\varepsilon}{dt} = \frac{1}{d}\frac{dh}{dt}, \tag{2}$$

where d is the diameter of the cylindrical indenter. The viscosity (η) can be determined at any time during the isothermal annealing by the following equation:

$$\eta = \frac{\sigma}{3}\left(\frac{d\varepsilon}{dt}\right)^{-1} = \frac{pd}{9}\left(\frac{dh}{dt}\right)^{-1}. \tag{3}$$

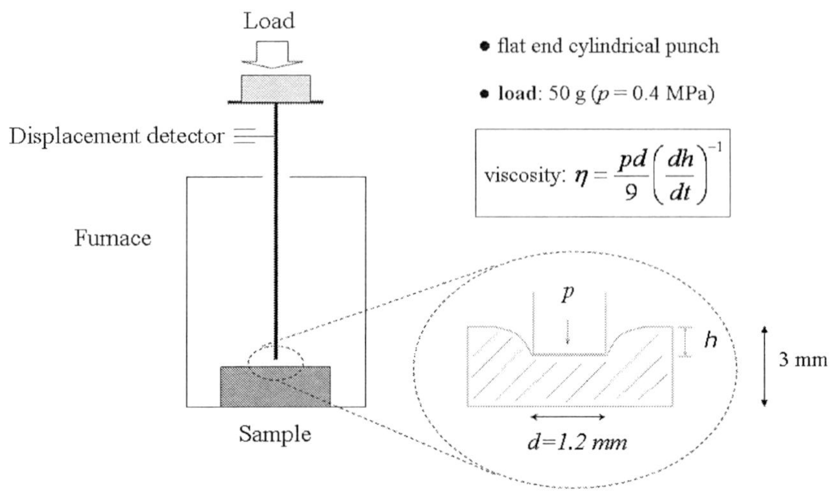

Figure 5. The experimental setup of the indentation creep device used in this study.

The indentation rate as a function of annealing time at 682 K is shown in Figure 6. The heat flow versus annealing time is also plotted in this figure. The plateau in the indentation rate at the beginning of the test indicates a steady state creep of the supercooled liquid. The vertical dotted line shows that the indentation rate started to decrease when the first exothermic peak evolved due to the formation of hard QC particles. Using the heat flow versus time and the indentation rate versus time data, the viscosity was calculated as a function of the heat released during the formation of QC phase normalized by the total heat, H/H_{total}. The latter quantity is called relative released heat (H_{rel}). The values of H_{rel} together with the heat flow are plotted as a function of annelaing time at 682 K in Figure 7. Complementary TEM and X-ray diffraction experiments have shown that the relative released heat obtained in the first DSC peak was close to the QC fraction as discussed in section 2. As a consequence, the values of H_{rel} can be regarded as a measure of the QC fraction. The effect of QC particles on the viscosity was studied at 677, 682 and 687 K up to $H_{rel} = 0.2$, as till this value other crystalline phases did not form according to X-ray diffraction experiments. The dot in the heat flow curve in Figure 7 represents the state corresponding to $H_{rel} = 0.2$ at 682 K.

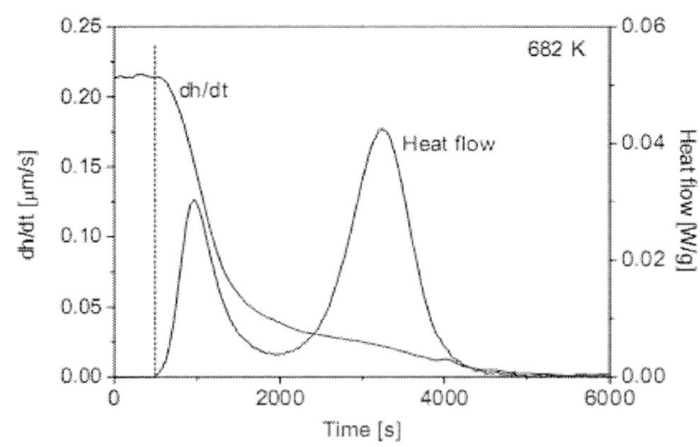

Figure 6. The heat flow and the indentation rate (dh/dt) as a function of time during isothermal annealing at 682 K.

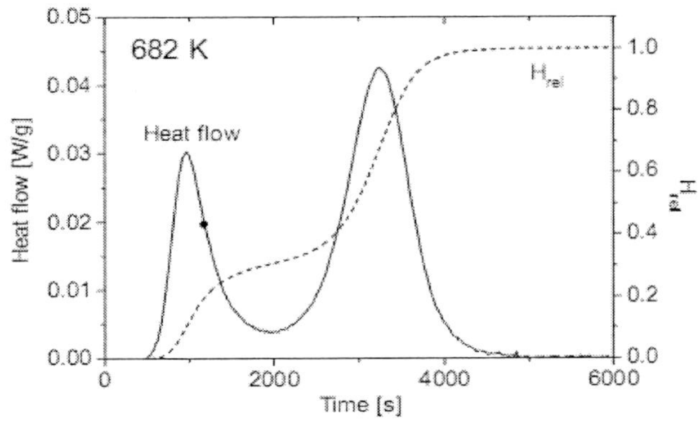

Figure 7. The heat flow and the corresponding relative released heat, H_{rel} as a function of time during isothermal annealing at 682 K. The dot in the heat flow curve represents the state related to $H_{rel} = 0.2$.

In Figure 8 the viscosity as a function of the relative released heat is plotted up to the value of $H_{rel} = 0.2$. As the volume fraction of the QC phase increases, the viscosity of the amorphous–nano-QC composite increases. It is noted that Yan and coworkers [Yan, 2004] have also found that nano-scale crystalline precipitates increase the viscosity of $Zr_{41.25}Ti_{13.75}Ni_{10}Cu_{12.5}Be_{22.5}$ bulk metallic glass.

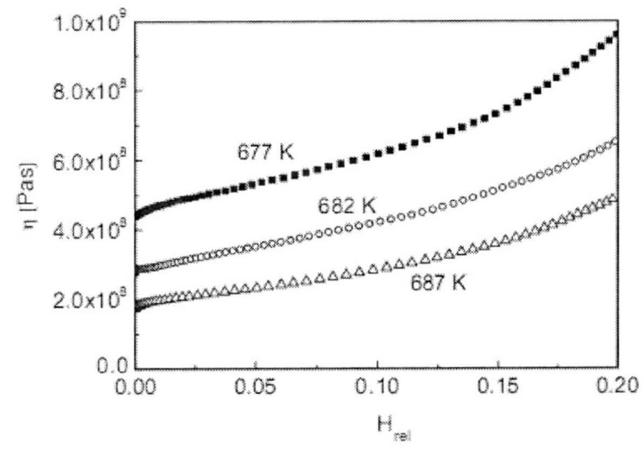

Figure 8. The viscosity as a function of the relative released heat during formation of the QC phase at 677, 682 and 687 K.

The annealed samples can be considered as a dilute suspension of undeformable spherical QC particles in a viscous liquid phase. A general relationship between the viscosity and the volume fraction of the undeformable spherical particles (V) in dilute suspensions was derived theoretically by other authors [Lundgren, 1972; Bedeaux, 1983; Beenakker, 1984; Hsueh, 2005]:

$$\frac{\eta}{\eta_L} = \frac{1}{1 - 2.5\,V}, \qquad (4)$$

where η_L is the viscosity of the liquid phase. In these calculations the suspension is treated as a mixture of two fluids, one fluid having an infinitely large viscosity (hard particles) and the other fluid having viscosity η_L (liquid). It has been shown that if all the hydrodynamic interactions between the spheres are taken into account, the ratio of η/η_L differs from eq.(4) by maximum 6% up to V=0.2 [Beenakker, 1984]. In order to study the validity of this relationship for our QC–supercooled liquid composite, the apparent viscosity was normalized by the viscosity of the liquid phase measured before starting the QC formation (η_{L0}). In Figure 9 the reciprocal of the normalized viscosity, η_{L0}/η is plotted as a function of H_{rel} which is taken to be equal with V.

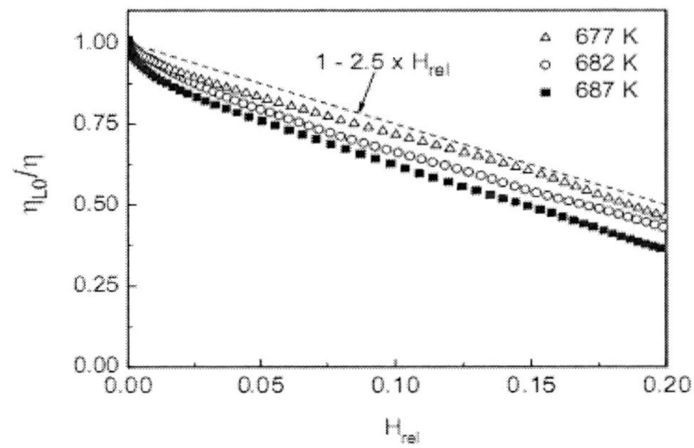

Figure 9. The reciprocal normalized viscosity in supercooled liquid–QC suspension as a function of the relative released heat at 677, 682 and 687 K.

Figure 9 shows that the η_{L0}/η versus H_{rel} relationship can be approximated by linear functions with slopes close to -2.5 for all the three temperatures. The slight deviation of the experimental data from the function 1 - 2.5 x H_{rel} (denoted by the dashed line in Figure 9) can be explained by the change of viscosity in the liquid phase i.e. by the deviation of the real values of η_L from η_{L0} when QC particles are formed. Consequently, these results show that at high temperature the QC–supercooled liquid composite deforms as a dilute suspension and at least up to about 20 % volume fraction of QC, η can be given by the following formula [Gubicza, 2008]

$$\eta = \eta_L(H_{rel}) \cdot f(H_{rel}), \tag{5}$$

where $f(H_{rel})$ accounts for the effect of the hard QC particles on the viscosity and it has the following form

$$f(H_{rel}) = \frac{1}{1 - 2.5H_{rel}}. \tag{6}$$

The viscosity of the liquid phase, η_L was determined as the ratio of the experimentally determined viscosity, η, and $f(H_{rel})$ given in eq.(6) and plotted as a function of H_{rel} at 677, 682 and 687 K in Figure 10. It can be seen that η_L increases slightly with increasing the QC fraction.

The temperature dependence of viscosity of the liquid phase was assumed to obey the following relationship:

$$\eta_L(H_{rel}) = \eta_0(H_{rel})\exp\left(\frac{Q(H_{rel})}{RT}\right), \qquad (7)$$

where η_0 is the pre-exponential factor in viscosity, Q is the activation energy, R is the universal gas constant ($R = 8.31\,\mathrm{J \cdot K^{-1} \cdot mole^{-1}}$) and T is the absolute temperature. The values of η_0 and Q were calculated from the slope of $\ln \eta_L$ vs. 1/T plot for different values of H_{rel} and plotted in Figs. 11a and b, respectively. The pre-exponential factor η_0 increased while the activation energy decreased with increasing fraction of the QC phase. The two effects are largely compensated by each other resulting in only a slight increase of the viscosity of liquid phase with increasing the QC fraction as it was shown in Figure 10. The increase of the pre-exponential factor and the decrease of the activation energy during the formation of the QC phase can be probably attributed to the changes in the chemical composition of the remaining liquid phase. In section 3 we have discussed that the liquid phase was largely depleted in Ti and enriched in Be during the formation of QC particles. It was shown previously that the increase of the fraction of smaller Be atoms in ZrTiCuNiBe bulk metallic glasses usually results in the decrease of the activation energy of diffusion [Macht, 2001]. It was also observed for other compositions that increasing the concentration of smaller atoms among the alloying elements in BMGs leads to smaller values both of the activation energy and the pre-exponential factor, D_0 of diffusion [Frank, 1994]. The smaller average atomic size in BMGs most probably results in an easier thermal activation of the atomic jump and also reduces the distance of a single jump thereby reducing the activation energy and D_0, respectively. The pre-exponential factor of the viscosity, η_0 is inversely proportional to D_0 therefore η_0 increases with increasing the fraction of smaller atoms in the liquid phase due to the formation of QC phase.

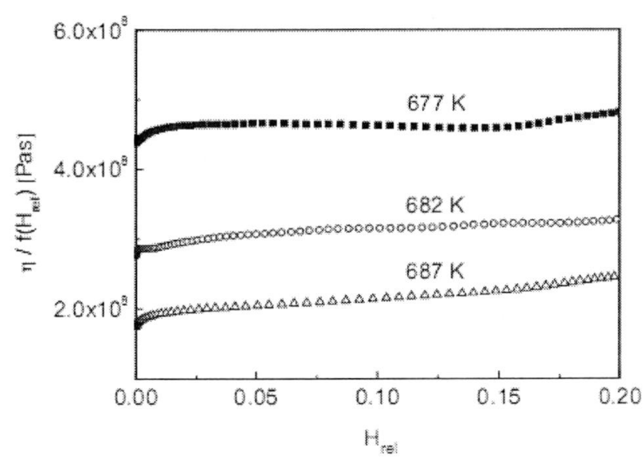

Figure 10. The viscosity of the liquid phase in supercooled liquid–QC suspension as a function of the relative released heat at 677, 682 and 687 K.

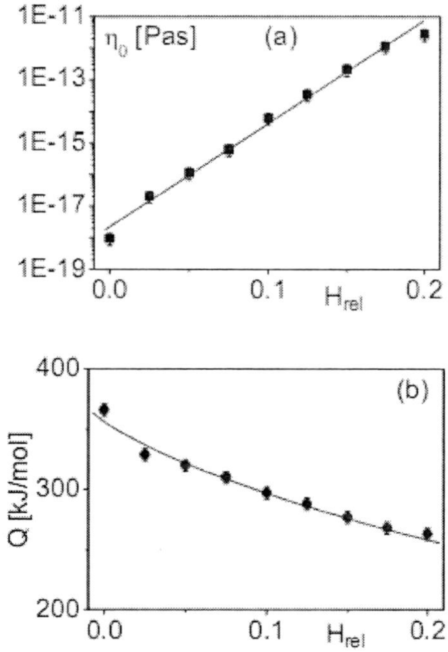

Figure 11. The changes of the pre-exponential factor, η_0 (a) and the activation energy, Q (b) of creep for the liquid phase in supercooled liquid–QC suspension as a function of the relative released heat.

5. Influence of Quasicrystalline Phase Formation on the Room Temperature Mechanical Behavior of BMGs

The effect of formation of QC phase on the deformation behavior of a $Zr_{44}Ti_{11}Cu_{10}Ni_{10}Be_{25}$ BMG at room temperature was studied by three-point bending tests on 14 mm long samples of 1.6 mm height and 2.3 mm width at a displacement rate of 0.005 mm s^{-1} [Lendvai, 2008]. Figure 12 shows the results of three-point bending tests carried out at room temperature on samples annealed for different times at 682 K. To permit direct comparison of specimens with slightly different cross-sections, the bending force normalized with the second moment of area (*F/I*) is plotted as a function of deflection. It can be seen that while the initial fully amorphous sample shows plasticity, the samples containing different fractions of QC phase failed already in the elastic deformation regime. The large decrease in strength can be explained mainly by a loss of free volume due to structural relaxation in the beginning of annealing [Suh, 2003]. The increase of viscosity in the amorphous phase caused by the changes of chemical composition due to the formation of QC particles may also contribute to the reduction of strength [Schuh, 2007]. The latter effect may make stress relaxation by means of viscoplastic flow more difficult (e.g. in a crack tip region) thereby decreasing the bending strength when QC particles are formed.

It is noted that in the case of a $Zr_{65}Al_{7.5}Ni_{10}Cu_{7.5}Pd_{10}$ BMG alloy annealed for 60 s at 705K, it was found that the QC phase increased the strength at RT [Inoue, 2001]. The QC phase was identified to have an icosahedral structure with an average particle size of about 30 nm and a high volume fraction of 80–90%. This bulk nanoquasicrystalline alloy shows higher compressive fracture strength (1820 MPa) combined with a significantly improved plastic elongation of 0.5%, as compared with those (1630 MPa and nearly 0%) for the corresponding amorphous single phase alloy with the same alloy composition. Similarly, improved mechanical properties were obtained for bulk $Zr_{58}Al_9Ni_9Cu_{14}Nb_{10}$ QC–glass composite processed by conventional copper mould casting [Qiang, 2007], where the QC phase was the majority phase with about 500 nm grain size. The room temperature compression stress–true strain curve exhibits a 2% elastic deformation up to failure, and a maximum fracture stress of 1850 MPa at a quasi-static loading rate of 4.4 x 10^{-4} s^{-1}. This mechanical behavior is superior to quasicrystal alloys developed earlier, and is comparable to Zr-based bulk metallic glasses and their nanocomposites which seems to be related to the existence of a glassy phase in between the large-grain QC phases. The synthesis of the in situ QC–glass composite suggests a new approach to develop promising QC materials for engineering applications.

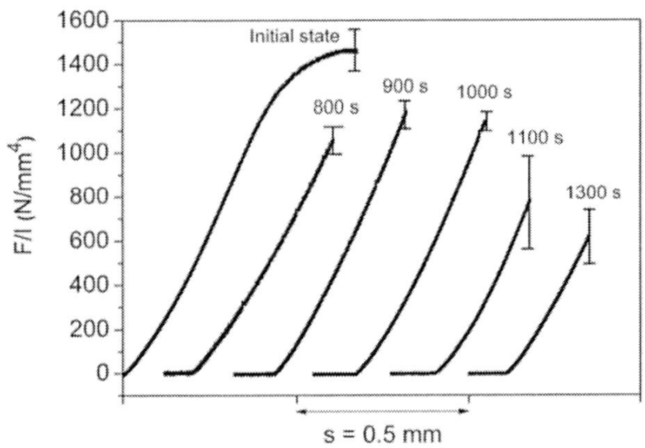

Figure 12. Results of three-point bending tests for samples annealed for different times at 682 K. *F/I* is the applied force normalized by the second moment of area and *s* is the deflection. The error bars indicate the uncertainties of maximum values of *F/I*.

6. CONCLUSION

The formation of quasicrystalline phase and its effect on the mechanical behavior was studied in a $Zr_{44}Ti_{11}Cu_{10}Ni_{10}Be_{25}$ BMG alloy. During annealing of this alloy, first metastable QC particles and in a second step stable crystalline phases are formed partly from the liquid phase and partly by transformation from the quasicrystals. The nanosized QC particles are spherical and have an internal grain structure. The QC particles formed in ZrTiCuNiBe BMGs during annealing have icosahedral structure and are enriched in Ti and Ni and depleted from Be and Cu.

The formation of the QC phase increases the viscosity of BMGs measured above the glass-transition temperature. The experimental values of viscosity basically follows the theoretical function describing dilute suspensions of hard particles, although the change of chemical composition in the liquid phase also affects the viscosity. As a result of increasing Be/Ti ratio in the liquid phase, the pre-exponential factor and the activation energy of viscosity increased and decreased, respectively, with increasing fraction of the QC phase.

The formation of the QC phase during annealing is accompanied by the decrease of room–temperature bending strength which can be attributed mainly to a loss of free volume and partially to an increase of viscosity in the amorphous phase as a consequence of the change of its chemical composition.

Acknowledgments

This work was supported by the Hungarian Scientific Research Funds, OTKA, Grant No. K-67692 and 81360. The authors are grateful to Dr. J.L. Lábár, Dr. Gy. Vörös, Mr. E. Agócs, Mr. D. Fátay and Mr. Z. Kuli for their assistance in experiments.

References

Bedeaux, D. Physica A 1983, 121, 345-361.

Beenakker, C. W. J. Physica A 1984, 128, 48-81.

Bletry, M.; Guyot, P.; Brechet, Y.; Blandin, J. J.; Soubeyroux, *J. L. Intermetallics* 2004, 12, 1051-1055.

Chang, H. J.; Kim, D. H.; Kim, Y. M.; Kim, Y. J.; Chattopadhyay, *K. Scripta Mater* 2006, 55, 509-512.

Chen, M. W.; Zhang, T.; Inoue, A.; Sakai, A.; Sakurai, *A. Appl. Phys. Lett.* 1999, 75, 1697-1699.

Cseh, G.; Chinh, N. Q.; Tasnádi, P.; Juhász, A.; *J. Mater Sci.* 1997, 32, 5107-5111.

de Hey, P.; Sietsma, J.; van den Beukel, *A. Acta Mater* 1998, 46, 5873-5882.

Eckert, J.; Mattern, N.; Zinkevitch, M.; Seidel, *M. Mater Trans* JIM 1998, 39, 623-634.

Fan, C.; Wilson, T. W.; Dmowski, W.; Choo, H.; Richardson, J. W.; Maxey, E. R.; Liaw, P. K. *Intermetallics* 2006, 14, 888-892.

Fátay, D.; Gubicza, J.; Szommer, P.; Lendvai, J.; Blétry, M.; Guyot, P. *Mater Sci. Eng. A* 2004, 387–389, 1001-1004.

Frank, F. C. *Proc. R. Soc.* London A 1952, 215, 43-46.

Frank, W.; Hörner, A.; Scharwaechter, P.; Kronmüller, H. *Mater Sci. Eng. A* 1994, 179-180, 36-40.

Galano, M.; Rubiolo, G. H. *Scripta Mater* 2003, 48, 617-622.

Gubicza, J.; Labar, J. L.; Agocs, E.; Fatay, D.; Lendvai, *J. Scripta Mater* 2008, 58, 291-294.

Han, W. T.; Tomozawa, M. *J. Am. Ceram. Soc.* 1990, 73, 3626-3632.

Heggen, M.; Spaepen, F.; Feuerbacher, *M. Mater Sci. Eng. A* 2004, 375-377, 1186-1190.

Heilmaier, M.; *J. Mater Proc. Techn.* 2001, 117, 374-380.

Holzer, J. C.; Kelton, K. F. *Acta Met. Mater* 1991, 39, 1833-1843.

Hsueh, C-H.; Becher, P. F. *J. Am. Ceram. Soc.* 2005, 88, 1046-1049.

Inoue, A.; Zhang, T.; Saida, J.; Matsushita, M.; Chen, M. W.; Sakurai, T. *Mater Trans.* JIM 1999, 40, 1181-1184.

Inoue, A. *Mater Sci. Eng. A* 2001, 304–306, 1-10.

Kelton, K. F. *Intermetallics* 2006, 14, 966-971.

Kelton, K. F. *J. Non-Cryst. Solids* 334-335 (2004) 253-258

Kelton, K. F.; Gangopadhyay, A. K.; Kim, T. H.; Lee, G. W. *J. Non-Cryst. Solids* 2006b, 352, 5318-5324

Kim, W. J.; Ma, D. S.; Jeong, H. G. *Scripta Mater* 2003, 49, 1067-1073.

Köster, U.; Meinhardt, J.; Roos, S.; Liebertz, H. *Appl. Phys. Lett.* 1996, 69, 179-181.

Kühn, U.; Eckert, J.; Mattern, N.; Schultz, L. *Appl. Phys. Lett.* 2000, 77, 3176-3178.

Kühn, U.; Eckert, J.; Mattern, N.; Schultz, L. *Phys. Stat. Sol.(a)* 2005, 202, 2436-2441.

Kühn, U.; Eymann, K.; Mattern, N.; Eckert, J.; Gebert, A.; Bartusch, B.; Schultz, L. *Acta Mater* 2006, 54, 4685-4692.

Lábár, J. L. *Ultramicroscopy* 2005, 103, 237–249.

Lee, K. S.; Ha, T. K.; Ahn, S.; Chang, Y. W. *J. Non-Cryst. Sol.* 2003, 317, 193-199.

Lendvai, J.; Fátay, D.; Gubicza J. *Mater Sci. Eng. A* 2008, 483–484, 607-610

Lendvai, J.; Gubicza, J.; Lábár, J. L.; Kuli, Z. *Int. J. Mater Res.* 2009, 100, 439-442.

Liu, L.; Chan, K. C. *Intermetallics* 2004, 12, 1143-1148.

Liu, L.; Chan, K. C. *J. Alloys. Comp.* 2004b, 364, 146-155.

Lundgren, T. S. *J. Fluid Mech.* 1972, 51, 273-299.

Macht, M.-P.; Naundorf, V.; Fielitz, P.; Rüsing, J.; Zumkley, Th.; Frohberg, G. *Mater Sci. Eng. A* 2001, 304-306, 646-649.

Mechler, S.; Wanderka, N.; Macht, M.-P. *Mater Sci. Eng. A* 2004, 375–377, 355–358.

Murty, B. S.; Ping, D. H.; Hono, K.; Inoue, A. *Appl. Phys. Lett.* 2000b, 76, 55-57.

Murty, B. S.; Ping, D. H.; Hono, K.; Inoue, A. *Scripta Mater* 2000, 43, 103-107.

Murty, B. S.; Hono, K. *Mater Sci. Eng. A* 2001, 312, 253-261.

Prakash, O.; Jones, D. R. H. *Acta Mater* 1996, 44, 891-897.

Qiang, J. B.; Zhang, W.; Xie, G.; Kimura, H.; Dong, C.; Inoue, A. *Intermetallics* 2007, 15, 1197-1201.

Révész, Á.; Henits, P.; Kovács, Z. J Alloys Comp 2010, doi:10.1016/j.jallcom.2009.10.175

Saida, J.; Matsushita, M.; Li, C. *Appl. Phys. Lett.* 2000, 76, 3558-3560.

Saida, J.; Inoue, A. *J. Non-Cryst. Solids* 2003b, 317, 97-105.

Saida, J.; Imafuku, M.; Sato, S.; Matsubara, E.; Inoue, A. *J. Non-Cryst. Solids* 2007, 353, 3704-3708.

Saida, J.; Matsushita, M.; Inoue A. *J. Appl. Phys.* 2001, 90, 4717-4719.

Saida, J.; Sherif El-Eskandarany, M.; Inoue, A. *Scripta Mater* 2003, 48, 1397-1401.

Schuh, C. A.; Hufnagel, T. C.; Ramamurty, U. *Acta Mater.* 2007, 55, 4067-4109.

Suh, D.; Dauskardt, R. H. *J. Non-Cryst. Solids* 2003, 317, 181-186.

Van de Moortéle, B.; Epicier, T.; Pelletier, J. M.; Soubeyroux, J. L. *J. Non-Cryst. Solids* 2004, 345-346, 169-172.

Wanderka, N.; Macht, M-P.; Seidel, M.; Mechler, S.; Stahl, K.; Jiang, J. Z. *Appl. Phys. Lett.* 2000, 77, 3935-3937.

Waniuk, T. A.; Busch, R.; Masuhr, A.; Johnson, W. L. *Acta Mater* 1998, 46, 5229-5236.

Wei, B. C.; Yu, G. S.; Löser, W.; Xia, L.; Roth, S.; Wang, W. H.; Eckert, J. *Mater Sci. Eng.* A 2004, 375–377, 1161-1164.

Wollgarten, M.; Mechler, S.; Davidov, E.; Wanderka, N.; Macht, M.-P. *Intermetallics* 2004, 12, 1251-1255.

Xing, L. Q.; Eckert, J.; Löser, W.; Schultz, L. *Appl. Phys. Lett.* 1998, 73, 2110-2112.

Xing, L. Q.; Hufnagel, T. C.; Eckert, J.; Loser, W.; Schultz, L. *Appl. Phys. Lett.* 2000, 77, 1970-1972.

Xing, L.-Q.; Li, Y.; Ramesh, K. T.; Li, J.; Hufnagel, T. C. *Phys. Rev. B* 2001, 64, 180201(R).

Yan, M.; Sun, J. F.; Shen, J. *J. Alloys Comp.* 2004, 381, 86-90.

Yu, E. C.; Li, J. C. M. *Phil. Mag.* 1977, 36, 811-825.

In: Quasicrystals: Types, Systems, and Techniques
Editor: B. E. Puckermann, pp. 163-173
ISBN 978-1-61761-123-0

Chapter 7

SURFACE STRUCTURE OF TWO-FOLD AL-NI-CO DECAGONAL QUASICRYSTAL: PERIODICITY, APERIODICITY, DEFECTS AND SECOND PHASE STRUCTURE

Jeong Young Park[a1]

Graduate School of EEWS (WCU Program), Korea Advanced Institute of Science and Technology (KAIST), Daejeon, Republic of Korea

ABSTRACT

The atomic structure of the 2-fold decagonal Al-Ni-Co quasicrystal surface has been investigated using scanning tunneling microscopy (STM). Decagonal quasicrystals are made of pairs of atomic planes with pentagonal symmetry periodically stacked along a 10-fold axis. It is, therefore, expected that the 2-fold surfaces exhibit a periodic direction along the 10-fold axis, and an aperiodic direction perpendicular to it. The surface shows rough and cluster-like structures at low annealing temperatures (T<1000K), whilst annealing to temperatures in excess of 1000K results in the formation of step-terrace structures. The surface consists of terraces separated by steps of heights 1.9, 4.7, 7.8, and 12.6 Å. Ratios of step heights can be properly assigned to different τ powers, suggesting a well defined quasiperiodic long-range order. At the annealing temperature (1100K < T < 1150K), atomically resolved STM images of the 2-fold plane reveal atomic rows along the 10-fold direction with a periodicity of 4 Å. The spacing between the parallel rows is aperiodic, with distances following a Fibonacci sequence. We found that the quasiperiodic order in the sequence of atomic rows is destroyed by the presence of phason defects. Above the heating temperature of 1200K, formation of second phase structures was observed. The formation of a second phase could be associated with the preferential evaporation of Al at the elevated temperature.

---

1 a)Author to whom correspondence should be addressed. Electronic mail: jeongypark @kaist.ac.kr.

1. INTRODUCTION

Quasicrystals are metallic alloys that exhibit atomic scale order, but not periodic order. (Shechtman, Blech et al. 1984; Janot 1992) Atomic scale properties of these materials are different from single crystalline material, for example, extraordinary mechanical properties (Dubois ; Park, Ogletree et al. 2006), electrical and thermal transport properties (Martin, Hebard et al. 1991; Zhang, Cao et al. 1991), and electronic structure (Rotenberg, Theis et al. 2000). A key question is how the quasiperiodic atomic arrangement is fundamentally associated with these material properties, and what is different from the crystalline arrangement. To address these issues, decagonal quasicrystals provide a desirable configuration where the structure is periodic along one dimension, and quasiperiodic in the two other dimensions (Steurer 2004). 2-fold decagonal quasicrystals are especially intriguing since the surface exhibits atomic scale coexistence of periodic and quasiperiodic ordering. This unique atomic arrangement allows us to study the influence of surface ordering upon surface properties, for example, friction (Park, Ogletree et al. 2005), transport, diffusion of adsorbates, etc.

The bulk atomic structure of Al–Co–Ni quasicrystals has been studied intensively using high-resolution transmission electron microscopy (TEM), high-angle dark-field scanning transmission electron microscopy (HAADF-STEM) (Abe, Saitoh et al. 2000), and x-ray and electron diffraction(Steurer 2004). The quasicrystalline phases have clearly shown an ordered arrangement of columnar atom clusters, which have a decagonal shape with pentagonal symmetry and a diameter of 2.0 nm. The bulk structure of the decagonal quasicrystal possesses a 10-fold symmetry axis along the periodic direction and two sets of five equivalent 2-fold symmetry axes with increments of 36° in the quasiperiodic plane.

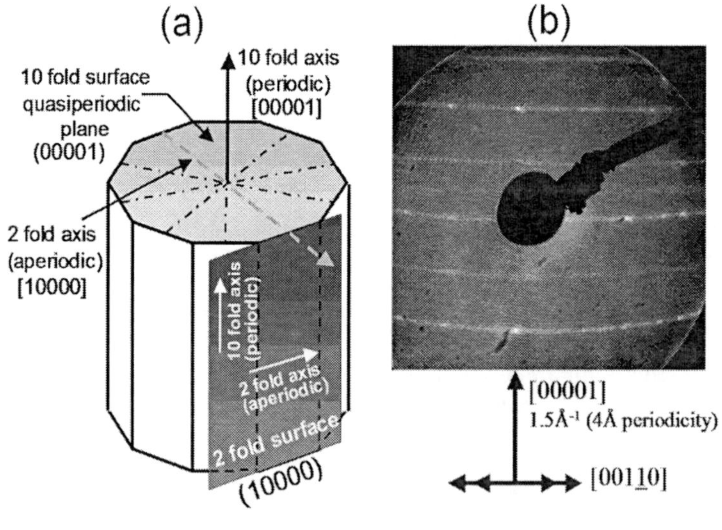

Figure 1. (a) A schematic of the bulk model of the decagonal Al-Ni-Co quasicrystal. According to the bulk model, the 2-fold (10000) plane shows periodicity along the [00001] direction, and aperiodicity along the [00110] direction. (b) LEED pattern of the 2-fold Al-Ni-Co decagonal quasicrystal surface at the electron energy of 65 eV, and reciprocal lattice basis vectors of the decagonal structure with 4 Å periodicity.

Recently, the surface structure of the 10-fold Al–Ni-Co decagonal quasicrystalline surface has also been studied (Cox, Ledieu et al. 2001; Kishida, Kamimura et al. 2002; Ferralis, Pussi et al. 2004; Sharma, Franke et al. 2004; Yuhara, Klikovits et al. 2004; Mader, Widmer et al. 2009). From low energy electron diffraction (LEED) and scanning tunneling microscopy (STM) studies, it is reported that there is only one type of monoatomic step on the 10-fold surface with a step height of 0.2 nm. The symmetry of each layer is not decagonal but pentagonal and two adjacent layers are related by inversion symmetry. Several important issues, such as surface atomic density, first-layer concentrations and their relation to the structure, similarity of surface atomic structure with the bulk model have been addressed in the 10-fold Al-Ni-Co surface based upon STM measurements, He-atom scattering (HAS), high resolution low energy electron diffraction (SPA-LEED), and low energy ion scattering spectroscopy (Cox, Ledieu et al. 2001; Kishida, Kamimura et al. 2002; Ferralis, Pussi et al. 2004; Sharma, Franke et al. 2004; Yuhara, Klikovits et al. 2004) .

The surface atomic structure of 2-fold Al-Ni-Co has been studied using scanning tunneling microscopy (STM) (Kishida, Kamimura et al. 2002; Park, Ogletree et al. 2005), low energy electron diffraction (LEED), and low energy He-atom scattering (HAS) (Sharma, Franke et al. 2004). The surface reveals both periodic and aperiodic ordering.

The schematic of the bulk structure of a decagonal quasicrystal is shown in Figure 1a where the indexing system by Steurer et al. (Steurer 2004) is used. Decagonal quasicrystals possess a 10-fold symmetry axis along the periodic direction, and two sets of five equivalent 2-fold symmetry axes rotated by 18° in the quasiperiodic plane. For clarity, round and square brackets denote surface orientation and reciprocal lattice vector, respectively. The indices correspond to the five basis vectors commonly used to describe the decagonal reciprocal lattice. The last index refers to the (00001) basis vector along the periodic direction, while the first four refer to the four basis vectors in the quasiperiodic plane. The bulk structure of the decagonal quasicrystal is composed of quasiperiodic planes of (00001) which are equally spaced with an interlayer spacing of 0.4 nm. In this schematic of bulk structure, the 2-fold surface, or (10000) plane, has periodicity along [00001], and quasi-periodicity along the 2-fold direction or [001   0].

In this Communication, we present the detailed atomic structures of the 2-fold Al-Ni-Co decagonal quasicrystal, and their dependence on sample preparation. Aperiodicity can be represented by a sequence of atomic rows following the Fibonacci sequence, and inflation symmetry with the Golden Mean (τ). Ratios of step heights can be properly assigned to different τ powers, suggesting a well defined quasiperiodic long-range order. We show that this aperiodic ordering can be destroyed in the proximity of any interlayer phason defect. At high annealing temperatures (>1200K), the atomic surface with a second phase structure was revealed.

2. Experimental

We prepared and characterized our samples in a UHV chamber with a base pressure of 1.0 x 10^{-10} torr. The chamber contained a commercial RHK atomic force microscope (AFM)/scanning tunneling microscopy (STM) mounted on a 6” flange(Park, Ogletree et al. 2004; Park, Ogletree et al. 2005). The sample could be translated from the AFM to a 3-axis manipulator with a heating stage. Cleaning was accomplished by a combination of electron beam bombardment and Ar^+ ion sputtering. Low energy electron diffraction (LEED) and Auger electron spectroscopy (AES) were used for surface analysis. Samples and cantilevers could also be transferred from air through a load-lock without breaking the vacuum, which makes measurements possible with various cantilevers with different spring constants and metal coatings.

By using conductive cantilevers, the electrical current between the tip and sample could be measured and used for feedback in scanning tunneling microscopy (STM) mode (Park, Sacha et al. 2005). We used two types of cantilevers coated with approximately 50 nm of either W_2C or TiN, and with spring constants of 48 or 90 N/m for STM mode. The high stiffness of the cantilevers suppresses the jump to contact instability found in soft cantilevers, thus ensuring stable tunneling. Using a field emission scanning electron microscope, the radii of the metal-coated tips were found to be 30-50 nm.

Single grain 2-fold decagonal $Al_{72.4}Ni_{10.4}Co_{17.2}$ quasicrystals were grown at Ames Laboratory at Iowa State University, and prepared in the form of samples of 1cm x 1cm x 1.5 mm (Fisher, Kramer et al. 1999). In the UHV chamber, they were cleaned by cycles of Ar^+ sputtering at 1 keV, and annealed for 1 to 2 hours by electron bombardment at a temperature between 1100 K and 1200 K, as monitored by an optical pyrometer. The details on the sample preparation and UHV cleaning are described elsewhere (Park, Ogletree et al. 2005; Park, Ogletree et al. 2005).

3. RESULTS AND DISCUSSION

3.A. Step Structure of the 2-Fold Al-Ni-Co Decagonal Quasicrystal Surface

In the previous LEED study of the 2-fold Al-Ni-Co surface, W. Theis et al. (Theis, Rotenberg et al. 2003) reported the observation of intense and sharp spots corresponding to a 4 Å period along the 10-fold axis, as well as streaks with broadened peaks reflecting an 8 Å periodicity. Along the 2-fold or quasiperiodic direction, they observed diffraction peaks at the bulk reciprocal lattice points $(n + m\tau)$ 0.6 $Å^{-1}$ where n, m are integers, and τ is the golden mean. The resulting LEED pattern is shown in Figure 1b along with reciprocal lattice basis vectors of the decagonal structure with 4 Å periodicity. The broad stripe pattern in the LEED pattern corresponds to 8 Å periodicity, suggesting the presence of a disordered region which is presumably associated with surface reconstruction. The 4 Å period corresponds to a periodic stacking of two inequivalent layers constituting the smallest possible periodicity in the decagonal phase.

An STM image taken after low temperature annealing (<1000K) shows the surface to be rough and cluster-like, as shown in Figure 2a. The surfaces exhibit the step-terrace structure with the higher heating temperature (> 1000K), as shown in Figure 2b. Similar STM observationw were reported by E. Cox et al. (Cox, Ledieu et al. 2001) who observed the rough and cluster-like features at low annealing temperatures for the 10-fold Al-Ni-Co surface. The step heights of this surface were described elsewhere (Park, Ogletree et al. 2005). Figure 3 shows the STM image of 2-fold Al-Ni-Co surface (Vs = 1.0V, It = 0.1nA), where terraces are separated by several types of atomic steps. From the height profile across the dashed line as shown in Figure 3b, the step heights are 1.9±0.1 Å (X_H), 4.7±0.1 Å (short,

S_H), 7.8±0.2 Å (long, L_H), and 12.6±0.3 Å (S_H+L_H). Of these, steps with heights of S_H, L_H, and L_H+S_H satisfy the ratio of the Golden Mean ($\tau \sim 1.618$), reflecting the aperiodic structure along the normal direction of the surface or [10000]. While steps with height of L and S were reported by Kishida and others (Kishida, Kamimura et al. 2002), the step with the height of 2 Å has not been reported.

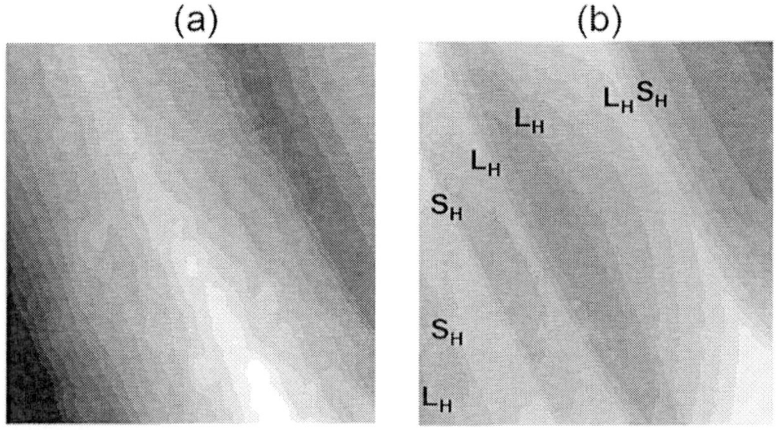

Figure 2. 100 nm x 100 nm STM image after annealing at (a) 950K for two hours, (b) 1130K for two hours. LH and SH are described in the text and Figure 3.

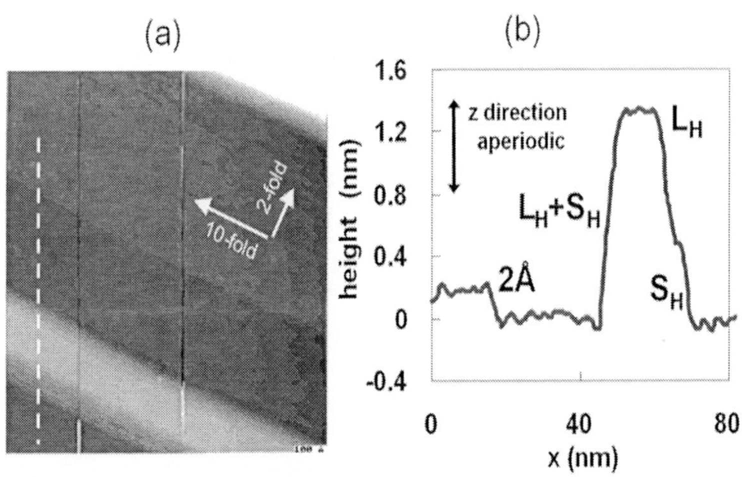

Figure 3. (a) STM image of 100 nm x 100 nm (Vs = 1.0V, It = 0.1nA) of 2-fold Al-Ni-Co surface. (b) The line profile of the STM image showing several types of steps with a height of 4.7±0.1 Å (short, S_H), 7.8±0.2 Å (long, L_H), and 12.6±0.3 Å height (S_H+L_H), and 1.9 Å. The ratio of L and S is approximately equal to Golden Mean (~ 1.62), suggesting the aperiodic structure along the vertical direction to the plane. (Reprinted figure with permission from (Park, Ogletree et al. 2005) .)

Additional evidence on the presence of a well-defined quasiperiodic long-range order can be given by analyzing the step heights distribution in terms of successive powers of the golden mean. Table 1 shows the ratios of the step heights of four types of steps reported by Park et al.

Table 1. Ratio of step heights of four types of atomic steps (1.9±0.1 Å(X_H), 4.7±0.1 Å (short, S_H), 7.8±0.2 Å (long, L_H), and 12.6±0.3 Å height (S_H+L_H)). XH was taken as the reference unit to derive the different ratios. τ is Golden Mean (=(1+√5)/2) ~ 1.618). Ratios between atomic steps are properly assigned as different τ powers within the experimental accuracy

	$L_H + S_H$	L_H	S_H	X_H
X_H	6.6 ± 0.5	4.1± 0.2	2.5± 0.2	1
S_H	2.7± 0.1	1.7± 0.1	1	
L_H	1.60 ± 0.08	1		
	$L_H + S_H$	L_H	S_H	X_H
X_H	τ^4 (6.854)	τ^3(4.236)	τ^2(2.618)	1
S_H	τ^2	τ	1	
L_H	τ	1		

The shortest step was taken as the reference unit (X_H =1.9±0.1Å) to derive the different ratios. As can be seen in the table, all of the ratios listed in the table can be properly assigned to different τ powers (within the experimental accuracy). It is noteworthy that the ratio involving higher steps provide the better accuracy, as expected in a system with a well-defined long-range order.

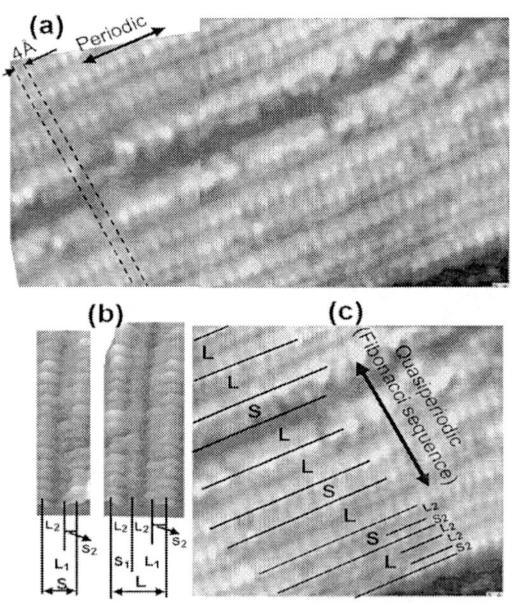

Figure 4. (a) Collage of STM images (145 Å x 90 Å). of two contiguous regions of the 2-fold Al-Ni-Co surface. The terraces are made of rows of periodically arranged atoms (4 Å) along the ten-fold direction and separated by distances L and S. (b) Expanded view showing the interior in the L and S sections. L contains two atomic rows, separated by L_2 and S_2 distances. S contains one row, at distances of L_2 and S_2 from the boundary. (c) The sequence of L and S spacings between rows follows a Fibonacci sequence. The trench in the center of (a) and (c) is due to a missing L+S section. (Reprinted figure with permission from (Park, Ogletree et al. 2005).)

Figure 4a shows a collage of two high resolution STM images of a terrace. Except for a defect in the form of a missing row (visible as a dark band), it consists of atomic rows of close but not exactly the same apparent height, with variations of ± 0.3 Å. Two different lengths, S = 7.7±0.3 Å and L = 12.5±0.4 Å, separate the rows and define the sides of pseudo unit cells. Secondary rows of lower apparent height are visible inside the cells, two within L and one within S, as shown in Figure 4b.

The spacing between these secondary rows are L_2 = 4.9±0.3 Å, and S_2 = 2.8±0.2 Å. As shown in the figure (4b), an intermediate-level partition can be considered with L_1 and S_1 separations, where $L_1 = L_2 + S_2$ and $S_1 = L_2$. The ratios L/S, L_1/S_1 or L_2/S_2 are all close to the Golden Mean.

The L and S distances form an LSLSLLSLLS sequence (Figure 4c), which corresponds to a Fibonacci sequence (a Fibonacci sequence is a progression of numbers that are sums of the previous two terms: f(n+1) = f(n) + f (n-1)[2]), for n = 6. If we substitute L and S by the subsections L_1, S_1, or further by L_2, S_2 we obtain: $L_1S_1L_1L_1S_1L_1L_1S_1L_1S_1L_1L_1S_1L_1S_1L_1$, and $S_2L_2L_2S_2L_2S_2L_2$ $L_2S_2L_2S_2L_2L_2S_2L_2L_2S_2L_2S_2L_2L_2S_2L_2L_2S_2L_2$ respectively. These sequences, visible in the STM image, correspond to Fibonacci sequences for n = 7 and 8. The process of increasing the number of units by subdividing the large units into smaller ones to create self-similar, but not identical patterns is called inflation. In the processes of inflation between n = 6 and 7, L is substituted by L_1S_1, and S by L_1. From n = 7 to 8 the process is inverted, i.e., $L_1 \rightarrow S_2L_2$, and $S_1 \rightarrow L_2$. This inversion does not alter the Fibonacci sequence but causes it to shift.

3.B. Atomic Scale Disordering on the 2 Fold Al-Ni-Co Surface

As shown before, the 2-fold Al-Ni-Co surface consists of atomic rows of atoms with 4 Å periodicity along the 10-fold direction and aperiodically spaced in the 2-fold direction. This quasiperiodic sequence on the surface can be perturbed by the presence of interlayer phason defects. The interlayer phason defect is the defect (Jeong and Steinhardt 1993) in which a phason flip occurs between the adjacent layers along the 10-fold direction. This is a local atomic arrangement which is related to the strain in the phason degree of freedom. Particularly, the number density of phason defects is of importance since it is associated with the fundamental question as to why the quasicrystalline structural order is realized in the material. One model regarding this issues the perfectly ordered quasiperiodic model, where the quasicrystal is assumed to be energetically stabilized (Steinhardt and Jeong 1996). The other is the random tiling model, in which the quasicrystal is assumed to be stabilized by a configurational entropy related to the phason disorder . Kishida and others (Kishida, Kamimura et al. 2002) found that the number density of phason defects in a 2-fold Al-Ni-Co surface is quite low (~ 2×10^{-4} nm^{-2}), suggesting the quasicrystal is likely to be energetically stabilized. In our study, the number density of the interlayer phason defect is also small (< 8×10^{-4} nm^{-2}), consistent with a perfectly ordered quasiperiodic model. Theoretically, it is predicted that the phason defect destroys the Fibonacci chain. For example, the conductance change due to the phason defect which gives rise to the irregular variation of the Fibonacci chain was studied by Moulopoulos and Roche (Moulopoulos and Roche 1996). Figure 5 shows STM images of regions containing such defects which destroy the Fibonacci sequence

SLLSLL. The atomic row in the left of the phason defect does not follow aperiodic ordering, and interestingly has 8Å periodicity.

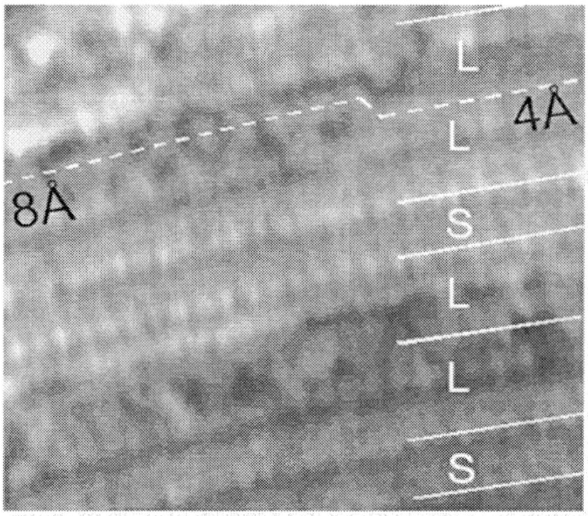

Figure 5. 90 Å x 90 Å atomic scale STM images (Vs =1.2V, I = 0.1nA) shows interlayer phason defects which break the Fibonacci sequence ordering.

3.C. Atomic Structure of Second Phase Structures

Atomic scale order-disorder phase transitions are a challenging question in structurally complex systems such as aperiodic quasicrystals. Structural phase transitions of decagonal Al-Ni-Co quasicrystals have been studied with positron annihilation spectroscopy, x-ray diffraction and neutron scattering (Baumgarte, Schreuer et al. 1997; Frey, Weidner et al. 2002; Sato, Baier et al. 2004). K. Sato et al. found that decagonal Al-Ni-Co quasicrystals undergo order-disorder transitions at 1140K, based on positron annihilation spectroscopy results (Sato, Baier et al. 2004). Also, Frey et al. reported that the diffuse layer with 8 Å periodicity vanishes above 1173K based on in-situ X-ray and neutron diffraction experiments.

After annealing the sample at higher temperatures (> 1200K), we observed a atomic structure that is clearly distinct from the quasicrystalline surface structure prepared with lower temperatures (< 1150K). Figures 6a and 6b show the STM images revealing two domains separated by the 2 Å step. The lower domain (Domain A) is composed of atomic rows with 4 Å periodicity. Domain B exhibits a more disordered structure but the aperiodicity remains along the 2-fold direction, as marked by the lines in Figure 6a. In the quasicrystalline phase, as shown in Figure 6, the lower terrace reveals atomic rows with 8 Å periodicity. It is an interesting point that this 8 Å periodicity is not visible after the phase transition (in Domain A), which is consistent with Frey et al's result that the diffuse layer vanishes after heating the sample at 1173K. The second order structure at the elevated temperature could be associated with the preferential evaporation of Al, followed by the diffusion of Al (Hocker and Gahler 2004) due to its higher mobility, leading to the disordered surface caused by the loss of aluminum atoms.

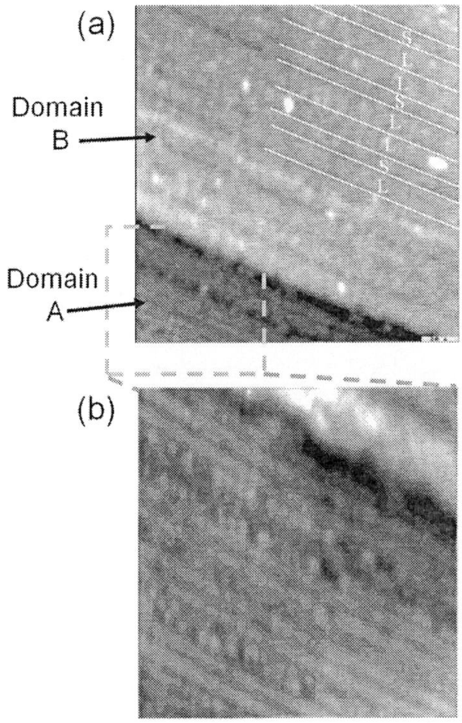

Figure 6 (a) 180 Å x 180 Å atomic scale STM images (Vs =1.2V, I = 0.1nA) of the surface after annealing at 1200K.(b) 90 Å x 90 Å STM image reveals two domains across 2 Å step.

4. Conclusion

The atomic structure of the 2-fold Al-Ni-Co surface shows quasiperiodic ordering (along the 10-fold direction) and periodic ordering with 4 Å periodicity (along the 2-fold direction), consistent with the bulk structure of a decagonal quasicrystal. After in-situ oxidation, the dark depressions are created on the surface, which is presumably attributed to chemisorbed oxygen molecules. We found that the surface exhibited 4 types of steps (X_H, S_H, L_H, S_H+L_H). We found ordering which obeys the Fibonacci sequence is partially disturbed near interlayer phason defects. Above heating temperatures of 1200K, formation of second phase structures were observed.

Acknowledgments

The Author acknowledges the valuable comments from Patricia Thiel and Miquel Salmeron and the support by WCU (31-2008-000-10055-0) program and NRF grant (No. 2010-0005390) through the National Research Foundation of Korea funded by the Ministry of Education, Science and Technology.

REFERENCES

Abe, E., K. Saitoh, et al. (2000). "Quasi-Unit-Cell Model for an Al-Ni-Co Ideal Quasicrystal based on Clusters with Broken Tenfold Symmetry." *Phys. Rev. Lett.* 84(20): 4609-4612.

Baumgarte, A., J. Schreuer, et al. (1997). "X-ray diffraction study of decaprismatic Al-Co-Ni crystals as a function of composition and temperature." *Philosophical Magazine a-Physics of Condensed Matter Structure Defects and Mechanical Properties* 75(6): 1665-1675.

Cox, E. J., J. Ledieu, et al. (2001). "The ten-fold surface of the decagonal Al72Ni11Co17 quasicrystal studied by leed, SPA-LEED, AES and STM." Materials Research Society Symposium Proceedings 643(Quasicrystals--Preparation, Properties and Applications): K11.13.11-K11.13.16.

Dubois, J.-M. "Useful Quasicrystals. World Scientific, New Jersey. 2005." Useful Quasicrystals. *World Scientific,* New Jersey. 2005.

Ferralis, N., K. Pussi, et al. (2004). "Structure of the tenfold d-Al-Ni-Co quasicrystal surface." Physical Review B: *Condensed Matter and Materials Physics* 69(15): 153404/153401-153404/153404.

Fisher, I. R., M. J. Kramer, et al. (1999). "On the growth of decagonal Al-Ni-Co quasicrystals from the ternary melt." *Phil. Mag.* 79(3): 425-434.

Frey, F., E. Weidner, et al. (2002). "Temperature dependence of the 8-angstrom superstructure in decagonal Al-Co-Ni." *Journal of Alloys and Compounds* 342(1-2): 57-64.

Henley, C. L. Quasicrystal: The State of the Art, edited by D. P. DiVincenzo and P. J. Steinhardt (*World Scientific*, Singapore, 1991), p. 111.

Hocker, S. and F. Gahler (2004). "Aluminium diffusion in decagonal quasicrystals." *Physical Review Letters* 93(7).

Janot, C. (1992). Quasicrystals: A Primer. Oxford, Clarendon Press.

Jeong, H. C. and P. J. Steinhardt (1993). "Finite-Temperature Elasticity Phase-Transition in Decagonal Quasi-Crystals." *Physical Review* B 48(13): 9394-9403.

Kishida, M., Y. Kamimura, et al. (2002). "Scanning tunneling microscopy of an Al-Ni-Co decagonal quasicrystal." *Physical Review* B 65(9).

Mader, R., R. Widmer, et al. (2009). "High-resolution scanning tunneling microscopy investigation of the (12110) and (10000) two-fold symmetric d-Al-Ni-Co quasicrystalline surfaces." *Physical Review* B 80(3).

Martin, S., A. F. Hebard, et al. (1991). "Transport-Properties of Al65cu15co20 and Al70ni15co15 Decagonal Quasi-Crystals." *Physical Review Letters* 67(6): 719-722.

Moulopoulos, K. and S. Roche (1996). "Role of phason defects on the conductance of a one-dimensional quasicrystal." *Physical Review* B 53(1): 212-220.

Park, J. Y., D. F. Ogletree, et al. (2004). "Friction and adhesion properties of clean and oxidized Al-Ni-Co decagonal quasicrystals: a UHV atomic force microscopy / scanning tunneling microscopy study." *Tribology Lett.* 17(3): 629-636.

Park, J. Y., D. F. Ogletree, et al. (2005). "Atomic scale coexistence of periodic and quasiperiodic order in a two-fold Al-Ni-Co decagonal quasicrystal surface." *Phys. Rev.* B 72: 220201.

Park, J. Y., D. F. Ogletree, et al. (2005). "Elastic and inelastic deformations of ethylene-passivated tenfold decagonal Al-Ni-Co quasicrystal surfaces." *Physical Review* B 71(14): 144203.

Park, J. Y., D. F. Ogletree, et al. (2005). "High frictional anisotropy of periodic and aperiodic directions on a quasicrystal surface." *Science* 309(5739): 1354-1356.

Park, J. Y., D. F. Ogletree, et al. (2006). "Adhesion properties of decagonal quasicrystals in ultrahigh vacuum." *Philosophical Magazine* 86(6-8): 945-950.

Park, J. Y., G. M. Sacha, et al. (2005). "Sensing dipole fields at atomic steps with combined scanning tunneling and force microscopy." *Physical Review Letters* 95(13): 136802.

Rotenberg, E., W. Theis, et al. (2000). "Quasicrystalline valence bands in decagonal AlNiCo." *Nature* (London) 406(6796): 602-605.

Sato, K., F. Baier, et al. (2004). "Study of an order-disorder phase transition on an atomic scale: The example of decagonal Al-Ni-Co quasicrystals." *Physical Review Letters* 92(12).

Sharma, H. R., K. J. Franke, et al. (2004). "Structure and morphology of the tenfold surface of decagonal Al71.8Ni14.8Co13.4 in its low-temperature random tiling type-I modification." Physical Review B: *Condensed Matter and Materials Physics* 70(23): 235409/235401-235409/235410.

Sharma, H. R., K. J. Franke, et al. (2004). "Investigation of the twofold decagonal Al71.8Ni14.8Co13.4(1 0 0 0 0) surface by SPA-LEED and He diffraction." *Surface Science* 561(2-3): 121-126.

Shechtman, D., I. Blech, et al. (1984). "Metallic Phase with Long-Range Orientational Order and No Translational Symmetry." *Phys. Rev. Lett.* 53: 1951.

Steinhardt, P. J. and H. C. Jeong (1996). "A simpler approach to Penrose tiling with implications for quasicrystal formation." *Nature* 382(6590): 431-433.

Steurer, W. (2004). "Twenty years of structure research on quasicrystals. Part I. Pentagonal, octagonal, decagonal and dodecagonal quasicrystals." *Zeitschrift fuer Kristallographie* 219(7): 391-446.

Theis, W., E. Rotenberg, et al. (2003). "Electronic valence bands in decagonal Al-Ni-Co." *Physical Review* B 68(10): 104205.

Yuhara, J., J. Klikovits, et al. (2004). "Atomic structure of an Al-Co-Ni decagonal quasicrystalline surface." *Physical Review* B: *Condensed Matter and Materials Physics* 70(2): 024203/024201-024203/024207.

Zhang, D. L., S. C. Cao, et al. (1991). "Anisotropic Thermal-Conductivity of the 2d Single Quasi-Crystals - Al65ni20co15 and Al62si3cu20co15." *Physical Review Letters* 66(21): 2778-2781.

In: Quasicrystals: Types, Systems, and Techniques
Editor: B. E. Puckermann, pp. 175-188
ISBN 978-1-61761-123-0

Chapter 8

BOUNDARY CONDITIONS FOR BEAM BENDING IN TWO-DIMENSIONAL QUASICRYSTALS

Yang Gao[1]
College of Science, China Agricultural University, Beijing 100083, People's Republic of China
Institute of Mechanics, University of Kassel, Kassel D-34125, Germany

ABSTRACT

For beam bending in two-dimensional orthorhombic quasicrystals, the reciprocal theorem and the general solution of plane elasticity of quasicrystals are applied in a novel way to obtain the appropriate boundary conditions accurate to all order for the beams of general edge geometry and loadings. By introducing two definitions for the decaying and regular states, the necessary conditions on the edge-data to induce only a decaying elastostatic are directly translated into the appropriate boundary conditions for the existence of a rapidly decaying solution within the beams. When stress and mixed conditions are imposed on the beam edge, these decaying state conditions for the case of bending deformation of quasicrystal beams are derived explicitly for the first time. They are then used for the correct formulation of boundary conditions for the beam theory solution.

1. INTRODUCTION

Since the discovery by Shechtman et al. [1], quasicrystals (QCs)—solids with a long-range orientational order and a long-range quasiperiodic translational order [2]—have become one of topics of intensive theoretical and experimental studies in the physics of condensed matter. The physical properties, such as the structure, electronic, magnetic, optical, thermal and mechanical properties of the material have been investigated intensively [3-6], which show their complex structure and unusual properties. Elasticity is one of important properties of QCs. Within the framework of the Landau–Lifshitz phenomenological theory, the elastic energies of QCs were formulated [7, 8]. In particular, the field of linear elastic theory of QCs

1 gaoyangg@gmail.com.

has been investigated for many years [9-12]. Great progress has been made in the fields of the mechanic involving the elasticity and defects, see review articles [13, 14] for detail.

Under external loads, the exact solution of linear elastostatic problems for slender and thin bodies consists of an interior component significant throughout the bodies and an outer (boundary layer) component in a decaying form. Near a lateral edge, the interior component is supplemented by boundary layer component which becomes insignificant away from the edge. The prescribed admissible boundary conditions can be satisfied only by a combination of these two components. However, the boundary layer solution, even just a leading term approximation, needed to fit the edge-data is rather intractable except for cases with simple geometries and load symmetries. This and the fact that the solution behavior near the edges is often not needed from practical viewpoint have driven people to take efforts over the years to formulate the interior solution, by assigning an appropriate portion of the prescribed edge-data to it, without any reference to the boundary layer solution.

Gregory and Wan [15-17] and Wan [18] developed a decay analysis technique determining the interior solution successfully and effectively, and provided the results for several plate problems. Through generalizing the method, a set of necessary conditions on the edge-data for the existence of a rapidly decaying solution is established, and various extensions have been found among elastic beams [19], one-dimensional (1D) hexagonal QC plates [20] and two-dimensional (2D) dodecagonal QC plates [21].

By generalizing the model and method for elastic beams or plates to QC beams and by invoking the general solution of 2D QCs, for the case of bending deformation, these decaying state conditions are obtained explicitly for the first time, when the stress and mixed edge-data are imposed on the beam edge. Boundary conditions at the end edge of the beams are generally satisfied by a combination of interior and decaying solution components. The results enable us to formulate the correct boundary conditions for QC beam theories with stress and mixed edge-data.

2. Plane Stress State of 2D QCs

For a 2D QCs referred to a Cartesian coordinate system (x, y, z), let x-y plane be the quasi-periodic plane and z be the periodic direction. For a narrow straight beam, the width in the y-direction is stress free. Therefore, it is plausible to set $\sigma_{yx} = \sigma_{yy} = \sigma_{yz} = H_{yx} = H_{yy} = H_{yz} = 0$. This is a plane stress assumption. We assume that the beam length in x-direction is denoted by l, the beam width in y-direction is set to be unit, and the beam height in z-direction is $2h$. The problem of beams may be decomposed into two fundamental problems: the extension of a beam and the bending of a beam. This chapter is concerned about the bending problem of a QC beam with a narrow rectangular cross-section.

For orthorhombic QCs, the point groups $2mm$, 222, mmm and $mm2$ belong to Laue class 4. Following Hu et al. [11], the general equations governing the plane stress state of 2D orthorhombic QCs in the absence of body forces can be written as:

$$\varepsilon_{\alpha\beta} = \frac{1}{2}\left(\partial_\beta u_\alpha + \partial_\alpha u_\beta\right), \qquad w_{x\beta} = \partial_\beta w_x, \tag{1}$$

$$\begin{aligned}&\partial_\beta \sigma_{\alpha\beta}=0,\\&\partial_\beta H_{x\beta}=0,\end{aligned}\tag{2}$$

$$\begin{aligned}\sigma_{xx}&=C_{11}\varepsilon_{xx}+C_{13}\varepsilon_{zz}+R_1 w_{xx},\\\sigma_{zz}&=C_{13}\varepsilon_{xx}+C_{33}\varepsilon_{zz}+R_3 w_{xx},\\\sigma_{zx}&=\sigma_{xz}=2C_{55}\varepsilon_{xz}+R_5 w_{xz},\\H_{xx}&=R_1\varepsilon_{xx}+R_3\varepsilon_{zz}+K_1 w_{xx},\\H_{xz}&=2R_5\varepsilon_{xz}+K_8 w_{xz},\end{aligned}\tag{3}$$

where α, $\beta = x$, z, u_α and w_x denote phonon and phason displacements in the physical and perpendicular spaces, respectively, $\sigma_{\alpha\beta}$ and $\varepsilon_{\alpha\beta}$ phonon stresses and strains, respectively, $H_{x\beta}$ and $w_{x\beta}$ phason stresses and strains, respectively. For plane stress problem, there are 9 independent constants in Eq. (3)

$$\begin{aligned}&C_{11}=\bar{C}_{11}+\bar{C}_{12}E_1+\bar{R}_7F_1,\ C_{13}=\bar{C}_{13}+\bar{C}_{12}E_2+\bar{R}_7F_2,\\&C_{33}=\bar{C}_{33}+\bar{C}_{23}E_2+\bar{R}_9F_2,\ C_{55}=\bar{C}_{55},\\&K_1=\bar{K}_1+\bar{R}_2E_3+\bar{K}_2F_3,\ K_8=\bar{K}_8,\\&R_1=\bar{R}_1+\bar{C}_{12}E_3+\bar{R}_7F_3,\ R_3=\bar{R}_3+\bar{C}_{23}E_3+\bar{R}_9F_3,\ R_5=\bar{R}_5,\\&E_1=\frac{\bar{C}_{12}\bar{K}_6-\bar{R}_7\bar{R}_8}{\bar{R}_8^2-\bar{C}_{22}\bar{K}_6},\ E_2=\frac{\bar{C}_{23}\bar{K}_6-\bar{R}_8\bar{R}_9}{\bar{R}_8^2-\bar{C}_{22}\bar{K}_6},\ E_3=\frac{\bar{R}_2\bar{K}_6-\bar{R}_8\bar{K}_2}{\bar{R}_8^2-\bar{C}_{22}\bar{K}_6},\\&F_1=\frac{\bar{C}_{22}\bar{R}_7-\bar{C}_{12}\bar{R}_8}{\bar{R}_8^2-\bar{C}_{22}\bar{K}_6},\ F_2=\frac{\bar{C}_{22}\bar{R}_9-\bar{C}_{23}\bar{R}_8}{\bar{R}_8^2-\bar{C}_{22}\bar{K}_6},\ F_3=\frac{\bar{C}_{22}\bar{K}_2-\bar{R}_2\bar{R}_8}{\bar{R}_8^2-\bar{C}_{22}\bar{K}_6},\end{aligned}\tag{4}$$

which are expressed by 7 elastic constants $\bar{C}_{ij}$ in phonon field, 4 constants $\bar{K}_i$ in phason field and 7 constants $\bar{R}_i$ in phonon-phason coupling field.

According to the general solution of plane elasticity of 2D QCs [22], the components of displacements take the form:

$$u_x=\delta_{1i}\partial_x\psi_i,\ \ u_z=m_i\partial_z\psi_i,\ \ w_x=l_i\partial_x\psi_i,\tag{5}$$

where $i = 1, 2, 3$, δ_{ij} is the Kronecker delta symbol, and the following summation convention has been used throughout this chapter: the Einstein summation over repeated lower case indices from 1 to 3 is applied, while upper case indices take on the same numbers as the corresponding lower case ones but are not summed. Besides, the potential functions ψ_i satisfy the equations

$$\nabla_I^2\psi_i=\partial_x^2\psi_i+\frac{1}{s_I^2}\partial_z^2\psi_i=0.\tag{6}$$

The values of m_i, l_i and s_i^2 are related by the following expressions:

$$\begin{aligned}&\frac{C_{33}m_i}{C_{13}+C_{55}+C_{55}m_i+(R_3+R_5)l_i}\\&=\frac{C_{55}+(C_{13}+C_{55})m_i+R_5l_i}{C_{11}+R_1l_i}\\&=\frac{R_5+(R_3+R_5)m_i+K_8l_i}{R_1+K_1l_i}=\frac{1}{s_i^2},\end{aligned} \tag{7}$$

where s_i^2 are three characteristic roots (or eigenvalues) of the following cubic algebra equation of s^2

$$as^6-bs^4+cs^2-d=0. \tag{8}$$

The constants in the preceding equations are

$$\begin{aligned}a&=C_{33}\left(C_{55}K_8-R_5^2\right),\\b&=C_{55}\left(C_{55}K_8-R_5^2\right)+C_{33}\left(C_{11}K_8+C_{55}K_1-2R_1R_5\right)\\&\quad+2R_5\left(C_{13}+C_{55}\right)\left(R_3+R_5\right)\\&\quad-K_8\left(C_{13}+C_{55}\right)^2-C_{55}\left(R_3+R_5\right)^2,\\c&=C_{33}\left(C_{11}K_1-R_1^2\right)+C_{55}\left(C_{11}K_8+C_{55}K_1-2R_1R_5\right)\\&\quad+2R_1\left(C_{13}+C_{55}\right)\left(R_3+R_5\right)\\&\quad-K_1\left(C_{13}+C_{55}\right)^2-C_{11}\left(R_3+R_5\right)^2,\\d&=C_{55}\left(C_{11}K_1-R_1^2\right).\end{aligned} \tag{9}$$

The appropriate boundary conditions for the beams will be given to the case of distinct eigenvalues s_i^2 in the following context, and the other boundary conditions for the case of equal eigenvalues can be obtained in a similar manner.

3. Necessary Conditions for a Decaying State

The top and bottom faces of the beams are taken to be traction free, so that

$$\sigma_{xz}=\sigma_{zz}=0,\ \ H_{xz}=0\ \ (z=\pm h). \tag{10}$$

The presence of any body or surface loads may be removed by a particular solution. On the curved edge of the beams, one of the following sets of edge data is prescribed

Case A:

$$\sigma_{xx}(0,z)=\bar{\sigma}_{xx}(z),\ \ \sigma_{xz}(0,z)=\bar{\sigma}_{xz}(z),\ \ H_{xx}(0,z)=\bar{H}_{xx}(z), \tag{11}$$

Case B:

$$u_x(0,z)=\bar{u}_x(z),\ \ \sigma_{xz}(0,z)=\bar{\sigma}_{xz}(z),\ \ w_x(0,z)=\bar{w}_x(z), \tag{12}$$

Case C:

$$\sigma_{xx}(0,z)=\bar{\sigma}_{xx}(z),\ \ u_z(0,z)=\bar{u}_z(z),\ \ H_{xx}(0,z)=\bar{H}_{xx}(z), \tag{13}$$

Case D:

$$u_x(0,z)=\bar{u}_x(z),\ \ u_z(0,z)=\bar{u}_z(z),\ \ w_x(0,z)=\bar{w}_x(z), \tag{14}$$

Case E:

$$\sigma_{xx}(0,z)=\bar{\sigma}_{xx}(z),\ \ \sigma_{xz}(0,z)=\bar{\sigma}_{xz}(z),\ \ w_x(0,z)=\bar{w}_x(z), \tag{15}$$

Case F:

$$u_x(0,z)=\bar{u}_x(z),\ \ \sigma_{xz}(0,z)=\bar{\sigma}_{xz}(z),\ \ H_{xx}(0,z)=\bar{H}_{xx}(z), \tag{16}$$

Case G:

$$\sigma_{xx}(0,z)=\bar{\sigma}_{xx}(z),\ \ u_z(0,z)=\bar{u}_z(z),\ \ w_x(0,z)=\bar{w}_x(z), \tag{17}$$

Case H:

$$u_x(0,z)=\bar{u}_x(z),\ \ u_z(0,z)=\bar{u}_z(z),\ \ H_{xx}(0,z)=\bar{H}_{xx}(z). \tag{18}$$

In generalization of analogous statements for elastic beams [19], two classes of exact states are investigated for the equations governing QC beams with free faces. One of these is designated as the interior state significant throughout the beams. The other complementary class corresponds to a boundary layer solution and is designated as the decaying state. An elastostatic state in the beams is said to be a *regular state*

$$\{u_\alpha,\sigma_{\alpha\beta},w_x,H_{x\beta}\}=O\left(M_1h^\lambda\right)\ \text{ as }\ h\to 0, \tag{19}$$

or a decaying state

$$\{u_{\alpha}, \sigma_{\alpha\beta}, w_x, H_{x\beta}\} = O\left(M_2 e^{-\gamma d/h}\right) \text{ as } h \to 0, \tag{20}$$

where M_1 and M_2 are the maximum modulus for the regular state and decaying state, respectively, d is the minimum distance of the observation point from the edge of the beams, and M_1, M_2, λ and γ are positive constants.

Supposing that the edge-data do give rise to the decaying state in the beams, we now apply the Betti-Rayleigh reciprocal theorem for QC media, which takes the form

$$\iint_S \left(\sigma_{\alpha\beta}^{(1)} u_{\beta}^{(2)} + H_{x\alpha}^{(1)} w_x^{(2)} - \sigma_{\alpha\beta}^{(2)} u_{\beta}^{(1)} - H_{x\alpha}^{(2)} w_x^{(1)}\right) n_{\alpha} \mathrm{d}S = 0, \tag{21}$$

where S is the surface of the beams which consists of two end planes and a lateral surface, n_{α} is the direction cosine of the outward normal to S. With the foregoing two definitions of elastostatic states in mind, now we take the state with a superscript "(1)" to be the exact solution of the beams, and the decaying state induced by the prescribed edge-data $\bar{\sigma}_{xx}$, $\bar{\sigma}_{xz}$, $\bar{H}_{xx}$, $\bar{u}_x$, $\bar{u}_z$ and $\bar{w}_x$. For the auxiliary state, denoted by superscript "(2)", we take any regular state which fulfills load-free conditions on S. Similar to the derivation of necessary conditions for a decaying state in QC plates [20, 21], generalizing Gregory and Wan's decay analysis technique to the QC beams, we finally obtain the necessary conditions for a decaying state on the end $x = 0$,

Case A:

$$\int_{-h}^{h} \left(\bar{\sigma}_{xx} u_x^{(2)} + \bar{\sigma}_{xz} u_z^{(2)} + \bar{H}_{xx} w_x^{(2)}\right) \mathrm{d}z = 0, \tag{22}$$

Case B:

$$\int_{-h}^{h} \left(\bar{u}_x \sigma_{xx}^{(2)} - \bar{\sigma}_{xz} u_z^{(2)} + \bar{w}_x H_{xx}^{(2)}\right) \mathrm{d}z = 0, \tag{23}$$

Case C:

$$\int_{-h}^{h} \left(\bar{\sigma}_{xx} u_x^{(2)} - \bar{u}_z \sigma_{xz}^{(2)} + \bar{H}_{xx} w_x^{(2)}\right) \mathrm{d}z = 0, \tag{24}$$

Case D:

$$\int_{-h}^{h} \left(\bar{u}_x \sigma_{xx}^{(2)} + \bar{u}_z \sigma_{xz}^{(2)} + \bar{w}_x H_{xx}^{(2)}\right) \mathrm{d}z = 0, \tag{25}$$

Case E:

$$\int_{-h}^{h} \left(\bar{\sigma}_{xx} u_x^{(2)} + \bar{\sigma}_{xz} u_z^{(2)} - \bar{w}_x H_{xx}^{(2)}\right) \mathrm{d}z = 0, \tag{26}$$

Case F:

$$\int_{-h}^{h}\left(\bar{u}_x\sigma_{xx}^{(2)} - \bar{\sigma}_{xz}u_z^{(2)} - \bar{H}_{xx}w_x^{(2)}\right)\mathrm{d}z = 0, \tag{27}$$

Case G:

$$\int_{-h}^{h}\left(\bar{\sigma}_{xx}u_x^{(2)} - \bar{u}_z\sigma_{xz}^{(2)} - \bar{w}_x H_{xx}^{(2)}\right)\mathrm{d}z = 0, \tag{28}$$

Case H:

$$\int_{-h}^{h}\left(\bar{u}_x\sigma_{xx}^{(2)} + \bar{u}_z\sigma_{xz}^{(2)} - \bar{H}_{xx}w_x^{(2)}\right)\mathrm{d}z = 0. \tag{29}$$

These necessary conditions (22)-(29) for the edge-data to induce only a decaying elastostatic state will be translated into the appropriate boundary conditions for the beam later in next section. The main difficulty in performing the preceding process lies in obtaining suitable regular states which fulfill load-free conditions on the surface of the beams.

4. THE AUXILIARY REGULAR STATES

Once a suitable regular state is constructed for the relevant edge-data, the translation is immediate. However, this is not the situation for general edge-data. Now our main task lies in obtaining accurate solutions for these regular states.

We can take a rigid body translation in the z-direction as the state 1, i.e.

$$u_x^{(2)} = 0,\ \ u_z^{(2)} = C,\ \ w_x^{(2)} = 0,\ \ \sigma_{\alpha\beta}^{(2)} = H_{x\beta}^{(2)} = 0, \tag{30}$$

where C is a constant.

The state 2 may be taken as a rigid body rotation,

$$\begin{aligned} &u_x^{(2)} = Cz,\ \ u_z^{(2)} = -Cx,\ \ w_x^{(2)} = 0,\\ &\sigma_{\alpha\beta}^{(2)} = H_{x\beta}^{(2)} = 0. \end{aligned} \tag{31}$$

Now, we look for the state 3 with the use of the general solution (5). The potential functions ψ_i are taken as

$$\psi_i = A_i xz, \tag{32}$$

where A_i are unknown constants to be determined later. After taking account of the expressions (7), the displacements and stresses obtained from Eqs. (3), (5) and (32) can be shown to be

$$u_x^{(2)} = \delta_{Ii} A_i z, \quad u_z^{(2)} = m_i A_i x, \quad w_x^{(2)} = l_i A_i z,$$
$$\sigma_{xx}^{(2)} = \sigma_{zz}^{(2)} = H_{xx}^{(2)} = 0, \quad \sigma_{xz}^{(2)} = \sigma_{zx}^{(2)} = \alpha_i A_i,$$
$$H_{xz}^{(2)} = \beta_i A_i, \tag{33}$$

where

$$\alpha_i = C_{55}\delta_{Ii} + C_{55} m_i + R_5 l_i, \quad \beta_i = R_5 \delta_{Ii} + R_5 m_i + K_8 l_i. \tag{34}$$

To obtain the state 4, according to the characteristics of bending deformation, by using Eq. (6) we assume

$$\psi_i = B_i\left(s_I^2 z^3 - 3x^2 z\right), \tag{35}$$

where B_i are unknown constants to be determined later. In virtue of the expressions (7), the displacements and stresses in Eqs. (3), (5) and (35) can be written as

$$u_x^{(2)} = -6\delta_{Ii} B_i xz, \quad u_z^{(2)} = 3m_i B_i\left(s_I^2 z^2 - x^2\right), \quad w_x^{(2)} = -6 l_i B_i xz,$$
$$\sigma_{xx}^{(2)} = -6\alpha_i s_I^2 B_i z, \quad \sigma_{zz}^{(2)} = 6\alpha_i B_i z, \quad \sigma_{xz}^{(2)} = \sigma_{zx}^{(2)} = -6\alpha_i B_i x, \tag{36}$$
$$H_{xx}^{(2)} = -6\beta_i s_I^2 B_i z, \quad H_{xz}^{(2)} = -6\beta_i B_i x,$$

To obtain the state 5, we take the potential functions ψ_i to be of the form

$$\psi_i = C_i\left(s_I^2 xz^3 - x^3 z\right) + D_i xz, \tag{37}$$

where C_i and D_i are unknown constants yet to be determined. In terms of the expressions (7), the displacements and stresses in Eqs. (3), (5) and (37) have the following form as

$$u_x^{(2)} = s_i^2 C_i z^3 - 3\delta_{Ii} C_i x^2 z + \delta_{Ii} D_i z,$$
$$u_z^{(2)} = m_i\left[C_i\left(3s_I^2 xz^2 - x^3\right) + D_i x\right],$$
$$w_x^{(2)} = l_i s_I^2 C_i z^3 - 3 l_i C_i x^2 z + l_i D_i z,$$
$$\sigma_{xx}^{(2)} = -6\alpha_i s_I^2 C_i xz, \tag{38}$$
$$\sigma_{zz}^{(2)} = 6\alpha_i C_i xz,$$
$$\sigma_{xz}^{(2)} = \sigma_{zx}^{(2)} = 3\alpha_i C_i\left(s_I^2 z^2 - x^2\right) + \alpha_i D_i,$$
$$H_{xx}^{(2)} = -6\beta_i s_I^2 C_i xz,$$
$$H_{xz}^{(2)} = 3\beta_i C_i\left(s_I^2 z^2 - x^2\right) + \beta_i D_i.$$

5. THE APPROPRIATE BOUNDARY CONDITIONS FOR THE DECAYING STATE

For the case of bending deformation of the QC beams, the appropriate boundary conditions can be explicitly determined as follows, at least for the edge-data in Cases A-C and E-G.

5.1. Case A

As the procedure in the preceding section indicates, any candidate for regular states must meet load-free conditions (10) and the requirements stipulated below

$$\sigma_{xz}=0,\quad \sigma_{xx}=0,\quad H_{xx}=0\quad (x=0). \tag{39}$$

Obviously, the state 1 satisfies the conditions (10) and (39), so the corresponding necessary condition are obtained from Eq. (22)

$$\int_{-h}^{h}\bar{\sigma}_{xz}\mathrm{d}z=0. \tag{40}$$

The second auxiliary regular state may be takes as the state 2, then the corresponding necessary condition is

$$\int_{-h}^{h}\bar{\sigma}_{xx}z\mathrm{d}z=0. \tag{41}$$

Selecting one state from the state 3 or 5 as the third auxiliary regular state, we obtain the third necessary condition for a decaying state when $\bar{H}_{xx}$ is prescribed

$$\int_{-h}^{h}\bar{H}_{xx}z\mathrm{d}z=0. \tag{42}$$

For 2D QC materials, beam theories are not previously known in the literature, but they can be obtained from elasticity theory of QCs by generalizing assumption technique of Euler-Bernoulli or Timoshenko to QC beams. Accordingly, the conventional stress boundary conditions of QC beam theories, similar to those of elastic beam theories [23], can be also obtained and consist of Eqs. (40)-(42), although they are formulated explicitly by an application of the reciprocal theorem and the general solution of 2D QCs.

5.3. Case B

Regular states must meet the conditions (10) and the requirements

$$\sigma_{xz}=0,\ \ u_x=0,\ \ w_x=0\ \ (x=0). \tag{43}$$

As in Case A, selecting a rigid body translations in the state 1 as the first auxiliary regular state, we certainly must have the corresponding necessary conditions (40).

We take the state 4 as the second auxiliary regular state. On substituting Eq. (36) into Eqs. (10) and (43), we obtain

$$\alpha_i B_i=0,\ \ \beta_i B_i=0. \tag{44}$$

From which we can determine the relationship among these unknown constants as

$$\frac{B_1}{\lambda_1}=\frac{B_2}{\lambda_2}=\frac{B_3}{\lambda_3},\ \ \lambda_i=e_{ijk}\alpha_j\beta_k, \tag{45}$$

where e_{ijk} is Levi-Civita permutation symbol with indices varying from 1 to 3. Inserting this auxiliary regular state (36) into Eq. (23), after taking account of the relationship (45), we obtain the second necessary condition for a decaying state when $\bar{u}_x$, $\bar{\sigma}_{xz}$ and $\bar{w}_x$ are prescribed

$$\int_{-h}^{h}\left(\bar{u}_x z+\frac{m_i s_I^2\lambda_i}{2\alpha_i s_I^2\lambda_i}\bar{\sigma}_{xz}z^2+\frac{\beta_i s_I^2\lambda_i}{\alpha_i s_I^2\lambda_i}\bar{w}_x z\right)\mathrm{d}z=0. \tag{46}$$

5.3. Case C

Consider the conditions on the end $x=0$,

$$\sigma_{xx}=0,\ \ u_z=0,\ \ H_{xx}=0\ \ (x=0). \tag{47}$$

In this case, the states 2 and 3 are chosen as the auxiliary regular states, then the first two necessary conditions are the conditions (41) and (42).

The third auxiliary regular state may be taken as the state 5. Substitution Eq. (38) into Eqs. (10) and (47) leads to

$$\begin{aligned}&\alpha_i C_i=0,\ \ 3\alpha_i s_I^2 C_i h^2+\alpha_i D_i=0,\\ &\beta_i C_i=0,\ \ 3\beta_i s_I^2 C_i h^2+\beta_i D_i=0.\end{aligned} \tag{48}$$

From which we have the relationship among these unknown constants

$$\frac{C_1}{\lambda_1}=\frac{C_2}{\lambda_2}=\frac{C_3}{\lambda_3},\ \ \alpha_i D_i=-3\alpha_i s_I^2 C_i h^2,\ \ \beta_i D_i=-3\beta_i s_I^2 C_i h^2. \tag{49}$$

On substituting Eq. (38) into Eq. (24), by using Eqs. (41), (42) and (49) we obtain,

$$\int_{-h}^{h}\left[\bar{u}_z\left(h^2-z^2\right)+\frac{s_i^2\lambda_i}{3\alpha_i s_I^2\lambda_i}\bar{\sigma}_{xx}z^3+\frac{l_i s_I^2\lambda_i}{3\alpha_i s_I^2\lambda_i}\bar{H}_{xx}z^3\right]\mathrm{d}z=0. \tag{50}$$

5.4. Case E

Regular states must meet the requirements on $x = 0$,

$$\sigma_{xz}=0,\ \ \sigma_{xx}=0,\ \ w_x=0\ \ (x=0). \tag{51}$$

If the states 1 and 2 are chosen as the auxiliary regular states, then the necessary conditions have the form as the conditions (40) and (41), respectively.

5.5. Case F

The conditions on $x = 0$ give

$$\sigma_{xz}=0,\ \ u_x=0,\ \ H_{xx}=0\ \ (x=0). \tag{52}$$

If the state 1 is chosen as the auxiliary regular state, then the necessary condition takes the form as the condition (40).

5.6. Case G

By noting that

$$\sigma_{xx}=0,\ \ u_z=0,\ \ w_x=0\ \ (x=0). \tag{53}$$

Once one state is chosen from the state 2, the necessary condition is the condition (41).

Up to here, attempts to derive similar results on boundary conditions for Cases D, H have not been successful, since we have not found any simple regular states suitable for these cases. This lack of success may be related to the fact that no suitable regular states needed for the application of the reciprocal theorem could be found for these cases, thus it is not likely that the desired results are forthcoming. More importantly, the appropriate boundary conditions for six sets of mixed edge-data are obtained for the first time.

6. The Appropriate Boundary Conditions for the Interior State

For each type of edge-data of bending deformation, these aforementioned necessary conditions for a decaying state (boundary layer solution) can then be converted into a set of boundary conditions appropriate for the interior solution or its various approximate beam theories, which do not involve the boundary layer solution components. As the preceding discussion in Introduction, the difference between the exact solution and the interior one is a decaying state. As an immediate consequence, $\bar{u}_\alpha$, $\bar{w}_x$ $\bar{\sigma}_{x\beta}$ and $\bar{H}_{xx}$ must satisfy the conditions

$$\begin{aligned}&\bar{u}_\alpha=\left[u_\alpha-u_\alpha^I\right]_{x=0},\quad \bar{w}_x=\left[w_x-w_x^I\right]_{x=0},\\&\bar{\sigma}_{x\beta}=\left[\sigma_{x\beta}-\sigma_{x\beta}^I\right]_{x=0},\quad \bar{H}_{xx}=\left[H_{xx}-H_{xx}^I\right]_{x=0},\end{aligned}\tag{54}$$

where u_α^I, w_x^I $\sigma_{x\beta}^I$ and H_{xx}^I are interior solutions. Therefore, the above necessary conditions apply to this difference evaluated at an edge of the beams. For the stress and mixed edge-data to induce only the interior state, the data must satisfy these conditions

$$\int_{-h}^{h}\hat{\sigma}_{xz}\mathrm{d}z=\int_{-h}^{h}\left[\sigma_{xz}^I\right]_{x=0}\mathrm{d}z,\tag{55}$$

$$\int_{-h}^{h}\hat{\sigma}_{xx}z\mathrm{d}z=\int_{-h}^{h}\left[\sigma_{xx}^I z\right]_{x=0}\mathrm{d}z,\tag{56}$$

$$\int_{-h}^{h}\hat{H}_{xx}z\mathrm{d}z=\int_{-h}^{h}\left[H_{xx}^I z\right]_{x=0}\mathrm{d}z,\tag{57}$$

$$\begin{aligned}&\int_{-h}^{h}\left[\hat{u}_x z+\frac{m_i s_I^2\lambda_i}{2\alpha_i s_I^2\lambda_i}\hat{\sigma}_{xz}z^2+\frac{\beta_i s_I^2\lambda_i}{\alpha_i s_I^2\lambda_i}\hat{w}_x z\right]\mathrm{d}z\\&=\int_{-h}^{h}\left[u_x^I z+\frac{m_i s_I^2\lambda_i}{2\alpha_i s_I^2\lambda_i}\sigma_{xz}^I z^2+\frac{\beta_i s_I^2\lambda_i}{\alpha_i s_I^2\lambda_i}w_x^I z\right]_{x=0}\mathrm{d}z.\end{aligned}\tag{58}$$

$$\begin{aligned}&\int_{-h}^{h}\left[\hat{u}_z\left(h^2-z^2\right)+\frac{s_i^2\lambda_i}{3\alpha_i s_I^2\lambda_i}\hat{\sigma}_{xx}z^3+\frac{l_i s_I^2\lambda_i}{3\alpha_i s_I^2\lambda_i}\hat{H}_{xx}z^3\right]\mathrm{d}z\\&=\int_{-h}^{h}\left[u_z^I\left(h^2-z^2\right)+\frac{s_i^2\lambda_i}{3\alpha_i s_I^2\lambda_i}\sigma_{xx}^I z^3+\frac{l_i s_I^2\lambda_i}{3\alpha_i s_I^2\lambda_i}H_{xx}^I z^3\right]_{x=0}\mathrm{d}z.\end{aligned}\tag{59}$$

where $\hat{u}_\alpha$, $\hat{w}_x$ $\hat{\sigma}_{x\beta}$ and $\hat{H}_{xx}$ are the actually prescribed edge-data. In the form, these conditions also conveniently provide the appropriate boundary conditions for slender or deep beam theories. To obtain the appropriate boundary conditions for a particular beam theory, we should expand in powers of h all terms in all necessary conditions and retain only a suitable number of terms in each expansion. The above results for transverse bending and in-plane extension illustrate the general method for deriving local necessary conditions for a decaying state and therewith appropriate boundary conditions for beam theories.

7. CONCLUSION

In this chapter we extend the model and method for elastic and QC plates to 2D QC beams in bending deformation, which enables us to formulate the correct boundary conditions of the QC beams with the mixed edge-data for the first time. However, attempts to derive the corresponding boundary conditions for displacement and other types of edge-data have not been successful. We have not found any simple auxiliary regular states suitable for these edge-data, but this does not mean that our approach is useless in these cases. It means that the required auxiliary states are themselves the solutions of certain particular boundary value problems, which, when solved once and for all, are to be used in the appropriate decaying state conditions.

ACKNOWLEDGMENTS

The work is supported by the National Natural Science Foundation of China (No. 10702077) and the Alexander von Humboldt Foundation in Germany.

REFERENCES

[1] Shechtman, D.; Blech, I.; Gratias, D.; Cahn, J. W. Metallic phase with long-range orientational order and no translational symmetry. *Phys. Rev. Lett.* 1984, 53, 1951–1953.

[2] Levine, D.; Steinhardt, P. J. Quasi-crystals: A new class of ordered structure. *Phys. Rev. Lett.* 1984, 53, 2477–2450.

[3] Wollgarten, M.; Beyss, M.; Urban, K.; Liebertz, H.; Koster, U. Direct evidence for plastic deformation of quasicrystals by means of a dislocation mechanism. *Phys. Rev. Lett.* 1993, 71, 549–552.

[4] Athanasiou, N. S.; Politis, C.; Spirlet, J. C.; Baskoutas, S.; Kapaklis, V. The significance of valence electron concentration on the formation mechanism of some ternary aluminum-based quasicrystals. *Int. J. Mod. Phys*. B. 2002, 16, 4665–4683.

[5] Park, J. Y.; Ogletree, D. F.; Salmeron, M.; Ribeiro, R. A.; Canfield, P. C.; Jenks, C. J.; Thiel, P. A.; High frictional anisotropy of periodic and aperiodic directions on a quasicrystal surface. *Science.* 2005, 309, 1354–1356.

[6] Park, J. Y.; Sacha, G. M.; Enachescu, M.; Ogletree, D. F.; Ribeiro, R. A.; Canfield, P. C.; Jenks, C. J.; Thiel, P. A.; Saenz, J. J.; Salmeron, M. Sensing dipole fields at atomic steps with combined scanning tunneling and force microscopy. *Phys. Rev. Lett.* 2005, 95, 136802.

[7] Bak, P.. Phenomenological theory of icosahedron incommensurate (quasiperiodic) order in Mn-Al alloys. *Phys. Rev. Lett.* 1985, 54, 1517–1519.
[8] Levine, D.; Steinhardt, P. J. Quasicrystals. I. definition and structure. *Phys. Rev.* B. 1986, 34, 596–616.
[9] Ding, D. H.; Yang, W. G.; Hu, C. Z.; Wang, R. H. Generalized elasticity theory of quasicrystals. *Phys. Rev.* B. 1993, 48, 7003–7010.
[10] Yang, W. G.; Wang, R. H.; Ding, D. H.; Hu, C. Z. Linear elasticity theory of cubic quasicrystals. *Phys. Rev.* B. 1993, 48, 6999–7002.
[11] Hu, C. Z.; Yang, W. G.; Wang, R. H.; Ding, D. H. Point groups and elastic properties of two-dimensional quasicrystals. *Acta Crystallogr*. 1996, 52, 251–256.
[12] Wang, R. H.; Yang, W. G.; Hu, C. Z.; Ding, D. H. Point and space groups and elastic behaviours of one-dimensional quasi-crystals. *J. Phys: Condens Matter*. 1997, 9, 2411–2422.
[13] Hu, C. Z.; Wang, R. H.; Ding, D. H. Symmetry groups, physical property tensors, elasticity and dislocations in quasicrystals. *Rep. Prog. Phys.* 2000, 63, 1–39.
[14] Fan, T. Y.; Mai Y. W. Elasticity theory, fracture mechanics, and some relevant thermal properties of quasi-crystalline materials. *Appl. Mech. Rev.* 2004, 57, 325–343.
[15] Gregory, R. D.; Wan, F. Y. M. Decaying states of plane strain in a semi-infinite strip and boundary conditions for plate theory. *J. Elast.* 1984, 14, 27–64.
[16] Gregory, R. D.; Wan, F. Y. M. On plate theories and Saint-Venant's principle. *Int. J. Solids Struct.* 1985, 21, 1005–1024.
[17] Gregory, R. D.; Wan, F. Y. M. On the interior solution for linearly elastic plate. ASME *J. Appl. Mech.* 1988, 55, 551-559.
[18] Wan, F. Y. M. Stress boundary conditions for plate bending. *Int. J. Solids Struct.* 2003, 40, 4107–4123.
[19] Gao, Y.; Xu, S. P.; Zhao, B. S. Boundary conditions for elastic beam bending. *C R Mecanique*. 2007, 335, 1–6.
[20] Gao, Y.; Xu, S. P.; Zhao, B. S. Boundary conditions for plate bending in one-dimensional hexagonal quasicrystals. *J. Elast.* 2007, **86, 221–233.**
[21] Gao, Y.; Xu, S. P.; Zhao, B. S. Stress and mixed boundary conditions for two-dimensional dodecagonal quasi-crystal plates. *Pramana-J. Phys.* 2007, 68, 803–817.
[22] Gao, Y. General solutions of plane elasticity of two-dimensional quasicrystals with crystal rotational symmetry. *Arch. Ration. Mech. Anal.* (submitted).
[23] Timoshenko, S. P.; Goodier, J.C. Theory of Elasticity. McGraw-Hill: New York, 1970.

In: Quasicrystals: Types, Systems, and Techniques
Editor: B. E. Puckermann, pp. 189-193
ISBN 978-1-61761-123-0

Chapter 9

MICROSTRUCTURAL STUDIES ON PLATE SHEETS OF AL-LI-CU-MG ALLOY REINFORCED WITH SICP METAL MATRIX COMPOSITES

***A. K. Srivastava*[1*1] *and Asim Bag*[2]**

[1]Electron Microscopy, Division of Materials Characterization
National Physical Laboratory, C.S.I.R., Dr. K.S. New Delhi, India
[2]Det Norske Veritas Pte Ltd, 10 Science Park Drive
DNV Technology Centre, Singapore 118224

ABSTRACT

The microstructural characteristics of a commercial quaternary AA8090 (Al-2%Li-1.2%Cu-0.8%Mg, by wt.%) alloy reinforced with 15 vol.% SiCp has been examined in detail. The composite material in the form of plate sheets with the thickness about 1600 μm was thinned to electron beam transparent (~ 20 nm thickness) using mechanical polishing and ion beam milling to carry out microscopy observations. In the alloy matrix (α - Al) the presence of δ'-precipitates ($L1_2$ structure, lattice parameter a = 0.401 nm) as tiny spheres of about 50 – 100 nm in size has been delineated. The presence of icosahedral quasicrystalline phase has also been observed in the matrix. In general, a lamellae structure of δ'-precipitate with the layer thickness of about 250 nm has been revealed on the grain boundaries. Adjacent to δ'-precipitate, a prominent region of precipitate free zones with a thickness between 65 – 85 nm is present at the boundaries. The distribution of SiCp in α-Al matrix is uniform with a clear interface exhibiting some dislocations.

PACS: *61.44.Br; 61.66.Dk; 68.37.-d.*

---

1 *Corresponding author. Address: Scientist, Electron Microscopy, Division of Materials Characterization, National Physical Laboratory, Council of Scientific and Industrial Research, Dr. K. S. Krishnan Road, New Delhi, 110 012, INDIA.Tel.: + 91 - 11 – 45609308; Fax: + 91-11-45609310; E-mail address: aks@nplindia.ernet.in (A. K. Srivastava).

Keywords: *Al-Li alloy; Quasicrystalline phase, Microstructure; Grain boundary.*

1. INTRODUCTION

Al-Li alloys have gained considerable interest in many industrial applications due to their low density, high strength and high modulus. Each wt.% addition of Li to Al, reduces the density by about 3% and enhances the elastic modulus by about 6% [1,2]. Further the addition of SiC particulates (SiCp) in such alloys enhances the properties of the composites in many fold [3,4]. These light weight and high strength reinforced SiCp, Al-Li-Cu-Mg – based metal matrix composites (MMCs) can be tailored to have optimized thermal and physical properties to meet the requirements of electronic packaging systems, for example cores, substrates, carriers and housings [4]. Many companies are dedicated in producing electronic grade components employing SiC reinforced aluminum. However a good microstructural homogeneity of various phases of the alloy and a fine and uniform distribution of SiC partculates in the alloy matrix is always a prerequisite to obtain the desired properties. Therefore the present investigations are devoted on a detailed investigations to understand the microstructural features of a commercial quaternary AA8090 (Al-2%Li-1.2%Cu-0.8%Mg, by wt.%) alloy reinforced with 15 vol.% SiCp.

2. EXPERIMENTAL

A commercial AA8090 (Al-2%Li-1.2%Cu-0.8%Mg, by wt.%) alloy reinforced with 15 vol.% SiCp, powder metallurgy processed, was available in the form of plate sheets with the thickness ~ 1600 μm. The material was thinned to electron beam transparent (~ 20 nm thickness) using tripode mechanical polishing and ion beam milling [5]. A transmission electron microscope (TEM, model Akashi EM-002B) was operated at the electron accelerating voltage of 200 kV to study the microstructural features constituting the quaternary alloy reinforced SiCp metal matrix composites.

3. RESULTS AND DISCUSSION

Microstructural investigations carried out on the powder metallurgy processed commercial AA8090 (Al-2%Li-1.2%Cu-0.8%Mg, by wt.%) alloy reinforced with 15 vol.% SiCp metal matrix composites has delineated many interesting features at micro - and nano – scale. Figure 1 shows a SiC particulate with a sharp tip embedded in the matrix. The boundary between interface at the SiCp and the matrix appeared clean except the presence of certain defects in the form of dislocations. The microstructure of these dislocations indicated that they nucleated at the interface and grown in the soft matrix of α-Al to a maximum vicinity of about 400 nm. The origin of such defects at the soft and ductile matrix (α-Al) and the hard particulate (SiCp) has been attributed to the difference in coefficient of thermal expansion and crystallographic geometrical constraints between the matrix (α-Al) and the reinforced (SiCp) phase. There is no porosity or any type of decohesion has been observed at the interface even at nano-scale.

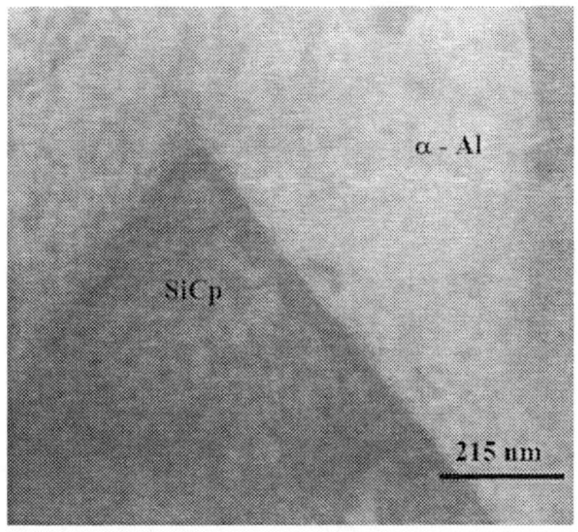

Figure 1. TEM bright micrograph showing a particulate of SiC embedded in α-Al.

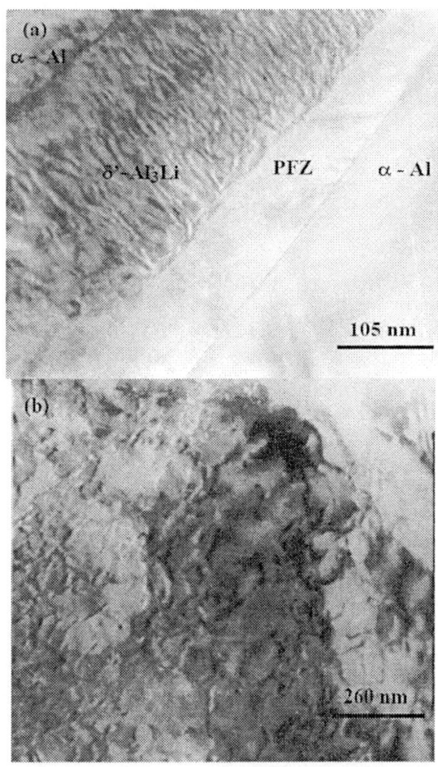

Figure 2. TEM bright field micrographs sowing (a) a layer of cellular structured δ'-Al3Li and precipitate free zone (PFZ) at the boundary of α-Al and (b) distribution spherical shaped δ'-Al3Li.

The investigations on matrix microstructure has revealed different morphologies of δ'-Al_3Li distributed in the matrix. At the grain boundaries, a fine distribution of lamellae / cellular structure of δ'- precipitates at the grain boundaries are diffracting strongly and forming a kind of layer (Figure 2 (a)). The layer thickness constituted of δ'- precipitates is about 250 nm. Occasionally the tiny spherical particles in addition to the lamellae structure of δ'- precipitates are also observed at the grain boundaries. A precipitate free zone (PFZ) appeared adjacent to the δ'- precipitates at the boundary. The thickness of PFZ varied between 65 – 85 nm. Such a significant feature of PFZ is stimulating in context of grain boundary migration, while external load applied during mechanical performance of these materials. Figure 2 (b) exhibits the presence of a spherical - shaped ultrafine precipitates of δ' with an average size of about (~ 50 – 100 nm) normally distributed in the matrix.

Selected area electron diffraction patterns have indicated a definite orientation relationship between the matrix (α-Al) and precipitate (δ'-Al_3Li). It is important to mention that both the Al (fcc, a = 0.404 nm) and δ' (ordered bcc, a = 0.401 nm) are cubic. A composite electron diffraction pattern from Al and δ' precipitate along [001] zone axis of a cubic crystal structure has been displayed in Figure 3. The crystallographic planes of δ' (010 and 100) and Al (020 and 200) are marked on the electron diffraction pattern.

A detailed study of the alloy matrix has further shown the presence of an icosahedral quasicrystalline (IQC) phase (Figure 4). These crystals are normally in spherical shape with an average size of about 600 nm. An inset in Figure4 is an electron diffraction pattern recorded along 2-fold orientation of an icosahedral quasicrystal.

CONCLUSION

SiC particulates (SiCp) reinforced Al-Li-Cu-Mg based metal matrix composites are important materials for electronic packaging systems. Electron microscopy evidences has been elucidated and discussed to interpret the evolution of different structures and microstructures while processing of these composites by a powder metallurgy route. A uniform distribution of SiCp has been noticed throughout in the alloy matrix without any porosity even at nano-scale. The orientation relationship between the δ'-Al_3Li and α-Al has been delineated. Such microstructural features are suitable for any future applications of these composite materials.

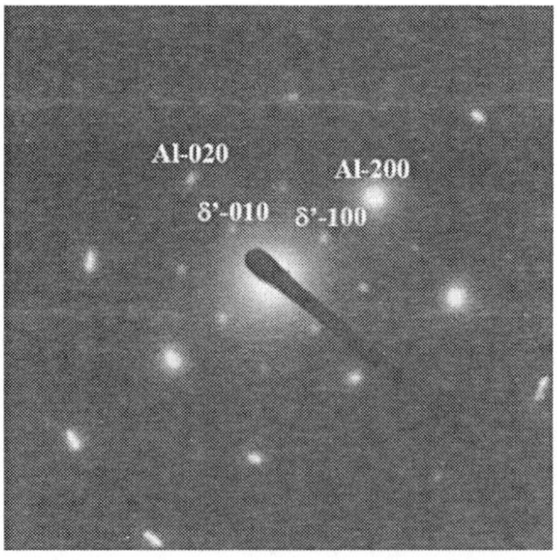

Figure 3. A composite selected area electron diffraction pattern from the δ'-Al3Li and α-Al along [001] zone axis of cubic crystal structure.

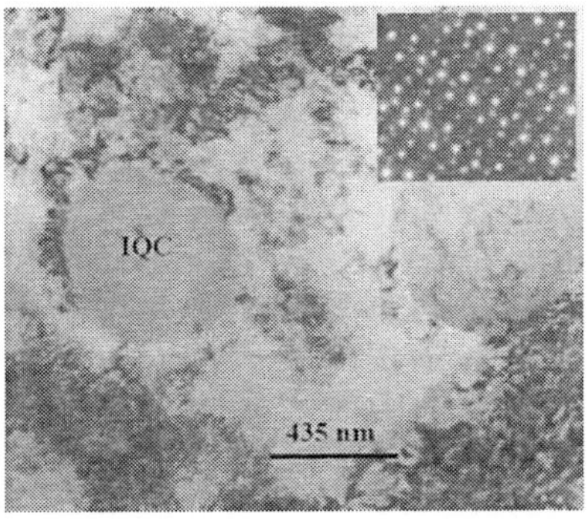

Figure 4. TEM bright micrograph showing the presence of an icosahedral quasicrystalline (IQC) phase in α-Al. Inset shows a corresponding selected area electron diffraction pattern from IQC.

ACKNOWLEDGMENTS

The authors are grateful to Professor E. S. Dwarakadasa (I.I.Sc. Bangalore) and Dr. S. Singh (N.P.L. New Delhi). AKS thanks to Professor C. Colliex (Orsay, France) for extending the facility of electron microscopy. AKS also acknowledges the BOYSCAST fellowship awarded by DST, Government of India.

REFERENCES

[1] A. Garg, A.K. Srivastava, T.R. Ramachandran, D.Bannerji, Science and Technology of Al-Li Alloys (editors: C.G. Krishnadas Nair, E.S. Dwarakadasa, Murli Saletore), Aluminium Association of India, Bangalore (1989) p. 53

[2] A.A. Csontos, E. A. Starke, *Int. J. Plasticity* 21 (2005) 1097

[3] P. Poza, J. Llorca, *Metall. Mater. Trans*. 30A (1999) 869

[4] K.K. Chawla, Structure and Properties of Composites, Materials Science and Technology (edited by R.W. Cahn, P. Hassen, E.J. Krammer) VCH Publishers Inc. New York, NY (USA), 1993, p. 122

[5] J. Ayache, P.H. Albarède, *Ultramicroscopy* 60 (1995) 195

In: Quasicrystals: Types, Systems, and Techniques
Editor: B. E. Puckermann, pp. 195-217
ISBN 978-1-61761-123-0

Chapter 10

MORPHOLOGIES OF ICOSAHEDRAL QUASICRYSTALS IN AL-MN-BE-(CU) ALLOYS

***Franc Zupanič*[1] *and Boštjan Markoli*[2]**
[1]Tonica Bončina, Niko Rozman, University of Maribor,
Faculty of Mechanical Engineering, Maribor, Slovenia
[2]University of Ljubljana, Faculty of Natural Sciences and Engineering,
Ljubljana, Slovenia

ABSTRACT

The shapes of icosahedral quasicrystalline (IQC) particles in Al-Mn-Be-(Cu) alloys were determined in samples subjected to very wide range of cooling rates: from around 10^6 K/s in very thin melt-spun ribbons down to below 100 K/s in permanent copper dies. Accordingly, the sizes of quasicrystalline particles ranged from few tenths of nanometres up to more than 100 μm. As a consequence, different methods were employed to properly characterize their shapes: projections of quasicrystalline particles using transmission electron microscopy (TEM), cross-sections of IQCs on metallographic polished surfaces, observation of deep etched samples and extracted particles in a scanning electron microscope (SEM). Despite of different sizes and shapes it was discovered that two the most important features are common to all of them:

- preferential growth in the three-fold directions
- tendency for faceting and adopting the shape of pentagonal dodecahedron.

The evolution of quasicrystalline shapes from apparently spherical particles to very large and highly branched dendrites is systematically presented. Special attention was devoted to the correct interpretation of quasicrystal shapes obtained from 2D-metallographic cross-sections.

INTRODUCTION

The point group symmetry of a crystal dictates its morphology [1]. Taking into account that the quasicrystals possess non-crystallographic point group symmetries, distinctive morphologies from those of periodic crystals may appear. It is to be stressed that two situations need to be distinguished. The equilibrium shape of a crystal is determined by a surface energy principle and can be different from the growth morphology which depends on the kinetics of the growth and processing conditions. Ideally at 0 K, the normal crystal should be bounded by flat surfaces, satisfying the condition of the minimum total surface energy. The effect of the temperature is to increase the disorder of the surface. The sharp edges and corners of the crystal at 0 K start rounding as the temperature is increased. The transition from a facetted plane to a smoothly curved one takes place at the roughening transition temperature. It is characteristic for each plane, being the highest for the close-packed plane. For a crystalline lattice, the roughening transition temperature generally scales with the lattice parameter. Thus a crystal with larger lattice parameter shows more faceting tendency. Since the quasicrystal can be considered as a periodic crystal with infinite periodicity, the roughening transition temperature should be infinite too; it should be faceted up to the melting temperature. This expectation was ruled out by the experimental evidence [2], since quasicrystals in the Al-Mn system were often found in the form of well-rounded dendrites. It indicates that the non-faceted growth of quasicrystals is due to dynamic roughening at high melt undercooling. This leads to a continuous growth and yields a rounded growth form. Nevertheless, the importance of growth steps (ledges) remains essential for the quasicrystals even in this case. Namely, using radiography Gastaldy et al. [3] clearly evidenced that a facetted growth proceeded by lateral motion of steps at the solid-melt interface and controlled by the interface kinetics. This indicated that quasicrystal growth is more comparable with the growth of both semiconductors and oxides than with that of pure metals.

Historically, polyhedral shapes are always associated with crystalline structures having periodic arrangement of atoms. Thus, it was not clear whether quasicrystals can have facets and definite polyhedral shapes. Ho et al. [4] adopted a bond oriented quasiglass model to determine whether facets are possible in such a case using only attractive potential. The results clearly indicate a strong tendency for faceting. However, they excluded pentagonal dodecahedron as a possible equilibrium shape. Afterwards, Ingersent and Steinhardt [5] made a detailed study of the shapes of quasicrystals by considering the combination of both attractive and repulsive interactions. They established that the equilibrium shapes of quasicrystals had some important differences from the crystal case. The main difference is that the shape obtained using only attractive interactions contained more than the minimal number of facets. The shape with the minimum number of facets consistent with the symmetry considered is obtained when higher neighbour repulsive interactions are included. This work also established firmly that the pentagonal dodecahedron can appear as an equilibrium shape when a more realistic situation of both attractive and repulsive potentials is taken into account.

Methods for Determining the Shapes of Quasicrystals

For determination of quasicrystal shapes several methods have been used. The first quasicrystalline phases were metastable and were obtained during rapid solidification. Rapid solidification of dilute Al-Mn alloys provide substantial undercooling promoting the nucleation of small quasicrystalline particles in the melt [6]. Later nucleates aluminium and grows very fast trapping the small quasicrystallites. This provides an excellent opportunity to study the equilibrium shape and the roughening behaviour of the quasicrystals. Due to small sizes of quasicrystalline particles, the basic shape of the small icosahedral quasicrystals in Al-Mn alloys was investigated by TEM. One of the initial difficulties was the problem of identification among the possible different shapes consistent with the icosahedral point group symmetry. It was found that the shapes are indistinguishable when projected along higher symmetry axes like threefold and fivefold axes [7]. However, as shown in Fig. 1, the projection along the twofold axes enable distinction between different shapes.

By developing polycrystalline stable quasicrystalline phases, as well as new aluminium alloys with increased quasicrystal-forming ability, determination of the quasicrystalline shape from the micrographs of the polished surface has become very important. Although the icosahedral quasicrystalline particles on the polished surface sometimes exhibited a pentagonal symmetry [8], it was not always clear, what was their exact shape, and what was the preferred growth direction. Namely, it appears to be a hard task to determine the 3D-shapes of microstructural constituents based on 2D-sections only. This is why Kral et al. [9] obtained 3D-shapes of phases in steels by using computer-aided visualization of 3D reconstructions from series of section images. They gradually removed approximately 0.2 μm of material in a step-wise manner and took photos of the section after each step. Nowadays, sequential cross-sectioning is done using focused ion beam FIB (usually a dual beam SEM/FIB system). In the systems equipped with EDS, 3D-elemental distribution can be obtained along with the particle shapes [10]. Nonetheless, such approach is feasible, but it is very time consuming, and it is unlikely suitable for regular routine investigations. It has not been applied for studying quasicrystals yet.

Direct observation of 3D-shapes of quasicrystals was done by decanting of the melt after growing single quasicrystals [11], using SEM observations on faceted microholes in slowly cooled icosahedral single quasicrystals [12] and inside microvoids caused by shrinkage during solidification [13]. 3D-shapes of quasicrystalline phases can also be obtained by deep etching and particles extraction techniques [14]. In both methods, the particles are isolated by dissolving the matrix using chemical and electrochemical methods, and characterized by the application of scanning electron microscopy (SEM). It is extremely important that a reagent does not dissolve quasicrystalline phases, and that they still exhibit their original shape after removing the matrix. An important disadvantage of the particle extraction technique is that all particles are mixed together, so no information can be obtained about their spatial distribution in the microstructure. The particle extraction also takes more time than deep etching offering more possibilities for the particles to be attacked by the etchant. Further, less material has to be removed during the deep etching. Satisfactory results have already been obtained when the etching depth approached the typical particle sizes. This also implies longer etching times when the particles are larger.

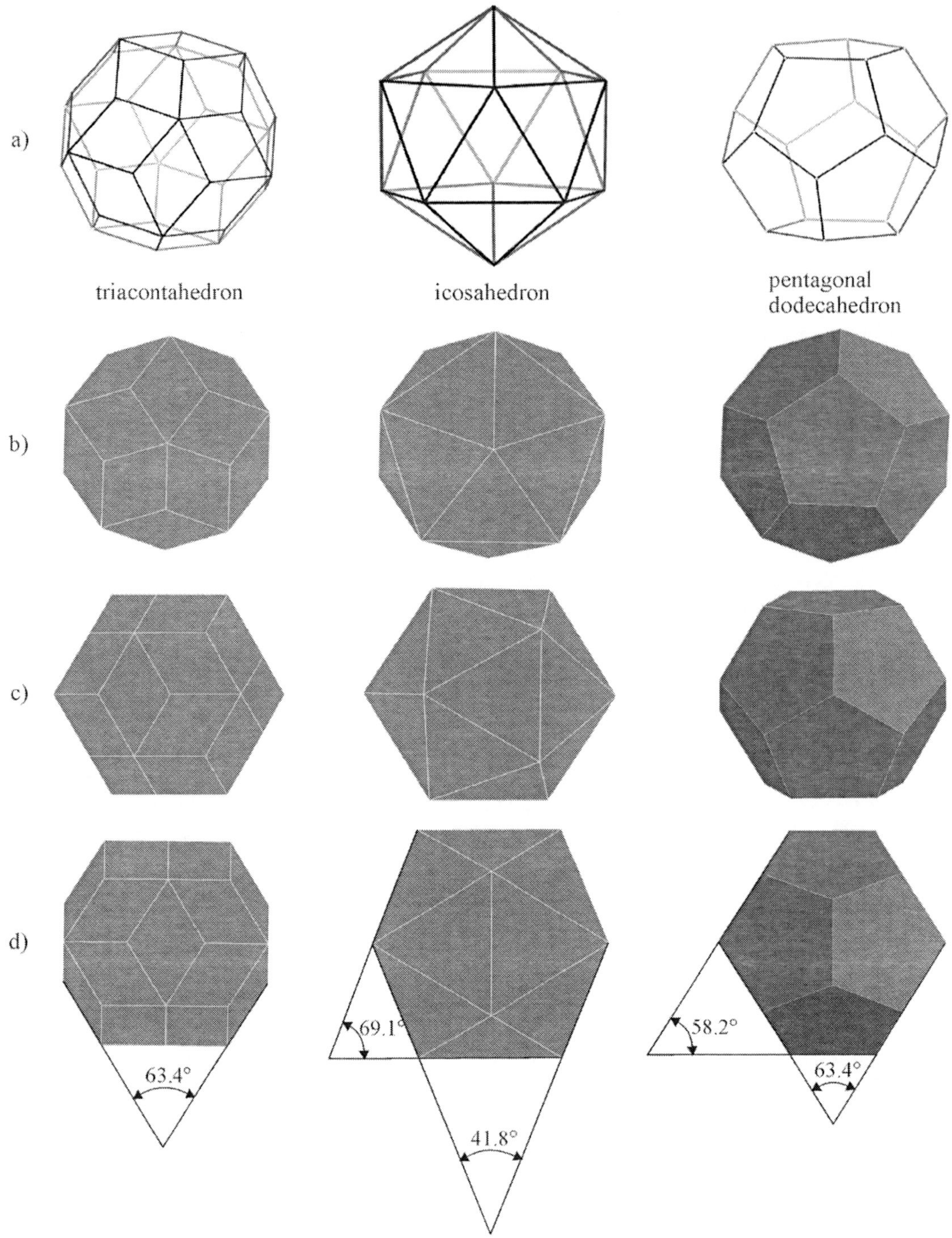

Figure 1. Shapes of a triacontahedron, icosahedron and pentagonal dodecahedron under different projections: a) general view, b) along fivefold axis, c) along threefold axis and d) along twofold axis. It can be seen that the shapes can be clearly distinguished in projections along the twofold axis.

Table 1. Morphologies of quasicrystalline phases in selected systems

System	type of quasicrystal	morphology	reference
Al15Mg50Zn35	IQC	a faceted dendritic growth with the fastest growth direction along a three-fold axes. The regular hexagonal-polygon icosahedral quasicrystal grains.	[23]
Al70Mn9Pd21	IQC	Archimedian polyhedron	[12]
Al71Pd21Mn8	IQC	pentagonal dodecahedron	[11]
Al72Ni12Co16	decagonal quasicrystal (DQC)	faceted morphology, lateral growth along the twofold axis normal to the plane (00002).	[24]
Al72Pd25Cr3	IQC	a planar growth with a growth direction parallel to a fivefold symmetrical axis	[25]
Al77.6Tc10Ir12.4	IQC (F-type structure)	a pentaprismatic growth morphology	[26]
Al-Mn	IQC	dendritic, preferred growth in three-fold directions	[2]
Al-Cu-Fe	IQC	pentagonal dodecahedron	[13]
Mg28Zn2Y	IQC	a perfect five-branch icosahedral quasicrystal, the icosahedral quasicrystal free growth with preferred growth direction of five-fold symmetry axes resulting in five-branch morphology.	[27]
Mg67.4Zn28.9Y3.7	IQC	primary IQC: petal-shaped with five and six branches, where each branch has faceted growth. Polygon-shaped IQC.	[28]
Mg-Zn-Er	IQC	eutectoid-lamellar morphology and the other with granular shape	[29]
Mg-Zn-Y	IQC	petal-like, polygon-like	[30]
Mg-Zn-Y	IQC	from petal-like morphology to spherical morphology.	[31]
Mg-Zn-Y	IQC	from petal-like morphology to spherical morphology.	[31]
Mg-Zn-Y	IQC	triacontahedral growth morphology.	[20]
Mg-Zn-Y-(Ti, Sb, Ce or C nanotubes)	IQC	petal-like to spherical	[32]
Mg-Zn-Y; Y-rich	IQC	petal-like with five branches, polygon-like morphology	[8]
Ti-Mn-Fe	IQC	dendritic, preferred growth in fivefold directions	[2]
Zn-Mg-RE (RE = Y, Tb, Dy, Ho or Er)	IQC	a pentagonal dodecahedral solidification morphology	[33]
Zn-Mg-Sc	IQC	a triacontahedral shape	[34]
Zr-Fe-Ni ternary metallic glass	IQC	a nearly spherical shape with a size of 5 to 20 nm	[35]
Sc12Zn88	IQC	pentagonal dodecahedra and rhombic triacontahedra	[17]

Experimentally Determined Quasicrystalline Shapes

Several investigations in the early stages were carried out to characterise the morphology of quasicrystals. Most of the initial work was carried out for Al-Mn and Al-Mn-Si icosahedral quasicrystals. Although one often obtains a dendritic structure with branches extending in the preferred threefold directions [2], Chattopadhyay et al. [15] indicated a well-facetted morphology for the quasicrystals. In the work of Thangaraj et al. [7] the pentagonal dodecahedron was established as the shape for the Al-Mn type of quasicrystal. For the most of the stable quasicrystals; e. g. in the Al-Cu-Fe system [13, 16], pentagonal dodecahedron was found as the equilibrium shape, in addition, quasicrystals with the icosidodecahedral morphology can be observed. Triacontahedral growth morphology of icosahedral quasicrystals was observed in several alloy system, such as: Sc12Zn88 [17], Al-Li-Cu [18], Al-Mg-Zn [19] and Zn-Mg-Y face-centred icosahedral alloys [20]. The preferred growth direction was mainly along treefold axes and occasionally also along fivefold axes [2].

The decagonal quasicrystal in Al-Cu-Co also exhibits well defined decagonal prism morphology [21], and there is also a report of a pencil shape of a one-dimensional quasicrystal [22].

The observed shapes of quasicrystals in selected systems are collected in Table 1. There is some confusion regarding quasicrystalline shapes, especially those determined from the 2D-sections.

Aims of the Present Work

The aim of this work is to determine the shapes of icosahedral quasicrystalline (IQC) particles in several alloys based on the Al-Mn system in samples subjected to very wide ranges of cooling rates. The intention is to systematically present the evolution of the shape, and much attention is given to procedures allowing correct interpretation of the quasicrystalline shape based on 2D-sections.

Experimental Work

The alloys were synthesised from pure Al and masters alloys AlMn20, AlBe5, AlCu10 and AlB3 provided by KBM Affilips B.V. using vacuum induction melting and casting into bars with 50 mm diameter. The compositions of the alloys (Table 2) were determined using ICP-AES (Inductively Coupled Plasma, Atomic Emission Spectroscopy). The bars were sectioned, remelted and cast into a rectangular copper die (100 mm × 10 mm × 1 mm) or melt-spun. The details are given elsewhere [36, 37].

Preparation of samples for the light-optical microscopy (LOM) and scanning electron microscopy (SEM) followed the standard mechanical metallographic procedures. In addition, considerable attention was given to deep etching and particles extraction techniques. The samples were observed under a light microscope Nikon Epiphot 300 and a scanning electron microscope Sirion 400 NC, FEI equipped with an EDS-analyser INCA 350, Oxford Instruments.

Table 2. Chemical compositions of the investigated alloys (ICP-AES)

alloy		Al	Mn	Be	Cu	B
Al-Mn	mass. %	79.8	21.2	-	-	-
	at. %	88.5	11.5	-	-	-
Al-Mn-Be	mass. %	90.6	5.4	4.0	-	-
	at. %	86.1	2.15	11.38	-	-
Al-Mn-Be-Cu	mass. %	90.64	4,24	0,68	4,44	-
	at. %	93.78	2.15	2.11	1.95	-
Al-Mn-Be-B	mass. %	92.33	3.93	0.77		2.97
	at. %	88.79	1.96	2.22		7.13

Transmission electron microscopy (TEM) was carried out in a FEI TITAN 80–300 and a JEOL 2000 FX. The TEM specimens for the TITAN were cut out at specific sites using the focussed ion beam (FIB) in an FEI Nova 200 Nanolab, and those for the JEOL 2000 FX specimens were prepared using the ion beam etching and polishing system GATAN PIPS 691 (3 keV, angle ± 2.5°).

Results and Discussion

In the first part only the characteristic of IQC in alloys based on the Al-Mn-Be system will be presented, followed by the characterisation of shapes in different conditions.

Icosahedral Quasicrystalline Phase in Al-Mn-Be-X Alloys

The microstructure of all alloys in melt-spun condition predominantly consisted of two phases IQC and α-Al, whereas mould castings contained additional phases. It is to be noted that in the Al-Mn alloy the IQC-phase was not present in the mould castings.

The aim of this section is only to provide some information regarding the characteristics of the IQC-phase that have some impact on the topics discussed in the next sections.

Analytical work in TEM was focussed on characterizing the i-phase. Fig. 2a shows the section through the two-phase (α-Al + i-phase) region, with corresponding diffraction patterns for the i-phase taken along two-, three- and fivefold axes. It is clear, that the diffraction patterns are not periodic and that the distances between the most important spots increase with the golden mean. The position of the most relevant diffraction spot (211111) in the twofold diffraction pattern indicated that the quasicrystalline phase possessed primitive icosahedral structure. The position of this spot is at g = 4.71 nm^{-1} in the reciprocal space, corresponding to the interplanar distance of d = 0.213 nm. The quasilattice constant a_R calculated using the following equation [39]

$$a_R = \frac{d\tau^3}{2} \qquad (0.1)$$

amounted to 0.45 nm, which is very close to a_R in the binary Al-Mn i-phase [40].

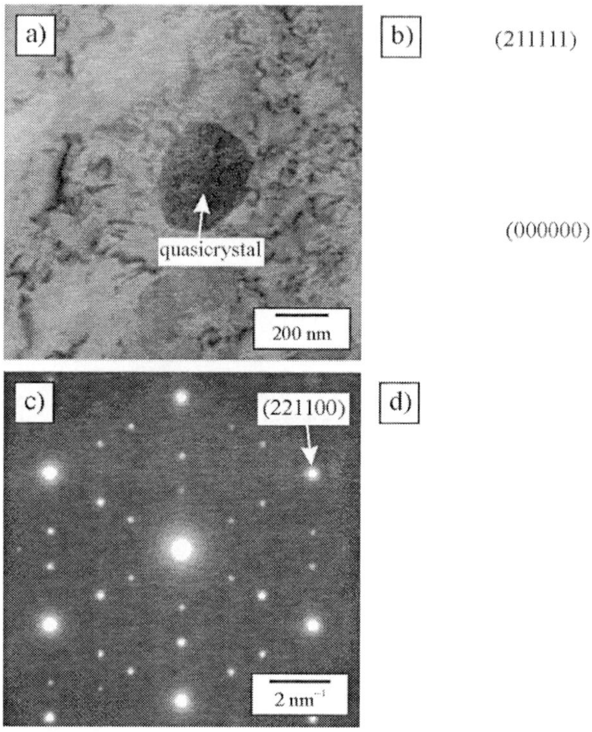

Figure 2. a) TEM micrograph of an individual quasicrystalline particle in Al-rich matrix in Al-Mn-Be. SAED-patterns taken along b) two-, c) three- and d) fivefold axis of the quasicrystalline particle. [38]

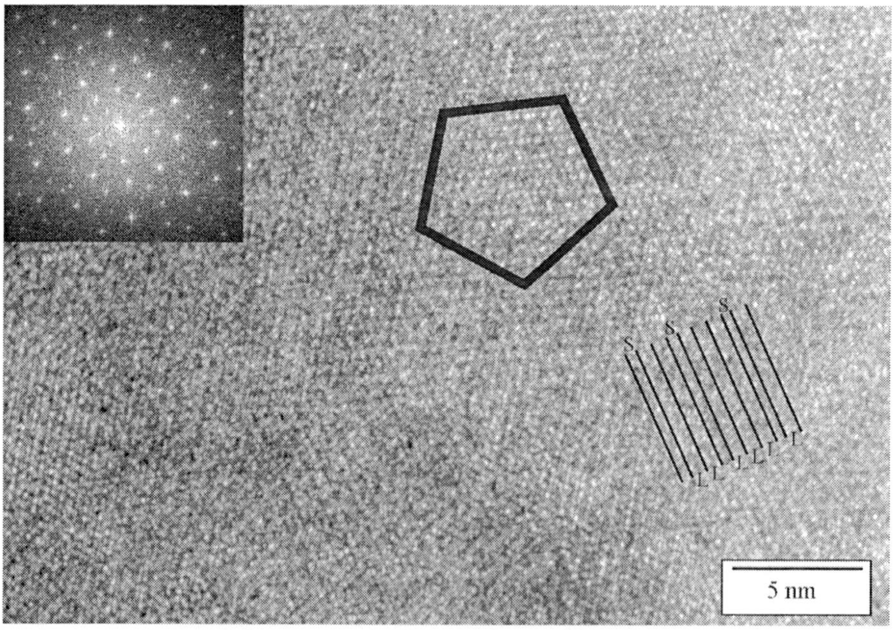

Figure 3. HRTEM of a quasicrystalline particle in a fivefold orientation with corresponding FFT insert (Fast Fourier Transform) in Al-Mn-Be alloy. [38]

A closer inspection of the diffraction patterns showed that the diffraction spots, especially the weaker ones, are deflected from their ideal positions. They did not lie along the same line, and the distances between them did not scale exactly with τ. Diffuse scattering can also be observed. This was a strong indication of disorder in the i-phase and the presence of so-called "phason strains". This is confirmed by a high-resolution electron microscopy (HRTEM) image taken along the fivefold orientation (Fig. 3). Fringes can be observed running parallel to five different directions related to the pentagonal symmetry. The spacing of the fringes was not periodic and can be related to the interplanar spacing observed in the diffraction patterns. The fringes did not continue for a long distance and local disturbances can be observed.

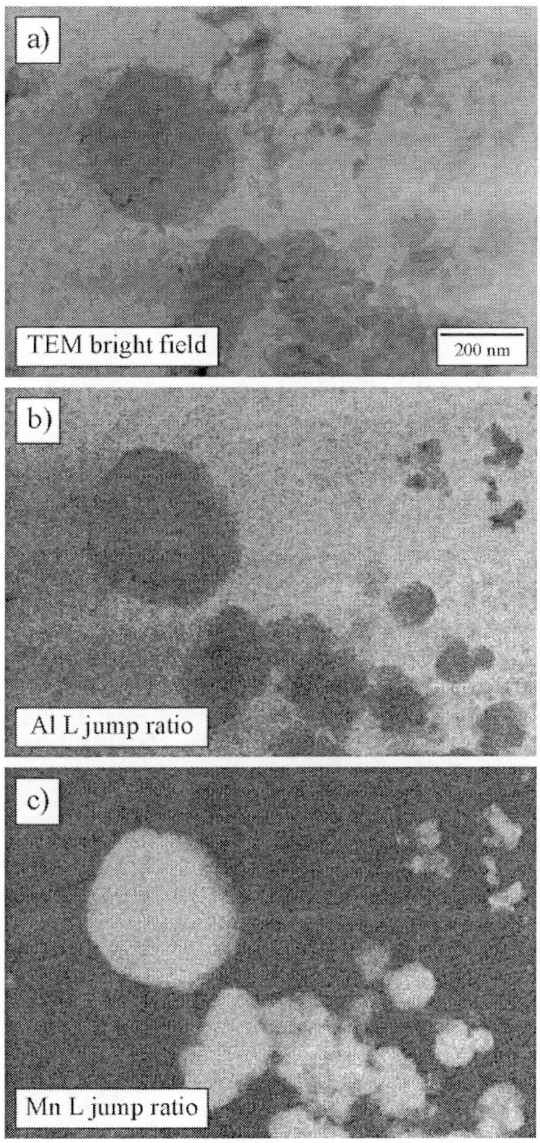

Figure 4. Distribution of elements using EFTEM in Al-Mn-Be alloy. a) TEM bright-field image, zero-loss filtered, b) elemental distribution of Al and c) elemental distribution of Mn (jump ratio images).[38]

For the determination of the chemical composition of the matrix and particles of the quasicrystalline phase, as well as for distribution of elements EDS (both in SEM and TEM), EFTEM, and AES were used. The results of EDS and EFTEM showed that manganese was concentrated in the quasicrystalline phase (Fig. 4), whereas both methods failed to detect beryllium. Using AES it was possible to perform both qualitative and quantitative analysis of IQC. It was found out that it contained in Al-Mn-Be alloy 14−16 at. % Mn and between 30−40 at. % Be, whereas the IQC in Al-Mn-Be-Cu also contained around 2 at. % Cu.

Morphology of the I-Phase in Melt-Spun Ribbons

Melt spinning provides the opportunity to study the equilibrium shape and the roughening behaviour of the quasicrystals because the rapid solidification of dilute Al-Mn alloys makes possible substantial undercooling promoting the nucleation of the quasicrystal in the melt. Later the Al nucleates and and its very fast grow very fast enable trapping the small quasicrystallites. It is to be noted that the shape reflects the shape at the trapping temperature.

Fig. 5 shows the microstructure typical for the 30–90 μm thick ribbons. These ribbons consisted of two regions indicated by A and B. Region A appeared in LOM micrographs and in lower SEM magnifications almost featureless. However, higher magnified SEM images (Fig. 5b) indicated a very uniform distribution of tiny spherical quasicrystalline particles (< 100 nm in diameter) in α–Al matrix. In the region B quasicrystalline particles were larger (up to 500 nm). They possessed both facetted shape, as well as the shape of dendrites with rounded arms.

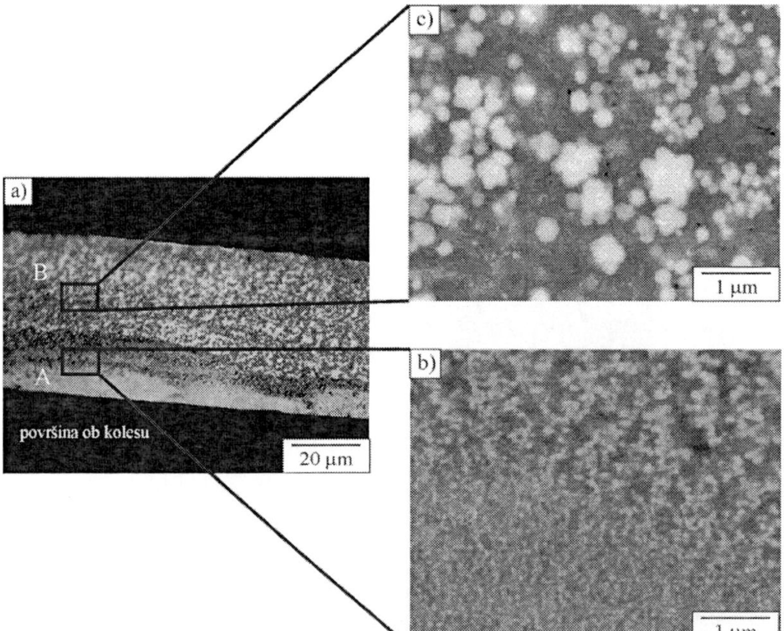

Figure 5. Longitudinal cross-section of the melt-spun ribbon in the alloy Al-Mn-Be. Microstructure typical for the 30–90 μm thick ribbons. a) LM-micrograph, b, c) enlarged images of the regions A and B (SEM, backscattered electron image) [37].

Typical transmission electron micrographs with the corresponding diffraction patterns are shown in Fig. 6. Extracted particles from the Al-Mn melt-spun ribbons had rounded edges with a size up to 100 nm (Fig. 6a). On the other hand, particles can often exhibit equiaxed form (Fig. 6b). Some of them had almost ideal spherical morphologies, whereas on others perturbations can be seen. Contrary to the first two cases, a particle in Fig. 6c possessed faceted morphology. In a TEM micrograph taken along a twofold axis the particle had a shape of a hexagon. The angles between edges were approximately 58° and 64°. As was discussed in the introduction, the triacontahedron, icosahedron and pentagonal dodecahedron can only be distinguished when a TEM-micrograph is taken in a twofold direction (Fig. 1d). When comparing Figs. 1d and 6c one can easily conclude that the observed particle had a shape of a pentagonal dodecahedron. This was already found in binary Al-Mn alloys [7], and ,obviously, is not affected by the addition of other elements (i.e., Be, B and Cu).

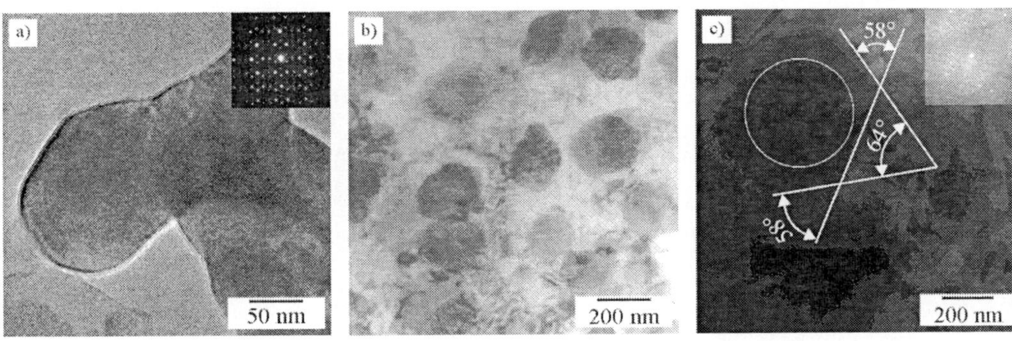

Figure 6. Particles in melt-spun ribbons. a) Extracted particles of i-phase in the alloy Al-Mn, the image was taken along twofold axis, b) bright-field micrograph of the alloy Al-Mn-Be and c) bright-field electron micrographs of the alloy Al-Mn-Be-B.

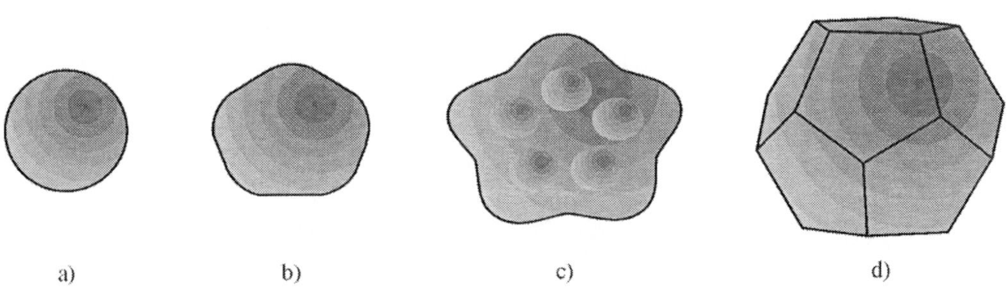

Figure 7. Morphologies of IQC observed in melt-spun ribbons. a) sphere, b) particle with rounded edges, c) particle with protuberances, d) pentagonal dodecahedron

The results can be related to the undercooling. The dynamic roughening is the most pronounced at large undercooling, therefore almost spherical particles form. With the decreasing cooling rate, the particles posses some flat faces with rounded corners. In some cases constitutional undercooling can lead to the morphological instability of the solid-liquid interface, resulting in formation of protuberances. Such development can lead to the formation of large dendrites with rounded arms [2]. At smaller undercooling the effect of dynamic roughening diminishes, resulting in formation of almost ideal pentagonal dodecahedrons. The schematic presentation of IQC-shapes found in melt-spun ribbons is shown in Fig. 7.

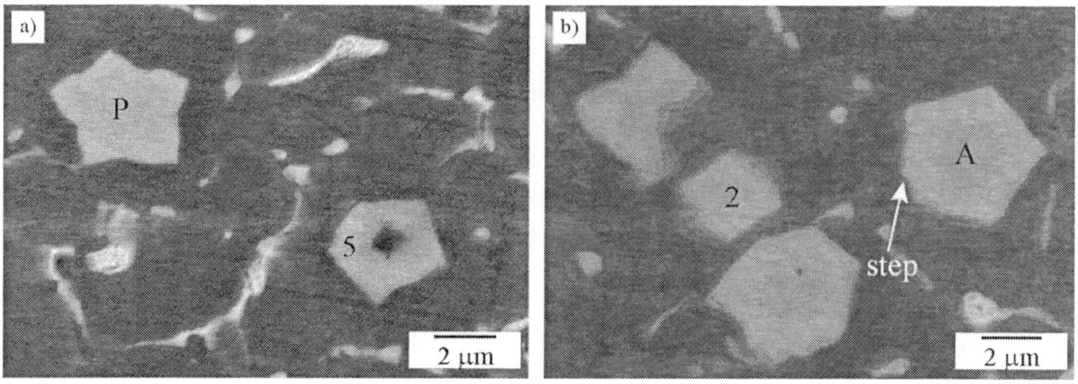

Figure 8. Shapes of IQC on metallographic specimens (back-scattered electron image, Al-Mn-Be-Cu alloy).

Shape of the Primary IQC in Alloys Cast Into a Copper Mould

The shape of the primary IQC was investigated in detail in the Al-Mn-Be-Cu alloy. In this alloy, the large quasicrystalline forming ability prevented the appearance of the concurrent phases that may influence the unconstrained growth of the IQC. Only Al_2Cu formed during terminal stages of solidification in the form of binary eutectic (α-Al + Al_2Cu). Thus, primary i-phase could form in the melt and grow freely in all directions. It exhibited several shapes from polyhedral to highly branched dendrites. The sizes of particles ranged from few micrometers up to several tenth micrometers. Accordingly, their 2D-shapes can already be recognized by observing the samples prepared for the light optical microscopy and scanning electron microscopy, without a necessity to use TEM. Metallographic cross-section in Fig. 8 shows some primary IQCs in the form of polygons and petal-like particles. In some cases almost regular pentagons and hexagons are present, whereas in others rectangles, trapezoids and other more irregular forms can be observed. The extracted particles often exhibited the shape of pentagonal dodecahedron. Some of them were almost ideally shaped, whereas others showed some protrusions on the vertices (Fig. 9).

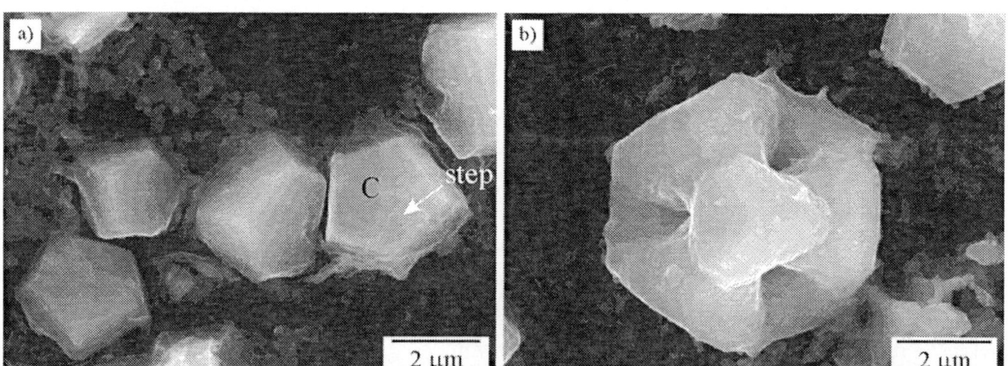

Figure 9. Shapes of primary IQC particles revealed by the particle extraction technique (secondary electron image, alloy Al-Mn-Be-Cu). a) Particles with a form of pentagonal dodecahedron and b) a particle with the predominant growth in the threefold directions

Let us assume that the primary IQC possesses the form of an ideal pentagonal dodecahedron. It would be of great value when we could determine the possible shapes that may be observed on the metallographic cross-sections. In order to reveal this let us make sections through a pentagonal dodecahedron using intersecting planes perpendicular to the principal symmetry axes of the pentagonal dodecahedron: fivefold, threefold and twofold (Fig. 10). When the section plane lies perpendicular to a fivefold axis, then the intersections normally have the form of the regular pentagon. However, when the intersection plane lies close to the equatorial plane of the pentagonal dodecahedron, also a decagon may appear. When the plane lies perpendicular to a threefold axis then the intersections have a form of triangles and hexagons. On the other hand, when an intersecting plane lies perpendicular to a twofold axis then in addition to hexagons, rectangles and octagons also appear. When the orientations of the intersecting planes deviates from the "ideal" orientations, then distorted polygons emerge (rectangles turn to trapezoids) or in case of orientations close to the fivefold axis decagons turn to nonagons and in the case of a twofold axis octagons turn to heptagons. It should be stressed that in the case of heptagons, octagons, nonagons and decagons particles have apparently the spherical shape. The form of intersections through the pentagonal dodecahedron allows determination of particle orientation. As was stated, pentagons appear only when the intersecting plane lies almost perpendicular to the fivefold axis. On this ground one can rather firmly state that the normal on the particle labelled with 5 in Fig. 8a lies almost parallel with the fivefold axis. On the other hand, trapezoidal shape of particle 2 indicates that the normal slightly deviates from the twofold axis; it is slightly tilted from the twofold axis towards a threefold axis.

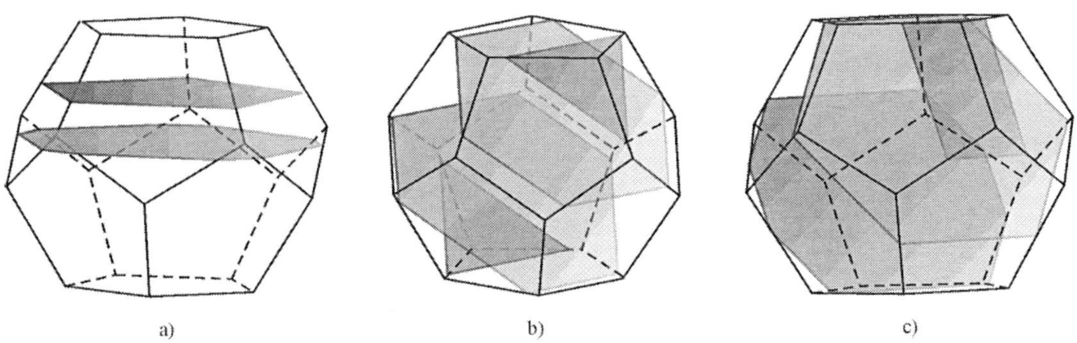

Figure 10. Sections through a pentagonal dodecahedron using intersecting planes perpendicular to a) fivefold, b) threefold and c) twofold axes.

Petal-like cross-sections cannot appear when particles have the shape of an ideal pentagonal dodecahedron, but only when the protrusions formed. These can form when accelerated growth occurs in particular directions. This is schematically depicted in Fig. 11 when a regular pentagon transform to a petal-like (star-like) shape as a result of the fastest growth rate in the vertex directions. In case of IQC the fastest growth takes place in the threefold directions. The conclusive evidence for this gives Fig. 9b. The particle is almost ideally oriented in a threefold orientation. The growth rate was the fastest in the threefold directions (vertices of pentagonal dodecahedron). One direction is perpendicular to the micrograph, and, the other three directions, are inclined for 41.8°. It can also be observed that growth rate is faster along the edges of the pentagonal dodecahedron than perpendicular to the fivefold faces. Some particles show evidence of sidewise growth of ledges on the fivefold faces (e. g. particle A in Fig. 8 and particle C in Fig. 9), which is the type of growth expected for IQC [1]. This growth is rather slow, therefore only small segments of fivefold faces can be observed at the vertices. Sections through such particles give rise to the five-petal and six-petal shapes found on the metallographic sections (e.g. the particle P in Fig 8a).

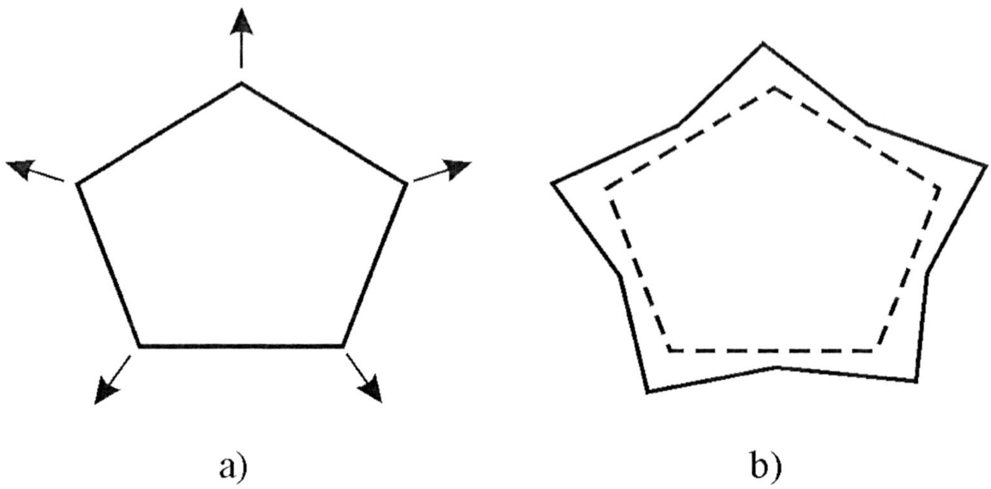

Figure 11. Transition of a regular pentagon to a five-petal shape due to preferred growth of the particle along vertices.

Fig. 12 shows additional shaped observed on metallographic specimens. They cannot be explained with sections through the pentagonal dodecahedron, even if they possessed small protuberances along threefold directions. The particles must have larger arms, as was revealed in extracted particle in Fig. 13a.

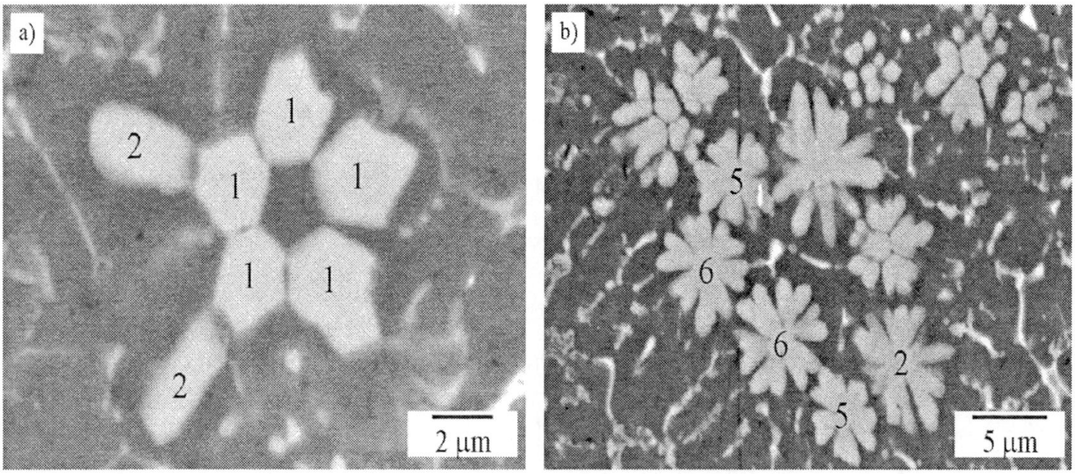

Figure 12. Different shapes of IQC on metallographic cross-sections (back-scattered electron image, alloy Al-Mn-Be-Cu). a) Simple dendrite, b) group of dendrites.

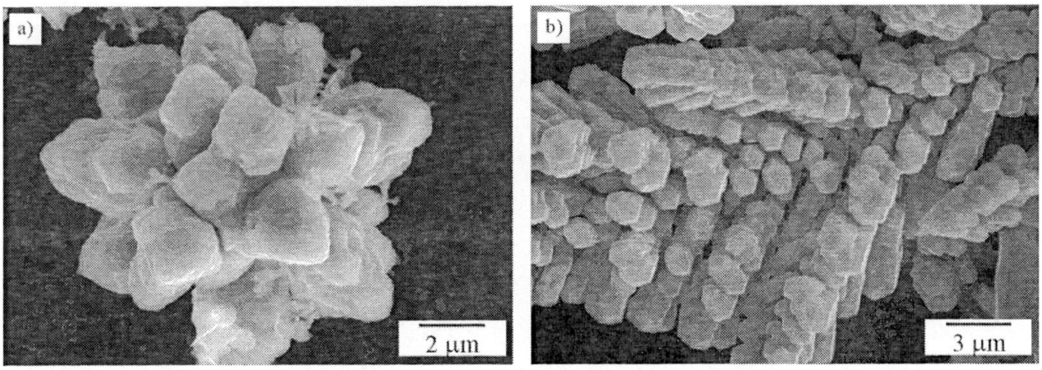

Figure 13. Shape of primary IQC particles, revealed by the particle extraction technique (secondary electron image, alloy Al-Mn-Be-Cu). a) Simple dendrite, b) highly branched dendrite.

Basic information regarding the mutual orientation of arms can be obtained by the use of stereographic projections along the principal axes of the icosahedral symmetry: fivefold, threefold and twofold (Fig. 14). Stereographic projection can also help us in determining the preferred growth directions of the dendritic arms. Let us show this for the stereographic projection when a fivefold axis points in the vertical direction. Let us assume the preferred growth in twofold, threefold and fivefold axes. When the preferred growth was in the fivefold directions, then section through the arms in the upper hemisphere would cut the central fivefold arm and in addition other five fivefold arms. Those arms are inclined relative to the central one for 63.43° (Fig. 15a). When the preferred growth direction was in the twofold directions, then ten arms would be cut, arranged in two circles around the twofold axis. The positions of the arms in the second circle would be shifted for 36° with reference to those in the first circle (Fig. 15b). When the threefold direction was the preferred growth direction then also ten threefold arms would be cut and arranged in two circles around the fivefold axis (Fig. 15c). In this case the arms in the second circle would lie directly behind those in the first circle. When comparing the shapes in Fig. 15 with Fig. 12a one can conclude that the preferred growth of the particle takes place along the threefold directions; this is typical for the most of icosahedral quasicrystals [2].

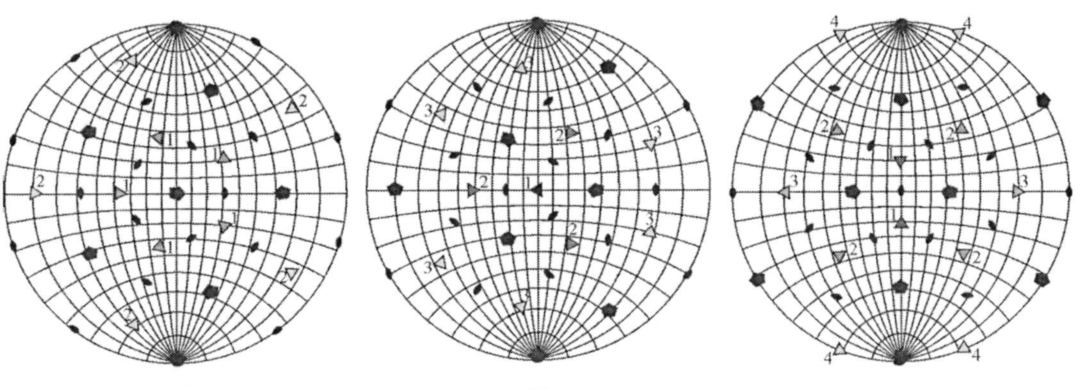

Figure 14. Stereographic projections of the upper hemisphere when the vertical lies parallel to the a) fivefold axis, b) threefold axis and c) twofold axis.

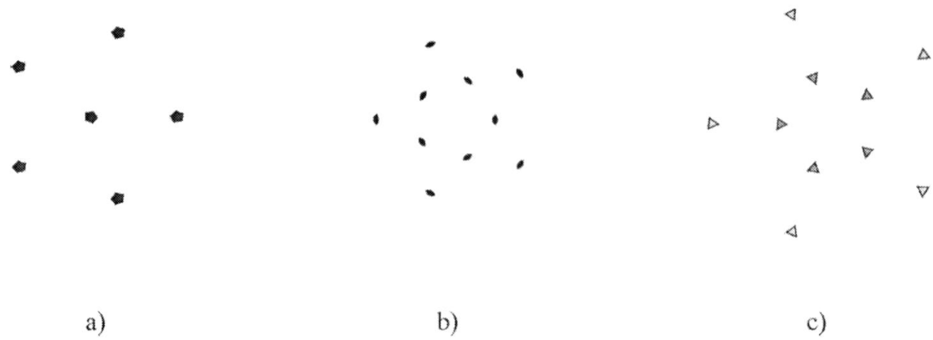

Figure 15. Schematic presentation of cuts through arms if the preferred growth of arms would take place along a) fivefold directions, b) threefold directions and c) twofold directions when the fivefold axis points in the vertical direction, like in Fig. 14a

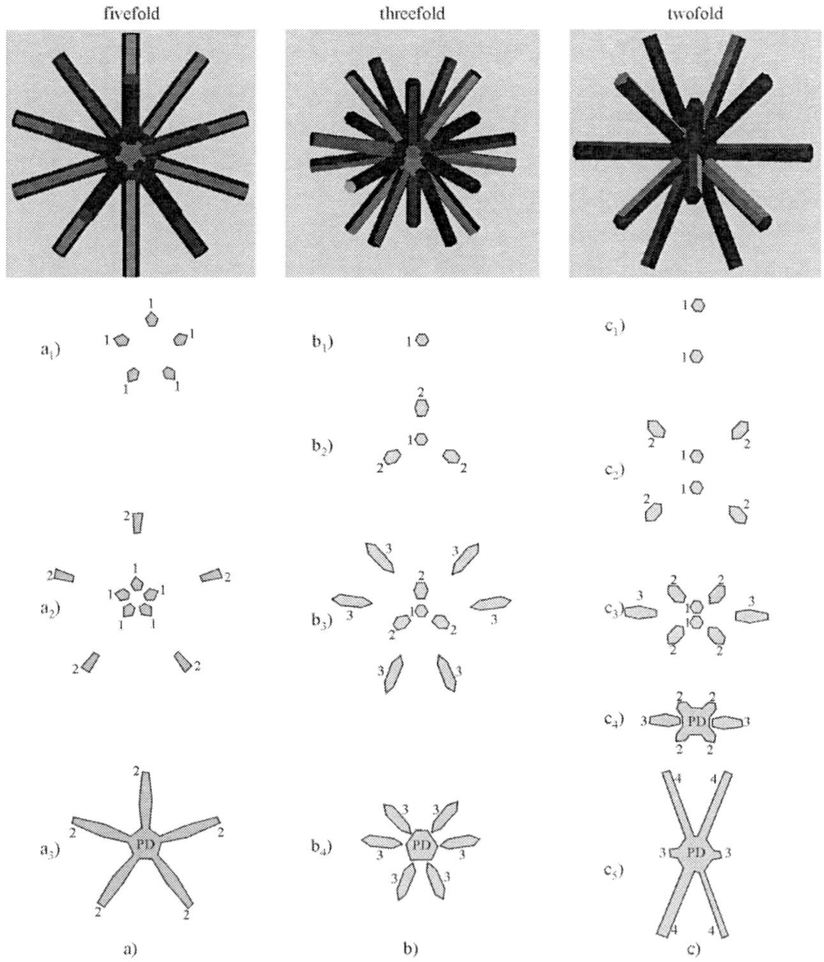

Figure 16. Sections through a pentagonal dodecahedron with arms projecting in the threefold directions. Sections are descending from the tips of the arms pointing in the vertical direction toward the centre of the pentagonal dodecahedron (PD). The intersecting plane is perpendicular to a) a fivefold axis (labels of the arms correspond to the labels of the poles in Fig. 14a) , b) a threefold axis (labels of the arms correspond to the labels of the poles in Fig. 14b) and c) a twofold axis (labels of the arms correspond to the labels of the poles in Fig. 14c)

Stereographic projections allow us to get the basic relationships, but it does not give us the exact shape of sections through the dendrites. For the sake of clarity, a model dendrite was made consisting of a central pentagonal dodecahedron, with dendritic arms extending in the threefold directions. All arms have the equal length and thickness. The sections through such dendrite were made using planes perpendicular to a fivefold, a threefold and a twofold axis. For each dendrite orientation several typical sections are shown in Fig. 16. Arms had rather small and uniform cross-sections therefore the distances between them are somewhat larger than in real dendrites (e.g. the dendrite in Fig. 13a). Closer observation of the dendrite in Fig. 12a revealed that its orientation lies close to the fivefold axis. Five cuts correspond to the arms indicated as 1 in Fig. 14a and Fig. $16a_2$, and the other two to those indicated by 2. The fact that only two of five possible arms indicated as 2 are visible gives rise to a conclusion that the normal does not lie exactly in the fivefold axis, but is slightly inclined towards a twofold direction.

Shapes of IQC particles in Fig. 12b can also be explained by sections through a dendrite, however in this case a good match is obtained if we assume that branching of arms occurred as schematically presented in Fig. 17. In this case particles indicated by 5 (Fig. 12b) lie almost exactly in the fivefold direction, those indicated by 6 in the threefold direction, and that indicated by 2 in the twofold direction. Fig. 13b shows the extreme case of highly branched dendrite.

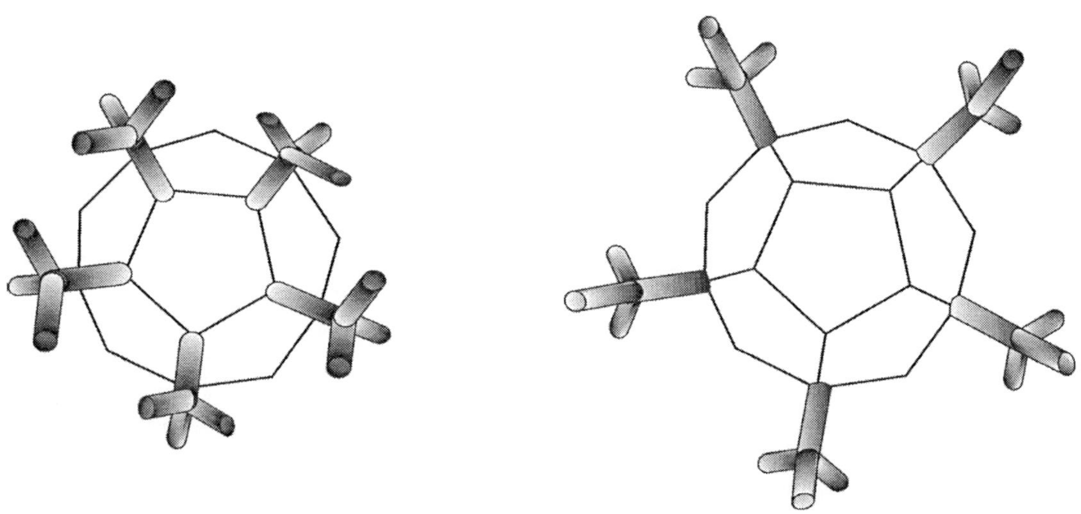

a) b)

Figure 17. Basic pentagonal dodecahedron with branched arms a) arms indicated by 1 in Fig. 14a and b) arms indicated by 2 in the same figure.

Eutectic IQC-Phase

Microstructures of Al-Mn-Be-(Cu) alloys subjected to moderate cooling rates during solidification contained in addition to the primary IQC, also IQC as a part of a two-phase microstructural constituent (α-Al + IQC), often referred as a binary eutectic. Fig. 18 shows two typical back-scattered electron micrographs. In Fig. 18a primary IQC particles (two are cut approximately in the fivefold symmetry and one in the threefold symmetry) seem to be completely surrounded with dendritic α-Al, and binary eutectic constituent being only in the interdendritic regions. It appears that the binary eutectic does not have any contact with the primary IQC particles. However, Fig. 18b indicates that there must be some correspondence with both since the orientations of the eutectic cells show clear relationship with the orientation of the IQC-dendrite. This is rather unambiguously confirmed in Fig. 19 in a deep-etched specimen, where eutectic rods started to grow from vertices of pentagonal dodecahedrons, preferentially in the threefold directions. This observation suggests that the preferred growth in threefold directions is typical for IQC in these alloys regardless whether IQC grows as a primary phase in the melt or mutually with the α-Al in the binary eutectic constituent. Growth of eutectic IQC starts either from the vertices of primary IQC (Fig. 19) or from a common centre (Fig. 20a) having frequently the form of a pentagonal dodecahedron. One can also observe that the branching starts at very small distances from the starting position (Fig. 19). This feature is most clearly evident in deep-etched specimens (Fig. 20). As the distance from the solidification centre increased, frequent branching occurred in order to keep the distance between branches as constant as possible. Detailed TEM analysis of the eutectic IQC showed that the whole rodlike structure is a monoquasicrystal because no sharp orientation changes were observed, not even at positions where branching took place [41]. This indicated that branching did not occur by e.g. twinning, which is typical for the Si phase in the Al-Si eutectic, but growth direction can be quickly changed to one of the other preferred directions due to the very high symmetry of the icosahedral phase. Furthermore, it was observed that orientation was changing uniformly along each rod. This kind of lattice rotation might indicate the presence of quasilattice defects, such as phason strains [42]. This was confirmed by peer examination of the diffraction patterns since many diffraction spots were deflected from their ideal positions (especially weaker ones), and the shapes of some spots showed strong anisotropy.

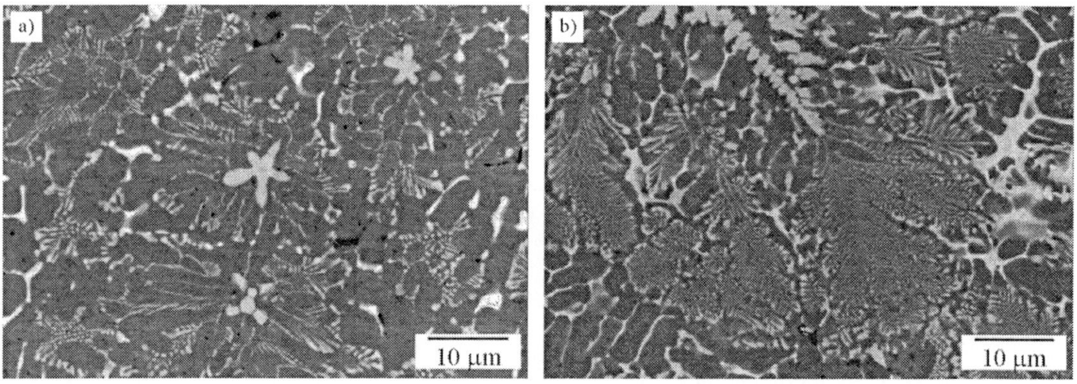

Figure 18. Back-scattered electron image of the areas with different amount of binary (IQC + α-Al) eutectic.

The eutectic IQC-phase has a rod-like appearance. The length of rods depends on the size of interdendritic regions and is typically in the order of ten micrometers, whilst their thickness is few 100 nm. The rods are non-faceted, and the branching occurs in threefold directions, the angle between branches is roughly 41°. The frequent branching enables the change of growth direction of the eutectic IQC in a very easy way and therefore allows the fast response to the ever changing growth conditions, i.e. the growth between dendritic arms. Eutectic can also grow in the cellular or dendritic manner, especially in the four-component Al-Mn-Be-Cu alloy due to the constitutional undercooling (Fig. 18b).

Figure 19. Secondary electron micrographs showing that the growth of the eutectic IQC-phase starts from the vertices of primary i-phases

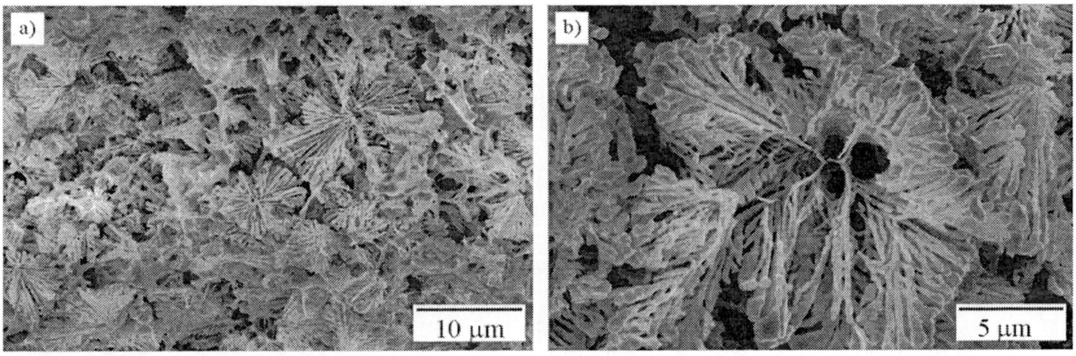

Figure 20. Morphology of eutectic IQC in deep-etched samples (secondary electron image)

The binary eutectic constituent could be classified as irregular rod-like eutectic, which is also typical for several eutectic systems: intermetallic phase – metallic solid solution and carbide – metallic solid solution (MC - γ_{Ni}) [43, 44]. The same kind of eutectic was also observed in Mg-Y-Zn-X alloys. Wang et al. [32] labelled it as a lamellar eutectic; however, this is not in accordance with the classification of eutectics.

Presence of Other Phases

The alloy Al-Mn-Be contained also other intermetallic phases besides IQC. Fig. 21 shows the presence of Be_4AlMn (dark phase) and Al_4Mn (platelike morphology). It can be inferred that Be_4AlMn-phase particles represented a suitable heterogeneous nucleation site for IQC. In fact, several IQC particles formed on each Be_4AlMn particle. It was discovered that even IQC in the form of dendrites (Fig. 22a) possessed a Be_4AlMn-particle in its centre. It seems that IQC also nucleates on hexagonal plates (Fig. 22b). The IQC particles appeared very often in the form of dendrites (Fig. 22a) and pentagonal dodecahedra. The most outstanding is the shape of the IQC dendrites (Fig. 22a), it looks like as though the branches were composed of a stack of pentagonal dodecahedra.

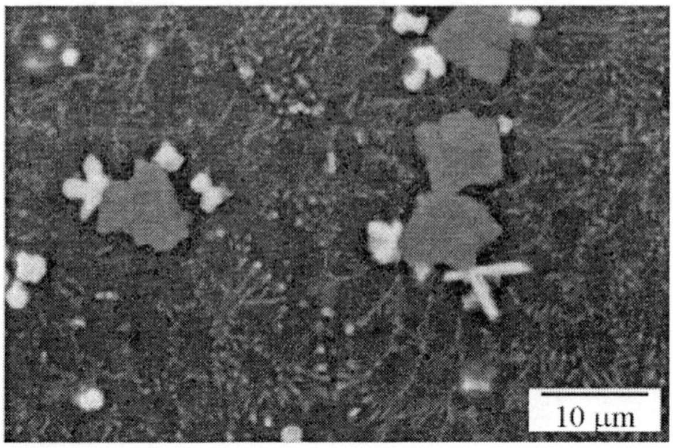

Figure 21. Backscattered electron image of the Al-Mn-Be alloy (mould casting)

These two shapes can be explained if we take into account that the growth rate of the i-phase is normally the fastest along its threefold directions and slowest along the fivefold directions, and, in addition, that the fivefold planes have the lowest surface energy and, therefore, the equilibrium shape of an icosahedral quasicrystal, represented the form of a pentagonal dodecahedron. After formation of the IQC at a particular undercooling, it grew very fast in the threefold directions, and the resulting surface was rather rough. When the solidification rate decreased the surface tension tended to stabilize the fivefold planes.

Figure 22. Extracted phases from the Al-Mn-Be mould casting (secondary electron image)

Both the i-phase and Be_4AlMn particles represented a heterogeneous nucleation site for α-Al. This is apparent from Fig. 21, showing several dendritic α-Al grains growing in a dendritic manner directly from the IQC- and Be_4AlMn- particles. It can be clearly seen that thin rods of i-phase also started to grow from IQC in (IQC + Be_4AlMn) agglomerates, dendritic and faceted IQC (Fig. 22). Also in this case eutectic growth occurred (IQC + α-Al).

It can be concluded that other phases promote the nucleation of IQC, but the growth characteristics remain almost unchanged: growth in the threefold directions and the shape of pentagonal dodecahedron.

Peculiar Dendritic Shape

Special form of IQC-dendrites was observed on the surface of some castings. Dendrites reach sizes of more than 100 μm, and those having special orientations have some interesting features. As an example, Fig. 23a shows a dendrite with the fivefold symmetry. In the middle of it almost regular pentagon appears, with dendritic branches growing from the vertices in the threefold directions. Edges of the pentagon appear as solid lines at smaller magnifications. However, higher magnifications revealed that the pentagon appeared at the positions with very high density of sectioned dendritic branches, and correspondingly with the very small distances between them; in the order of 1 μm and less. The branches in the threefold directions are not in one piece, but are composed of several parts. This morphology can be explained by assuming the highly branched dendrite, with distances between the parallel branches in the order of 5 μm. When dendrite is sectioned with the plane perpendicular to the fivefold direction, then the shape as shown in Fig. 23a can appear. In this case the highest lengths possessed the threefold branches inclined for 79.2° from the fivefold direction, those indicated with 2 in Fig. 14.

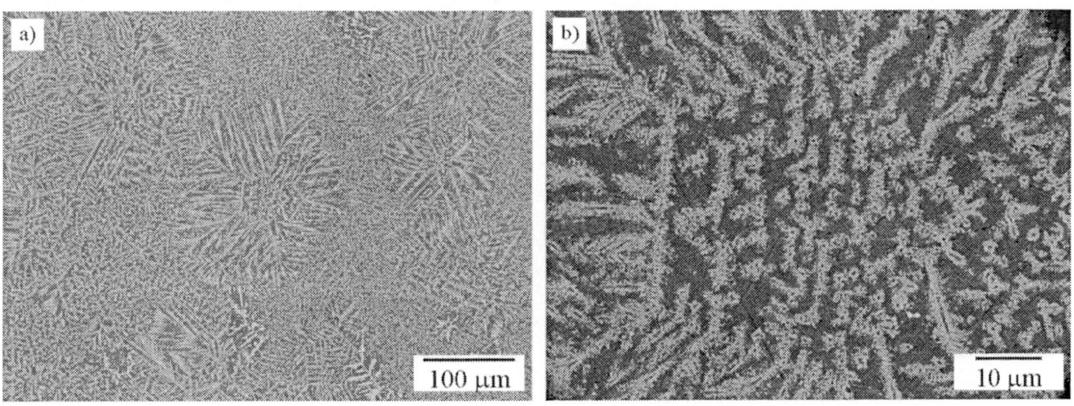

Figure 23. Peculiar dendritic shape in Al-Mn-Be-Cu mould casting (backscattered electron image)

CONCLUSIONS

The shapes of icosahedral quasicrystalline (IQC) particles in Al-Mn-Be-(Cu) alloys were determined in samples subjected to very wide ranges of cooling rates: from around 10^6 K/s in very thin melt-spun ribbons down to below 100 K/s in permanent copper dies. According to the results the following conclusions can be drawn.

During the melt spinning the following shapes of IQC were observed: spheres, particles with rounded edges, particles with protuberances and particles with the shape of pentagonal dodecahedron. The sizes of these particles were up to 500 nm.

Upon solidification in a copper mould the primary IQC and eutectic IQC were formed. The primary IQC possessed the shape of pentagonal dodecahedron either in the shape of almost ideal pentagonal dodecahedron or with small protuberances at the vertices. Very often dendritic IQC was present, with the arms in the threefold directions. The arms were either non-faceted, or faceted with the pentagonal facets. The eutectic IQC had the shape of rodlike eutectic. The rods grew predominantly in the threefold directions, exhibiting the high tendency for branching. They can very easily adjust to variable solidification conditions.

Other intermetallic phases found in the investigated alloys (Be_4AlMn, Al_4Mn) provided heterogeneous nucleation sites for IQC, however, the characteristics of IQC remained the same, with preferred growth in the threefold directions and tendency for faceting with pentagonal faces.

REFERENCES

[1] K. Chattopadhyay, N. Ravishankar, R. Goswami *Progress in Crystal Growth and Characterization of Materials* 1997, *34*, 237-249

[2] K. F. Kelton *International Materials Reviews* 1993, *38*, 105-137

[3] J. Gastaldi, G. Reinhart, H. Nguyen-Thi, N. Mangelinck-Noel, B. Billia, T. Schenk, J. Hartwig, B. Grushko, H. Klein, A. Buffet, J. Baruchel, H. Jung, P. Pino, B. Przepiarzynskik *Philos. Mag.* 2007, *87*, 3079-3087

[4] T. L. Ho, J. A. Jaszczak, Y. H. Li, W. F. Saam *Physical Review Letters* 1987, *59*, 1116-1119

[5] K. Ingersent, P. J. Steinhardt *Phys. Rev. B* 1989, *39*, 980-992

[6] D. Shechtman, I. Blech, D. Gratias, J. W. Cahn *Physical Review Letters* 1984, *53*, 1951-1953

[7] N. Thangaraj, G. N. Subbanna, S. Ranganathan, K. Chattopadhyay *J. Microsc.-Oxf.* 1987, *146*, 287-302

[8] D. Q. Wan, G. Yang, S. Chen, M. Zhu, Y. H. Zhou *Rare Metals* 2007, *26*, 435-439

[9] M. V. Kral, M. A. Mangan, G. Spanos, R. O. Rosenberg *Mater. Charact.* 2000, *45*, 17-23

[10] F. Lasagni, A. Lasagni, E. Marks, C. Holzapfel, F. Mucklich, H. P. Degischer *Acta Mater.* 2007, *55*, 3875-3882

[11] I. R. Fisher, M. J. Kramer, T. A. Wiener, Z. Islam, A. R. Ross, T. A. Lograsso, A. Kracher, A. I. Goldman, P. C. Canfield *Philos. Mag. B-Phys. Condens. Matter Stat. Mech. Electron. Opt. Magn. Prop.* 1999, *79*, 1673-1684

[12] C. Beeli, H. U. Nissen *Philos. Mag. B-Phys. Condens. Matter Stat. Mech. Electron. Opt. Magn. Prop.* 1993, *68*, 487-512

[13] T. Boncina, B. Markoli, I. Anzel, F. Zupanic *Materiali in Tehnologije* 2007, *41*, 271-277

[14] T. Boncina, B. Markoli, I. Anzel, F. Zupanic *Z. Kristall.* 2008, *223*, 747-750

[15] K. Chattopadhyay, S. Ranganathan, G. N. Subbanna, N. Thangaraj *Scripta Metallurgica* 1985, *19*, 767-771

[16] A. P. Tsai, A. Inoue, T. Masumoto *J. Mater. Sci. Lett.* 1987, *6*, 1403-1405

[17] P. C. Canfield, M. L. Caudle, C. S. Ho, A. Kreyssig, S. Nandi, M. G. Kim, X. Lin, A. Kracher, K. W. Dennis, R. W. McCallum, A. I. Goldman *Phys. Rev. B* 2010, *81*, 4

[18] B. Dubost, J. M. Lang, M. Tanaka, P. Sainfort, M. Audier *Nature* 1986, *324*, 48-50

[19] K. Sato, H. Uchiyama, Y. Takahashi, I. Kanazawa, R. Suzuki, T. Ohdaira, T. Takeuchi, T. Mizuno, U. Mizutani *Phys. Rev. B* 2001, *64*, 8

[20] N. Tamura, M. Beyss, K. Urban *Philos. Mag. Lett.* 1996, *74*, 89-97

[21] L. X. He, Y. K. Wu, X. M. Meng, K. H. Kuo *Philos. Mag. Lett.* 1990, *61*, 15-19

[22] A. P. Tsai, T. Masumoto, A. Yamamoto *Philos. Mag. Lett.* 1992, *66*, 203-208
[23] M. Zhu, G. C. Yang, F. Liu, D. Q. Wan, S. L. Cheng, Y. H. Zhou *Rare Metal Mat. Eng.* 2008, *37*, 1179-1182
[24] Y. C. Liu, G. C. Yang, Y. H. Zhou *Mater. Res. Bull.* 2000, *35*, 857-863
[25] A. P. Tsai, H. S. Chen, A. Inoue, T. Masumoto *Jpn. J. Appl. Phys. Part 2 - Lett.* 1991, *30*, L1132-L1135
[26] N. S. Athanasiou *Mod. Phys. Lett. B* 1997, *11*, 367-377
[27] D. Q. Wan, G. C. Yang, Y. H. Zhou *Rare Metal Mat. Eng.* 2007, *36*, 2166-2168
[28] M. Zhu, G. C. Yang, D. Q. Wan, S. L. Cheng, Y. H. Zhou *J. Mater. Sci. Technol.* 2009, *25*, 445-448
[29] J. H. Li, W. B. Du, S. B. Li, Z. H. Wang *Rare Metals* 2009, *28*, 297-301
[30] [30] D. Q. Wan, G. C. Yang, M. Zhu, Y. H. Zhou *Rare Metal Mat. Eng.* 2006, *35*, 1404-1407
[31] J. S. Zhang, H. W. Du, W. Liang, T. B. Li, C. X. Xu, H. X. Wang *Rare Metal Mat. Eng.* 2007, *36*, 381-385
[32] Z. F. Wang, W. M. Zhao, H. P. Li, J. Ding, Y. Y. Li, C. Y. Liang *J. Mater. Sci. Technol.* 2010, *26*, 27-32
[33] A. Niikura, A. P. Tsai, A. Inoue, T. Masumoto *Philos. Mag. Lett.* 1994, *69*, 351-355
[34] Y. Kaneko, T. Ishimasa In *Formation of icosahedral quasicrystal in Zn-Mg-Se alloy*; S. Hanada, Z. Zhong, S. W. Nam and R. N. Wright; Eds.; Japan Inst Metals: Sendai, Japan, 2001; pp 2523-2526
[35] J. Saida, M. Matsushita, A. Inoue *Mater. Trans. JIM* 2000, *41*, 543-546
[36] T. Boncina, B. Markoli, F. Zupanic *J. Microsc.-Oxf.* 2009, *233*, 364-371
[37] F. Zupanic, T. Boncina, B. Sustarsic, I. Anzel, B. Markoli *Materials Characterization* 2008, *59*, 1245-1251
[38] F. Zupanic, T. Boncina, A. Krizman, W. Grogger, C. Gspan, B. Markoli, S. Spaic *J. Alloy. Compd.* 2008, *452*, 343-347
[39] N. K. Mukhopadhyay, H. J. Chang, J. Y. Lee, D. H. Kim *Scr. Mater.* 2008, *59*, 1119-1122
[40] V. Elser *Phys. Rev. B* 1985, *32*, 4892-4898
[41] F. Zupanic, T. Boncina, N. Rozman, I. Anzel, W. Grogger, C. Gspan, F. Hofer, B. Markoli *Z. Kristall.* 2008, *223*, 735-738
[42] A. P. Tsai, A. Inoue, T. Masumoto *Progress in Crystal Growth and Characterization of Materials* 1997, *34*, 221-236
[43] F. Zupanic, T. Boncina, A. Krizman, B. Markoli, S. Spaic *Scr. Mater.* 2002, *46*, 667-672
[44] F. Zupanic, T. Boncina, A. Krizman, F. D. Tichelaar *J. Alloy. Compd.* 2001, *329*, 290-297

INDEX

A

B

C

D

E

F

G

H

I

Q

R

S